Luciano Floridi's Philosophy of Technology

Philosophy of Engineering and Technology

VOLUME 8

For further volumes:
http://www.springer.com/series/8657

Hilmi Demir
Editor

Luciano Floridi's Philosophy of Technology

Critical Reflections

Springer

Editor
Hilmi Demir
Philosophy Department
Bilkent University
Bilkent, Ankara, Turkey

ISSN 1879-7202 ISSN 1879-7210 (electronic)
ISBN 978-94-007-9197-8 ISBN 978-94-007-4292-5 (eBook)
DOI 10.1007/978-94-007-4292-5
Springer Dordrecht Heidelberg New York London

Printed on acid-free paper

Springer is part of Springer Science+Business Media (www.springer.com)

Preface

The ultimate aim of this volume is to further the philosophical reflection on technology within the context of Luciano Floridi's philosophy of technology. Philosophical reflection on technology is as old as philosophy itself, dating back to the Ancient Greek philosophers. The themes that have dominated the philosophical discourse on technology since then can be roughly categorized into three: (i) the social, cultural, and political impacts of technological developments; (ii) the epistemological status of technological knowledge, especially in relation to scientific knowledge; and (iii) the ontological status of the products of technology, i.e., technological artifacts. Luciano Floridi's philosophy of technology, which is based on his philosophy of information, has something to say about each of these themes. Not only that, his philosophical analysis of new technologies leads to a novel metaphysical framework in which our understanding of the ultimate nature of reality shifts from a materialist one to an informational one, in which all entities, be they natural or artificial, are analyzed as informational entities (Floridi 2010). This is the main rationale behind the choosing of his philosophy of technology as the topic of this volume.

There is no doubt that the information and communication technologies of the twentieth century have had a significant impact on our daily lives. They have brought new opportunities as well as new challenges for human development. According to Floridi, however, this is not the whole story. He claims that these new technologies have led to a revolutionary shift in our understanding of humanity's nature and its role in the universe. By referring to an earlier categorization, he calls this the "fourth revolution." The Copernican revolution was the first, leading to the understanding that we as humans are not at the center of the universe. The second revolution was the Darwinian realization that we are not unnaturally distinct or different from the rest of the animal world. The third was the Freudian revolution, which taught us that we are not as transparent to ourselves as we once thought. With the fourth revolution, says Floridi, "we are now slowly accepting the idea that we might be informational organisms among many agents …, *inforgs* not so dramatically different from clever, engineered artefacts, but sharing with them a global environment that is ultimately made of information, the infosphere. The information revolution [the fourth revolution] is not about extending ourselves, but about re-interpreting who we are" (Floridi 2008a).

This radical claim forms the basis of Floridi's philosophy of technology. Given this basis, philosophical reflection on technology is not only valuable in and of itself, but also brings a completely new framework of analysis for philosophy. In other words, philosophical reflection on technology takes a central role in philosophical analysis. To give an example, Floridi's analysis of object-oriented programming methodology (Floridi 2002), which relies on a method borrowed from a branch of theoretical computer science called Formal Methods, paves the way for defining a new macroethical theory, i.e., Information Ethics. The method he borrows from Formal Methods is the method of levels of abstraction. By using this method, Floridi claims that the moral value of human actions is not different in kind than the moral evaluation of other informational objects. The idea behind the method of levels of abstraction is quite simple and straightforward: the reality can be viewed from different levels. The roots of this simple idea go back to Eddington's work in the early decades of the twentieth century (Eddington 1928). Let me give a brief example in Floridi's own words:

> Suppose, for example, that we interpret p as Mary (p=Mary). Depending on the LoA and the corresponding set of observables, p=Mary can be analyzed as the unique individual person called Mary, as a woman, as a human being, as an animal, as a form of life, as a physical body, and so forth. The higher the LoA, the more impoverished is the set of observables, and the more extended is the scope of the analysis (Floridi 2002).

Perhaps the most crucial feature of the method of levels of abstraction is that the identification relation between two variables (or observables) is never absolute. Rather, the identification is always contextual and the context is a function of the level of abstraction chosen for the required analysis (Floridi and Sanders 2004a).

Floridi utilized his method not only in Information Ethics but also in several other subfields of philosophy. The following quote from his *Minds and Machines* article (2008), in which he responded to some objections raised against the method of levels of abstraction, provides a list of the areas in which the method has been used.

> Jeff Sanders and I were forced to develop the method of abstraction when we encountered the problem of defining the nature of agents (natural, human, and artificial) in Floridi and Sanders (2004b). Since then, we have been applying it to some long-standing philosophical problems in different areas. I have used it in computer ethics, to argue in favour of the minimal intrinsic value of informational objects (Floridi 2003); in epistemology, to prove that the Gettier problem is not solvable (Floridi 2004c); in the philosophy of mind, to show how an agent provided with a mind may know that she has one and hence answer Dretske's question "how do you know you are not a zombie?" (Floridi 2005a); in the philosophy of science, to propose and defend an informational approach to structural realism that reconciles forms of ontological and epistemological structural realism (Floridi 2004b); and in the philosophy of AI, to provide a new model of telepresence (Floridi 2005b). In each case, the method of abstraction has been shown to provide a flexible and fruitful approach (Floridi 2008c).

The jury is still out as to the truth value of the claim stated in the last sentence of this quote. One thing, however, is certain. Floridi's method borrowed from the Formal Methods branch of theoretical computer science and its applications have led to prolific and novel discussions in many different areas of philosophy. For the purposes of this volume, one of the most important applications of the method is in computer ethics. As mentioned above, Floridi claims that his Information

Ethics is a macroethical theory that provides a foundation for computer ethics. His Information Ethics consists of two main theses: (i) information objects *qua* information objects can be moral agents; and (ii) information objects *qua* information objects can have an intrinsic moral value, although possibly quite minimal, and hence they can be moral patients, subject to some equally minimal degree of moral respect (Floridi 2002).

The contributions in Part I of this volume are mainly centered on Floridi's Information Ethics and the method of levels of abstraction. These are Gordana Dodig-Crnkovic's "Floridi's Information Ethics as Macro-Ethics and Info-Computational Agent-Based Models," M.J. Wolf, F.S. Grodzinsky, and K.W. Miller's "Artificial Agents, Cloud Computing, and Quantum Computing: Applying Floridi's Method of Levels of Abstraction," Richard Lucas' "Levels of Abstraction and Morality," and Federica Russo's "The *Homo Poieticus* and the Bridge Between *Physis* and *Techne*."

Dodig-Crnkovic's ultimate aim in her chapter is to provide a general framework for the distribution of moral responsibility in multi-agent systems, which include humans as well as technological artifacts. In order to lay the groundwork for achieving this aim, she starts by providing her own interpretation of Floridi's Information Ethics, which she has been developing since 2006. Her interpretation, called the Info-Computationalist interpretation, is characterized by a recursive self-sustaining loop in which "the bottom-up construction of informational structures gives rise to top-down information re-structuring." In other words, the aggregate of the bottom-level elements forms a collective state that has emergent properties that are not reducible to the properties of the bottom-level informational structures. These emergent properties in turn influence the behavior of all bottom-level structures. Dodig-Crnkovic's interpretation is, to say the least, a novel one, because it allows a structured interaction between different levels of abstraction. In addition to her novel interpretation, she also states the similarities between Floridi's Information Ethics and the pragmatic approach to moral responsibility. The classical analysis of moral responsibility requires an agent with free will, and thus limits the domain of moral responsibility only to humans. In contrast, in the pragmatic approach, moral responsibility is not a result of an individual's duty; rather, it is a role defined by the externalist pragmatist norms of a group. Dodig-Crnkovic claims that Floridi's Information Ethics falls under the category of the pragmatic approach, and in that respect has the potential of providing the foundation for a moral framework in which technological artifacts can be assigned moral responsibility. Armed with these two preliminary explanations, i.e., the Info-Computationalist interpretation and the pragmatic character of Information Ethics, she uses Information Ethics to construct an artificial morality framework in which moral responsibility in intelligent systems is distributed across all agents, including technological artifacts. In her artificial morality framework, moral responsibility is handled as a regulatory mechanism that assures the desirable future behavior of intelligent systems.

Wolf et al.'s chapter, in a sense, is a continuation of an earlier article of theirs that appeared in *Ethics and Information Technology* (2009). In that article, they use two different levels of abstraction for analyzing the ethics of designing artificial agents. Their first level of abstraction, LoA1, is the user's view of an "autonomous system" such as a software package. The second level is the designer's perception of the system. Their ultimate conclusion in that paper is that the ethical responsibilities of a software designer significantly increase with the development of artificial agents because of the more intricate relationship between LoA1 and LoA2. In their contribution to this volume, they extend their original analysis by introducing a third level of abstraction, LoAS, the level that refers to society's perspective. This is important because new artificial agents not only have effects on individuals but also on the whole society that comprises those individuals. With this new addition, they test the applicability of Floridi's Information Ethics and the method of levels of abstraction to two new computing paradigms: cloud computing and quantum computing. Their overall conclusion is a positive one. They claim that although there are new challenges for Information Ethics in these two computational paradigms, Information Ethics has the potential of successfully meeting those challenges. It should be noted that their chapter also provides a nice and brief overview of the fundamental concepts of quantum computing.

Lucas' chapter is an extensive and detailed criticism of Information Ethics. He criticizes three notions that form the fundamentals of Floridi's theory, which are interactivity, autonomy, and adaptability. Lucas' ultimate conclusion is that Information Ethics, mainly because of being only formally defined, is too artificial and too simple for a natural characterization of morality. Although Floridi thinks that Lucas' understanding of Information Ethics is based on serious misunderstandings and that Lucas' chapter is beyond repair (please see Floridi's reply at the end of this volume), the chapter paves the way for a closer scrutiny of some of the arguments that Floridi has provided in defense of Information Ethics. An example might be helpful at this point. The essential motivation of Information Ethics is to be able to count artificial agents as moral agents. It should be noted that this essential motivation is somewhat different than the motivation behind the earlier characterizations of computer and information ethics. Moor (1985) is a good example of the classic treatment of the subject. In one of their earlier characterizations of Information Ethics, Floridi and Sanders consider a set of possible objections to their main claim about the moral value of artificial agents. These are the teleological objection, the intentional objection, the freedom objection, and the responsibility objection. They then provide counterarguments against those objections. Lucas thinks that none of these counterarguments sufficiently overcome the four possible objections that Floridi and Sanders consider. Of course, whether Lucas is right in his assessment or not is a matter of debate, but Lucas' reasoning urges us to reevaluate the fundamental arguments provided for the philosophical value of Information Ethics. In that respect, it is a valuable contribution to this volume.

Russo, in her chapter, focuses on one particular aspect of Floridi's Information Ethics, the reconciliation of *physis* and *techne* in a constructionist manner. According to Floridi, traditional macroethical theories take the situation which is bound to moral evaluation as given, but this traditional approach ignores the poietic nature of humans as ethical agents. Ignoring the poietic nature of humans is the ultimate basis of the dichotomy between *physis* and *techne* (Floridi and Sanders 2003). The demarcation line between these two has been disappearing because of digital technologies. Russo agrees with Floridi's analysis and attempts to take the analysis one step further. For Russo, the gradual disappearance of the demarcation line between *physis* and *techne* is not just a result of the new digital technologies; rather, it is dominated by new technologies in general. These new technologies include biotechnology and nanotechnology, which allow us to be "creating altogether new environments that pose new challenges for the understanding of us in the world." Floridi's Information Ethics, according to Russo, successfully accounts for the ethical implications of these new technologies, but, she continues, the epistemological implications are also at least equally important and need to be analyzed. This is what she aims to achieve in her chapter. In that respect, it would not be wrong to say that Russo takes Floridi's original analysis of digital technologies and applies it to a wider domain.

The two chapters in Part II provide novel ways of categorizing scientific and technological advancements on the basis of metrics different than Floridi's metric, which is based on introverted effects of scientific changes on the way we understand human nature. These are Anthony F. Beavers' "In the Beginning Was the Word and Then Four Revolutions in the History of Information" and Valeria Giardino's "I Mean It! (And I Cannot Help It): Cognition and (Semantic) Information."

Beavers, in his chapter, gives us a different categorization of the technological revolutions that mankind has experienced in its entire history. As mentioned above, Floridi's categorization of the information revolution as the fourth revolution is based on the metric of the way scientific developments change our understanding of ourselves. Thus, according to this metric, scientific developments that have led to a reassessment of humanity's fundamental nature and role in the universe are counted as revolutionary. Of course, as Floridi himself states, other metrics are also possible. In his chapter, Beavers offers a different metric that is not supposed to be an alternative to Floridi's metric, but rather complementary. The suggested metric is the history of information flow itself. In other words, technological and scientific advancements are categorized according to "the kind of information that can be stored and transmitted, the speed of information transmission, its preservation, and its reach." This metric also gives us four revolutions: the Epigraphic Revolution, the Printing Revolution, the Multimedia Revolution, and the Digital Revolution. The last one, which corresponds to Floridi's fourth revolution, is characterized by the introduction of automated information processing. There are two interesting features of Beavers' categorization that I would like to mention in this short preface. The first is that in his categorization, the Digital Revolution is not considered as a discontinuity from the previous revolutions, because information transmission and coding were also present, albeit in different forms, in the previous revolutions. What the Digital Revolution has brought to the table is new and revolutionary technological affordances that are made possible by automated information processing. This interesting feature, perhaps, is what fundamentally differentiates Beavers' categorization from Floridi's categorization. The second point is that the trajectory of the history of information flow is not characterized merely by the evolution of particular technologies, but also by the evolution of the informational networks that those particular technologies enable. After establishing his new categorization, Beavers situates the role of Philosophy, in particular the role of Philosophy of Information, in the historical context of the categorization by providing both valuable historical insights for the evolution of philosophical analysis and crucial questions that will help in the advancement of the Philosophy of Information as a new *philosophia prima*.

Giardino, in her chapter, also provides a different categorization of technological revolutions. Giardino argues that what Floridi calls the fourth revolution is in fact the second information revolution. The underlying reason for this difference is her analysis of information from a cognitive perspective. She thinks that we have been living in an informational environment all along, and that the infosphere includes all informational cognitive agents and cognitive tools. Given this understanding, any artifact that aids symbolic activities becomes an informational cognitive tool. Humans have been living in an informational environment since the time of the invention of the first tool that aided symbolic thinking. For Giardino, the correct characterization of information revolution(s) should be based on how information is transmitted across generations. Thus, the first information revolution is characterized by the transformation of the transmission from sequences of DNA to cultural transmission. The second information revolution, i.e., Floridi's fourth revolution, is characterized by the switch from cultural transmission to online transmission, according to Giardino. One of the valuable features of this chapter, among many others, is its interdisciplinary character. Giardino nicely brings together the literature on Philosophy of Information with the literature on Developmental Psychology and the literature on Cognitive Science.

The contributions in Part III take Floridi's Philosophy of Technology and Philosophy of Information as their basis and apply them to different domains: Elena Pasquinelli's "What Happens to Infoteachers and Infostudents After the Information Turn?" to education, Raphael Cohen-Almagor's "Content Net Neutrality: A Critic" to the regulation of freedom of speech on the Internet, and Armando Malheiro da Silva and Fernanda Ribeiro's "Information Science and Philosophy of Information: Approaches and Differences" to Information Science.

With the changes brought about by the information revolution, we humans have become inforgs that live in the infosphere, according to Floridi. The information revolution has led to a reontologization of our ordinary environment, where the divide between online and off-line has been disappearing. Our environment, the infosphere, "will become increasingly synchronized (time), delocalised (space) and correlated (interactions)," says Floridi. Pasquinelli, in her chapter, in light of Floridi's description of the infosphere, analyzes the past and possible future effects of the information revolution on educational institutions, practices, and actors. She starts her chapter with a diagnosis: the information revolution has not yet revolutionized education. For her, the main reason for this is the reluctance of educational institutions and actors in adopting the new tools, approaches, and paradigms that are made possible by information and computational technologies, especially in comparison to the institutions and actors of other domains. She then compares two different ways of changing the educational institutions and practices. The first is the top-down approach mostly adopted by policy makers. She cites the "One Laptop per Child"

(OLPC) program as an example of the top-down approach and shows the difficulties involved in changing educational practices in this way. According to Pasquinelli, change from the top is difficult, mainly because of the sheer size of educational institutions and the long tradition of educational paradigms and practices. A second reason, which is clearly seen in the OLPC case, is that top-down changes usually do not include students, who are the ultimate users of education, in the design of changing programs. Then she proceeds to give an example of a bottom-up approach that she claims to be more promising. Her fascinating example is the experience of Math on MXit from South Africa. With this example, she urges educational institutions and actors to implement the new technologies from the bottom up. The ultimate goal of such changes, for her, is to turn students into infostudents and teachers into infoteachers. During this transformation, which will be slow and gradual, she says, the old paradigms of education will be challenged because of the new tools and approaches of the information revolution. As the dominant example of the old educational paradigms, she gives the Victorian school, which was defined by the following three characteristics: (i) a dedicated and separated space for learning, (ii) a dedicated time for learning, and (iii) well-defined roles for the learner and the teacher. With the Internet, mobile phones, and digital media, she says, learning could occur anywhere and anytime. Moreover, the demarcation line between the student and the teacher will be blurred to the point of disappearance. In short, Pasquinelli's chapter is an informative and fascinating one in which she urges us to reontologize and reconceptualize our environment for education.

Cohen-Almagor, in his chapter, uses Floridi's Information Ethics in order to identify the democratic regulative principles of freedom of speech on the Internet and the responsibility of Internet Service Providers and Web Hosting Services. He starts his analysis by distinguishing three different senses of "net neutrality": (i) net neutrality as a nonexclusionary business practice; (ii) net neutrality as an engineering principle, allowing traffic on the Internet in a nondiscriminatory manner; and (iii) net neutrality as content nondiscrimination. He calls the third sense Content Net Neutrality. Although he accepts the first two senses as the fundamental principles that should underlie Internet regulation, he rejects Content Net Neutrality. Following Floridi's proactive approach to Ethics, which states that the ethical obligation in the information age is not limited to ethical behaviors in the infosphere but needs to extend to actively shaping the infosphere for the betterment of the humanity, Cohen-Almagor urges us to regulate the available content on the Internet. He argues that content that is morally repugnant and/or at odds with democratic ideals should not be made available on the Internet, and that the primary responsibility for this lies with Internet Service Providers and Web Hosting Services. Throughout his discussion, he uses several striking examples that seem to support his position.

As Silva and Ribeiro point out, Information Science as an autonomous field of study that appeared in the late 1950s. Since then, this new field of inquiry, which could be seen as a continuation of the library sciences, has seen an immense and rapid growth. Despite this rapid growth, however, its nature has not yet been precisely defined. This is perhaps due to the inherently interdisciplinary character of the field. Most interdisciplinary fields, for example Cognitive Science, have gone through a similar stage of development. Silva and Ribeiro, in their chapter, provide an all-encompassing framework for the nature and identity of Information Science. In their framework, Information Science is "a unitary yet transdisciplinary field of knowledge, included in the overarching area of the human and social sciences, which gives theoretical support to some applied disciplines such as Librarianship, Archivistics, Documentation and some aspects of Technological Information Systems." After providing their framework, they turn to Floridi's Philosophy of Information with the aim of finding a firm philosophical grounding for Information Science. While doing that, they state their own definition of information, which implies the following properties: structured by an action, integrated dynamical, has potentiality, quantifiable, reproducible, and transmissible. Their definition of information has some differences from Floridi's definition of semantic information. Perhaps one of the crucial differences is their distinction between informational data and noninformational data. The analysis of the differences and similarities between their definition of information and Floridi's semantic information is by itself valuable. Moreover, along the way they also bring together different threads of discussions, ranging from the French philosopher Ruyer's work on visual sensation to Søren Brier's Cybersemiotics. Given their analysis of Information Science and the connections they identify between Information Science and Philosophy of Information, it is plausible to conclude that Information Science could be understood as applied Philosophy of Information.

The main focus in Part IV is the epistemic and ontic aspects of Floridi's Philosophy of Information. The contributions here are Eric T. Kerr and Duncan Pritchard's "Skepticism and Information," Joseph E. Brenner's "Levels of Abstraction; Levels of Reality," and Steve T. McKinlay's "The Floridian Notion of the Information Object."

It is almost a truism to say that information should be "adequately created, processed, managed and used" (Floridi 2010). The bombardment of information that we all face in this day and age requires proper information management. As rightly pointed out by Kerr and Pritchard, proper information management requires paying attention to the connection between information and knowledge. After all, information is valuable as long as it paves the way for the acquiring of knowledge. In their chapter, Kerr and Pritchard focus on this important issue, i.e., the epistemic value of information. One of the milestones in the literature on the epistemic value of information is Dretske's book *Knowledge and the Flow Information*, in which a comprehensive epistemology based on information is provided. One of the controversial features of Dretske's framework is its denial of the principle of epistemic closure, which simply states that if an agent knows a proposition and knows that the proposition in question implies another one, then the agent also knows the implied proposition. Dretske's main reason behind the denial of closure is that, for him, information about appearances can never completely rule out skeptical doubts. Kerr and Pritchard claim that Dretske is wrong and that there are ways in which information could address skeptical doubts. They examine two such ways in their chapter: Ram Neta's contextual approach and John McDowell's disjunctivism. Kerr and Pritchard's chapter is valuable in and of itself. Moreover, it opens doors for a different approach to the epistemic value of information. Dretske's epistemological analysis is done in a hybrid context of doxastic and informational concepts. Kerr and Pritchard's analysis of the closure principle may also be understood as showing a need for moving to a purely informational context of analysis for knowledge, and this is exactly what Floridi does in his Philosophy of Information.

In his chapter, Brenner provides an extensive comparison of his logico-ontological theory, which is called Logic in Reality, and Floridi's Philosophy of Information. According to Brenner, "the broad theory of information proposed by Floridi requires an understanding of the properties and role of information at all levels of reality, in all entities." In other words, a complete theory of information should clarify the relevant ontological properties of information. Given the Kantian spirit of his theory, however, Floridi is quite cautious in making any ontological commitment about reality and entities. The method of levels of abstraction is proposed as a more inter-subjective, socially constructible (hence possibly conventional), dynamic, and flexible way to further Kant's approach. This method, claims Floridi, needs to be seen as a step away from internal realism, but this does not imply that it is a step toward external realism (Floridi 2008b). Thus, according to Brenner, in its current status Floridi's Philosophy of Information seems to be incomplete. Brenner claims that his Logic in Reality remedies this problem and complements Floridi's theory, and he discusses this at length in his chapter. To put it briefly, Logic in Reality is an extension of logic to complex real processes, providing a framework for analyzing and making inferences about complex real world entities and processes at all levels of reality, including biological, cognitive, and social levels. It is obvious from this nutshell definition that the processes that Logic in Reality aims to address include information production and transfer, as well. Some of the philosophically interesting features of Brenner's Logic in Reality are as follows. First, the proposed logic is nonpropositional and non-truth-functional. Second, it is grounded in a fundamental dualism, dynamic opposition, that is claimed to be inherent in energy and present in all real phenomena. In other words, real complex phenomena are in a contradictional relation between themselves and with their opposites. Third, the dynamic opposition in energy is accompanied by the law of the included middle, and thus there is no room for the principle of noncontradiction. Fourth, Logic in Reality neither requires nor commits to abstract categorical structures that separate different aspects of reality. Thus, most of the absolute distinctions of the traditional philosophical analysis, such as the one between epistemology and ontology, disappear in the framework of Logic in Reality. Fifth, Logic in Reality is based on a process-ontological view of reality, which means that the ontological inventory of the world is composed of processes at different levels of complex real phenomena. A direct result of this fifth feature is that Brenner's Logic in Reality implies an ontological levelism. As clearly stated in his defense of Informational Structural Realism (Floridi 2008b), Floridi is committed to the epistemological levelism that his method of levels of abstraction implies, but, as a result of his Kantian general framework, he finds ontological levelism untenable (Floridi 2008c). Brenner states that the ontological levelism that Floridi finds untenable is a result of the misconception of reality as seen through the glasses of classical logic and the traditional object-based ontological approach. In other words, according to Brenner, any ontological levelism that is based on an absolute distinction between epistemology and ontology is untenable, as Floridi rightly argues, but once the epistemology/ontology of Logic in Reality is adopted, then the ontological levelism becomes tenable and compatible with Floridi's Philosophy of Information.

As the ontological basis of his Philosophy of Technology, Floridi defends a form of structural realism which he calls Informational Structural Realism. In this particular version of structural realism, objects are considered as structural entities which are nothing but a collection of data clusters. This gives rise to Floridi's notion of informational objects as the fundamental ontological entities. As a side remark, it should be noted that although Floridi's analysis of objects in informational terms is quite novel, the history of including information as a fundamental entity in the metaphysics of the world dates back to Wiener's work on *Cybernetics* (Wiener 1948). In order to establish his notion of informational object, Floridi heavily relies on the lessons that he draws from object-oriented programming (OOP), both in terms of the methodology and of the ontology of OOP in constructing his Philosophy of Information and therefore his Philosophy of Technology. McKinlay's chapter focuses on the similarities and differences between OOP and Floridi's Philosophy of Information with respect to their ontology. McKinlay claims that the objects of OOP cannot be the informational objects that Floridi needs in his ontology simply because the objects of OOP are referents, whereas Floridi's informational objects are supposed to be ontologically primitive. McKinlay's claim is almost a direct result of his nominalism about conceptual objects such as OOP classes. His defense of nominalism heavily draws upon Quine's ideas. In addition to its philosophical value in terms of calling our attention to the ontological issues surrounding information and artifacts, McKinlay's chapter also provides a nice introduction to object-oriented programming.

The last chapter of the volume, "The Road to the Philosophy of Information," is Floridi's reply chapter in which each of the contributions are critically evaluated.

This volume, in my humble opinion, is quite promising in terms of achieving its ultimate aim, which is to further the philosophical reflection on technology.

References

Eddington, Arthur. 1928. *The nature of the physical world*. Cambridge: Cambridge University Press.

Floridi, Luciano. 2002. On the intrinsic value of information objects and the infosphere. *Ethics and Information Technology* 4: 287–304.

Floridi, Luciano. 2003. On the intrinsic value of information objects and the infosphere. *Ethics and Information Technology* 4(4): 287–304.

Floridi, Luciano. 2004b. The informational approach to structural realism. final draft available as IEG – Research Report 22.11.04. http://www.wolfson.ox.ac.uk/~floridi/pdf/latmoa.pdf

Floridi, Luciano. 2004c. On the logical unsolvability of the gettier problem. *Synthese* 142(1): 61–79.

Floridi, Luciano, and Jeffry W. Sanders. 2004a. On the morality of artificial agents. *Minds and Machines* 14(3): 349–379.

Floridi, Luciano, and Jeffry W. Sanders. 2004b. On the morality of artificial agents. *Minds and Machines* 14(3): 349–379.

Floridi, Luciano. 2005a. Consciousness, agents and the knowledge game. *Minds and Machines*. 15(3–4): 415–444.

Floridi, Luciano. 2005b. Presence: From epistemic failure to successful observability. *Presence: teleoperators and virtual environments* 14(6): 656–667.

Floridi, Luciano. 2008a. Artificial intelligence's new frontier: Artificial companions and the fourth revolution. *Metaphilosophy* 39(4/5): 652–654.

Floridi, Luciano. 2008b. A defence of informational structural realism. *Synthese* 161(2): 219–253.

Floridi, Luciano. 2008c. The method of levels of abstraction. *Minds and Machines* 18: 303–329.

Floridi, Luciano. 2010. *The philosophy of information*. Oxford: Oxford University Press.

Floridi, Luciano, and Jeffry W. Sanders. 2003. Internet ethics: The constructionist values of *Homo poieticus*. In *The impact of the internet on our moral lives,* ed. R. Cavalier, 195–214. New York: SUNY.

Moor, James. 1985. What is computer ethics? *Metaphilosophy* 16(4): 266–275.

Wiener, Norbert. 1948. *Cybernetics: or control and communication in the animal and the machine*. New York: Technology Press/Wiley.

Wolf, M.J., K. Miller, and F.S. Grodzinsky. 2009. On the meaning of free software. *Ethics and Information Technology* 11(4): 279–286.

Contents

Part I
Information Ethics and the Method of Levels of Abstraction

Chapter 1
Floridi's Information Ethics as Macro-ethics and Info-computational Agent-Based Models

Gordana Dodig-Crnkovic

1.1 Introduction

> *There are, however, "correct accounts" that may complement and reinforce each other, like stones in an arch.* Floridi (2008a, b, c, d)

Ten years after the introduction of Information Ethics (IE) which is an integral part the Philosophy of Information (PI) (Floridi 1999, 2002), Floridi's contribution to the subsequent production of knowledge in several research fields has been reviewed. Among others, two recent special journal issues dedicated to Floridi's work, *Ethics and Information Technology*, Vol. 10, No. 2–3, 2008 edited by Charles Ess and *Metaphilosophy*, Vol. 41, No. 3, 2010 edited by Patrick Allo witness the vitality of his research program of PI. It is far from a closed chapter in the history of philosophy. Contrariwise, it is of great interest for many researchers today, and its development can be expected to contribute to the elucidation of a number of central issues introduced or enhanced by Information and Communication Technologies, ICT.

For IE, moral action is an information processing pattern. It focuses on the fundamentally informational character of reality (Floridi 2008a) and our interactions with it. According to Floridi, ICTs create our new informational habitat "*constituted by all informational entities (such as informational agents, their properties, interactions, processes and mutual relations)*" which is an abstract equivalent of an eco-system. IE is thus a generalization of environmental ethics towards a:

- less anthropocentric concept of agent, including non-human (artificial) and distributed (networked) entities

G. Dodig-Crnkovic (✉)
School of Innovation, Design and Engineering, Computer Science Laboratory,
Mälardalen University, Västerås, Sweden
e-mail: gordana.dodig-crnkovic@mdh.se

H. Demir (ed.), *Luciano Floridi's Philosophy of Technology: Critical Reflections*,
Philosophy of Engineering and Technology 8, DOI 10.1007/978-94-007-4292-5_1,

- less biologically biased concept of patient as a 'centre of ethical worth' in any form of existence.
- more inclusive conception of environment that encompasses both natural and artificial eco-systems.

As moral judgments vitally depend on the information about what the case is and what is understood to be a desirable state of affairs, the macro-ethical behavior of networks of agents depends on mechanisms of information processing and communication. Moral responsibility increases for an agent who gets better informed. Information streams in the Infosphere can both enrich and pollute the informational environment for an agent. Those informational processes are essential in the analysis of behaviors of networks of agents, biological and artificial.

Classical ethics approaches typically look at individual (e.g. Virtue Ethics) or group behavior (e.g. the Ethics of Rights) while IE gives a framework for an agent-based approach. It is important to notice that Floridi's Philosophy of Information with Information Ethics is a research program and not a single theory. As a macro-ethics, applicable to networks of communicating agents and at the same time giving a fundamental-level view of information patterns and processes, IE can help identify general mechanisms and understand their workings. The insight into the underlying informational machinery helps to improve our analysis of ICT-enhanced systems. It is now possible to study the effects of different types of information communication, and their influence on informational networks, including the role of misinformation, disinformation, censorship of information (lack of information) and similar.

1.2 Info-computationalist Perspective on Some Basic Ideas of Information Ethics

In what follows, I will present examples of agent-based analysis of IE in socio-technological systems, elucidating ethical issues of IE within the Info-Computationalist framework as defined in Dodig-Crnkovic (2006a, 2009, 2010) and Dodig-Crnkovic and Müller (2010). That will say I will try to emphasize the diversity of existing ethical approaches, their mutual relations and the role IE plays in a deeper understanding of ethical conditions, based on dual-aspect ontology with information as a structure and computation as a process. In this reading, the contribution of IE is primarily within meta-ethics, but it sheds new light even on normative and descriptive ethics as well as on applied ethics.

IE provides a conceptual space and analytic tools for addressing the dynamic/cybernetics character of relationships between information objects. This approach helps establishing links between information, knowledge and practices of ethics. The proposed Info-Computationalist interpretation reveals a recursive self-sustaining loop: bottom-up construction of informational structures giving rise to top-down information re-structuring (emergent property). Bottom level information elements produce – through mutual interactions – a collective state that in its turn influences the behavior of each of the bottom-state elements. It should be emphasized that this mechanism, though exhibiting circularity, does not produce "vicious circles" as it stands in a continuous interaction with the environment which provides variation.[1]

The explication of the role of IE is based on the following Info-Computational elements:

1. Ontology is informational; the fabric of reality is (proto) information. (Informational Structural Realism, (Floridi 2008a))
2. Being is process of (natural) computation = Being is information processing, based on natural computing, which is both digital and analog. (Pancomputationalism 2009)
3. Information (structure) and computation (process) are two basic complementary concepts that constitute dual-aspect ontology.
4. Informational structures are physical; there is no information without physical implementation.
5. Based on physical laws, informational structures interact, evolve, and build more and more complex constellations, especially in intelligent living organisms that use "raw information"/(proto) information from the world to construct knowledge and form decisions. (Info-Computational Naturalized Epistemology (Dodig-Crnkovic 2008))
6. Ethical norms are among mechanisms that humans have developed in order to provide guidance in decision making and conduct. They can be understood as a result of successive evolution of info-computational structures in goal-driven living organisms.
7. Informational structures constitute complex systems which can be analyzed on different levels of organization/levels of description/levels of abstraction. IE is the first ethical approach focused on the fundamental level of information.

The above is based on the following fundamental principles, defined in Dodig-Crnkovic and Müller (2010)

(IC1) The ontologically fundamental entities of the physical reality are information (structure) and computation (change).
(IC2) Properties of a complex physical system cannot be derived solely from the properties of its components. Emergent properties must be taken into account.
(IC3) Change of informational structures is governed by laws.
(IC4) The observer is a part of the system observed.

[1] Among physical systems, living organisms are known to use this type of mechanisms in diverse contexts, such as metabolism, reproduction, growth end alike. On a theoretical level, Computing with Computer Science as its subset presents a rich source of examples of self-referential, circular systems that are not vicious, but perform intelligible functions like e.g. program loops, fractals and other recursive functions.

The idea of Levels of Abstraction is central to PI and even to IE, so in what follows I will try to frame the Info-Computational reading of the role of LoA in PI.

1.2.1 On the Concept of Levels of Abstraction

One of the most important insights of PI and IE is their explicit addressing of different Levels of Abstraction/Levels of Organization/Levels of Description of analysis:

> *LoAs are teleological, or goal-oriented. Thus, when observing a building, which LoA one should adopt -architectural, emotional, financial, historical, legal, and so forth – depends on the goal of the analysis. There is no "right" LoA independently of the purpose for which it is adopted, in the same sense in which there is no right tool independently of the job that needs to be done.* (Floridi 2008a, b, c, d)

Epistemologically LoA depends on the *type of interaction* between the cognizing agent and the object of the study. The type of interaction is in its turn defined by the teleological nature of knowledge production/acquisition.

Historically, research fields have typically addressed one level of abstraction/organization/description of reality. There are microscopes and there are telescopes, and visible with the help of those instruments/research tools are their specific worlds. In a microscope, no stars are visible, and in a telescope, no atomic structures. Why is it not common for a framework to encompass several levels of description e.g. to start with a very basic level of organization and encompass all levels up to macroscopic ones? For each of the layers, emergent properties show up as a result of systemic organizational phenomena. The difference between information and knowledge is not the difference in stuff but the difference in organization (structure). Looking into knowledge with fine resolution, one will only find information. Likewise, looking at the world through informational spectacles one will only see information in different constellations. Looking at a human with fine resolution, one will find only atoms, which again are known to us as information.

Every level of organization/level of complexity/level of abstraction has its own "rules of the game" and every new one emerges from the previous ones. The classical ethical discourse uses conceptual repertoire based in everyday human experience. The following passage from Hongladarom (2008) addresses the movement from the level with maximum abstraction of PI towards the detail-rich world of everyday life in the analysis of individual's right to (informational) privacy.

> *And here we are descending from the level of abstraction toward the greater specificity of everyday reality. Even if we believe that ontology is constituted by information, since reality can be described in more and more details and at deeper levels of abstraction, thus necessitating the need for more information, the need to protect privacy would not be affected because there being the Infosphere as basic reality does not mean that all information should be in the hands of the political authority. The question about Infosphere and privacy is designed to illustrate a challenge of the anti-naturalist who emphasizes the putative possibility of the individual against the ontology, but the two need not be in conflict with each other.*

Some critics feel uneasy with the Levels of Abstraction in fear of ethical relativism, but the fear is unfounded. Defining the Level of Abstraction adds to our understanding of a model. An analogy with natural sciences is instructive. Physics has specific models of the world on many different Levels of Abstraction: from elementary particles, atoms, molecules, solid state, classical mechanics and fluid dynamics, astrophysics to cosmological level. There is also a remarkable emerging field of complex systems which is not only about phenomena on specific levels of organization, but also deals with interactions among different levels. As a result, a complex system as a whole exhibits properties that are distinct from the properties of its individual parts. PI uncovers similar complex structures in epistemology and ontology while IE does the same for ethics. This makes IE a promising research program, and its practical applications are already many and will surely increase in number and importance.

1.2.2 On the Idea of Good in Information Ethics

One of frequent misunderstandings of IE is related to the intrinsic value of informational objects, which in its turn is connected to the understanding of the Levels of Abstraction of a model. A common misconception that follows this confusion is that IE will provide machinery for automatization of ethical decision-making. However, being on a fundamental level, IE will in the first place help us understand basic structures and underlying mechanisms. IE in relation to traditional ethical approaches is like molecular biology in relation to classical biology. We do not expect molecular biology to give us all answers on questions of the living world, but it provides a solid underpinning for the rest of biology. As in other research fields, the diversity of ethical approaches is still equally valuable, and it presupposes human judgment and interaction among theoretical structures.

Informational objects are a priori valuable. *If nothing else is known, we are advised not to destroy or distort informational structures*. On the higher levels of organization, such as the human one, it might well be that we must clean our mail inboxes or hard discs and that is of course not ethically problematic. Respect for information is grounded in respect for nature. One should not destroy natural objects without good reason. Nonetheless, that does not imply that we are not allowed to change anything in the world.

Hongladarom (2008) finds parallels of Floridi's IE with Spinoza's ethics in their ethical naturalism, and concludes that variety of approaches is after all inevitable. On the level of everyday practices, unity in diversity is naturally achieved through interactions:

> *What this translates to the contemporary situation of information ethics is that there are always bound to be many different ways of conceptualizing one and the same reality, and it is the people's needs, goals and desires that often dictate how the conceptualizing is done. However, when different groups of people interact, these systems become calibrated with one another. This is possible because they already belong to the same reality.*

Among the criticisms of IE, Capurro (2008) focus on the intrinsic value of informational objects, Brey (2008) makes a proposal to modify IE from a value-based into a respect-based theory in order to agree with the received view that *"inanimate things in the world deserve moral respect, not because of intrinsic value, but because of their (potential) extrinsic, instrumental or emotional value for persons"*, while Søraker (2007) proposes attribution of relational value to informational objects making the distinction between intrinsic, relational, and instrumental value. All critique points towards humans as a nexus of our ethical interest, which PI is from the outset constructed to avoid:

> *IE adopts this informational ontology (or better: the corresponding LoA) as a minimal common denominator that unifies all entities.* (Floridi 2008a, b, c, d)

This move towards connecting PI's decentralized universal perspective with classical ethical and human-centered approaches is, however, justified and necessary. We as a civilization are (still) "only" humans and our way of cognizing the world is (still) "only" human, so even if we at times adopt a fundamental level of informational structures and processes, it is in the first place in an attempt to understand the basic underlying mechanisms.

Even in a future anticipated hybrid world of humans and intelligent artifacts the relationships between different ethical frameworks and levels of description is necessary. In words of Hongladarom (2008): "*The individual cannot extricate herself from her own specific and fine-grained details of her social and physical environment.*"

A similar conclusion comes from Grodzinsky et al. (2008) who also seek to connect LoA of PI with those more everyday ethical issues one is used to: "*at levels of abstraction that are more concrete (i.e., where implementation details are visible)*". This recurring wish for providing more specific examples of connections between IE and classical ethical approaches is the evidence of interest in applying IE analysis.

Focusing on a fundamental level of organization and radically rethinking our relationships with each other and with the world, IE essentially contributes to our ability to understand underlying mechanisms of ethical behavior in networks of humans and intelligent artifacts. The observed progress towards increased distribution of cognitive functions in such systems (Magnani 2007) necessitates application of PI.

1.2.3 On the Artificial Agency and Morality

This article concerns systems of humans and intelligent adaptive artifacts and in the first place the problem of (moral) responsibility distribution. It argues that for all practical purposes, moral responsibility in autonomous intelligent systems is best handled as a regulatory mechanism, with the aim to assure their desirable behavior. "Responsibility" is thus ascribed an intelligent artifact in much the same way as "intelligence" and it is considered to be a matter of degree. We will expect a (morally) responsible artifactual intelligent agent to behave in a way that is traditionally thought to require human (moral) responsibility.

In order to make the point about artificial moral agency (Grodzinsky et al. 2008) adopt concept of Levels of Abstraction and discuss the difference between artificial agents whose behavior is completely defined by their designers, and agents able to learn and adapt, changing their own programs autonomously. They conclude that designers and other concerned stakeholders must maintain responsibility for those artifacts, no matter how autonomous they may be. Actually this conclusion shall not come as a surprise. The question Grodzinsky, Miller and Wolf ask: "*Can an artificial agent that changes its own programming become so autonomous that the original designer is no longer responsible for the behavior of the artificial agent?*" in the perspective of distributed responsibility discussed in detail later on, gets an obvious answer. Such an artificial agent with an artifactual equivalent of "free will" can not be more autonomous than a human within a techno-social system. Even though humans have free will and autonomy, there is a distribution of responsibility in a system.[2]

> *Again: the idea of building moral responsibility into artificial agents is not meant as leaving those agents outside of the techno-sociological control.*

One of the central concepts in this context is the concept of agent. Unlike (Himma 2009) who concludes his essay by the claim that artificial moral agency is possible if it is possible for ICTs to be conscious, in the field of Agent Based Modeling (http://www.scholarpedia.org/article/Agent_based_modeling) agents are supposed to include even much simpler entities. Agent-Based Modeling (ABM) is an individual-based modeling of a phenomenon as a system of interacting agents (actors) such that agents have internal states.[3] Humans may in this context be seen as highly complex agents.

Agents in general may be as simple as cellular automata but may also have random-access memory, i.e. they can interact with the environment beyond concurrent state communication by using memory to save representations of the environment. Members of an agent society can share information and knowledge. Such agents are dynamically incoherent as their next state is not only dependent on the previous state but also on their memory (which keeps the same value until it is accessed). Agent interactions can be local, global or intermediate (small-world network). The system evolves over time, and since agents behave individually in parallel, interactions are generally asynchronous.[4] ABMs are powerful modeling tools which relate Artificial Life, Game Theory and Artificial Intelligence and in this context are useful in the studying of ethics in IE applications.

---

[2] As long as artifacts are under human control, such as GPS devices, we have no problem to follow their command. But what kind of assurance do we need when artifacts with superior cognitive capacities have their own agenda? I believe that we will get successively better insights into that issue as we enhance our own cognitive capacities through distributed cognition in networks of biological and artificial agents.

[3] Internal states are represented by discrete or continuous variables.

[4] In ABM, both time and space can be discrete or continuous.

1.2.4 IE's Constructive/Generative Nature

Enabling computational modeling in IE resembles adding a microscope to medical diagnostic tools. It will not replace a doctor's usual examination of a patient, but will provide a useful complement. The result of the investigation of a patient's health naturally depends on the diagnostic method. So on one level of analysis, the problem might be identified as a high level of leukocytes in the blood. On a higher granularity level, the same problem may appear as an infectious disease. On an even higher, social level, the problem may be characterized as an epidemic and a health-care problem.

Instead of being worried by the fact that different levels of abstraction show different views of the world and give different answers to questions, such as: what is wrong? (the leukocyte number is too high, the patient has an infectious disease, there is a threat of pandemic, etc.), we should be happy with the fact that we finally make explicit a variable always present in every analysis, a variable which otherwise is hidden and often the source of misunderstanding in ethical debate when two parts in a dialogue discuss the problem on different levels of abstraction, without even recognizing that.

In other words, information-centric IE is a complementary and not an alternative approach to traditional ethics. As already pointed out by Floridi (2008a, b, c, d), there is a plurality of possible approaches which may "*complement and reinforce each other, like stones in an arch.*"

The strongest side of IE is its focus on the understanding of mechanisms of ethical behavior on a conceptually more fundamental level than what conventional ethical approaches usually provide. Instead of assuming that an agent is perfectly well informed and perfectly rational, the modeling of ethical agent systems on the informational level permits studying effects of information communication and processing in networks of agents. It includes effects of imperfect information transmission and how the global behavior of a system changes when agents get distorted information or no information at all, or when an agent itself is not a perfectly rational human but maybe a less cognitively equipped machine/program. The grounds for normativity in such an info-computational system can be studied by simulation models as well.

Rather than providing an automaton for ethical norm generation from available information, IE presents a valuable tool for studying the effects of plurality of ethical choices and network configurations reflecting the macro-ethical character of IE.[5] As in general with different levels of abstraction, the answers from micro- and macro-perspectives are not necessarily identical.

Being especially suitable for the analysis of artificial multi-agent systems (MAS), Information Ethics may be expected to be at least a useful framework in supporting

[5] Unlike the micro-ethical level where one considers what an individual should do, at the macro-ethical level the question is what macro-systems, such as political institutions, corporations or professional organizations, should do.

generative studies and modeling ethics of techno-social systems as MAS models used for sociological and economic phenomena (Gilbert 2008; Epstein 2004).

1.3 Info-computational Models of Intelligent Agent | Systems – A Pragmatic Approach to Moral Responsibility

1.3.1 Ethics and Future Intelligent Agents

Engineering can be seen as a long-term, large-scale social experiment since the design, production and employment of engineered artifacts can be expected to have long-range effects (Martin and Schinzinger 1996). Especially interesting consequences might be anticipated if the engineered artifacts are intelligent, adaptive and autonomous. Such multi-agent systems present very suitable objects for testing the usefulness of Information Ethics in practice. On a more specific level of abstraction, intelligent artifacts are the focus of interest of Roboethics, a new field of applied ethics, which has brought about many interesting novel insights (Veruggio and Operto 2008; Roboethics). Ethical challenges addressed within Roboethics include the use of robots, ubiquitous sensing systems and ambient intelligence, direct neural interfaces and invasive nano-devices, intelligent softbots, robots aimed at warfare, and similar which actualize ethical issues such as values, responsibility, liability, accountability, control, privacy, self, (human) rights, and similar (Dodig-Crnkovic 2006b; Dodig-Crnkovic and Persson 2008).

If the prediction of Veruggio comes true, and the "internet of things" becomes reality "*the net will be not only a network of computers, but of robots, and it will have eyes, ears and hands, it will be itself a robot*" (Roboethics). This envisaged robot-net will indeed involve unprecedented ethical challenges. In accordance with the precautionary principle (Hansson 1997, 1999; Montague 1998) we have not only rights but also clear moral obligation to elucidate the ethical consequences of the possible paths of development. Concerned voices[6] ask: Are we in danger to become just objects of artificial intelligence? (Crutzen 2006)

> *In shaping responsibility ascription policies one has to take into account the fact that robots and softbots – by combining learning with autonomy, pro-activity, reasoning, and planning – can enter cognitive interactions that human beings have not experienced with any other non-human system* (Marino and Tamburrini 2006).

When predicting global development we have to take into account that while we are changing technology, technology in its turn is changing us (Becker 2006; Russell and Norvig 2003). The next question is what happens when cognitive capabilities of

[6] Vol. 6, IRIE, 2006 *International Review of Information Ethics* dedicated to Ethics of Robotics, see http://www.i-r-i-e.net/archive.htm

autonomous intelligent artifacts surpass those of humans. Are we going to have any need or indeed any means to control such systems? It is good to address those issues as we are developing new intelligent and autonomous learning technologies and anticipating their future advance.

While Roboetics has its focus on the phenomena on the level of traditional applied ethics, relying on already existing insights in a sense close to Deborah Johnson's views, Floridi's Information Ethics allows analysis beyond traditional approaches. Based on an informational level, IE makes possible search for underlying mechanisms – patterns and processes – that in a network of agents results in a certain behavior. As already mentioned while discussing agent-based models, IE is not only applicable for the modeling of artificial agent networks, but even includes the possibility of modeling human behavior. Artificial agents are just the first and simplest application which may be made in a straightforward way.

Since IE uncovers an underlying layer of reality, its goal in ethical praxis may be seen not as excluding existing ethical theory and practice, but helping us to understand the fine structure of phenomena. Next I will try to establish the relationship between IE, Roboethics and other classical applied ethics. Especially, I will focus on the issue of trust and responsibility as understood in different frameworks.

1.4 Moral Responsibility, Classical vs. Pragmatic Approaches

1.4.1 Classical Approach to Moral Responsibility, Causality and Free Will

A common approach to the question of moral responsibility is presented in Stanford Encyclopedia of Philosophy, according to which "*A person who is a morally responsible agent is not merely a person who is able to do moral right or wrong. Beyond this, she is accountable for her morally significant conduct. Hence, she is an apt target of moral praise or blame, as well as reward or punishment*" (Eshleman 2004; Siponen 2004).

In order to decide whether an agent is morally responsible for an action, it is believed to be necessary to consider two parts of the action: causal responsibility and mental state (Nissenbaum 1994). In this view, the mental state aspect of a moral action is what distinguishes morally responsible agents. *Traditionally only humans are considered to be capable of moral agency.* The basis of the human capability of action is intention, and this internal state is seen as the origin of an act that, depending on the effects it causes, can imply moral responsibility (Johnson 2006; Dennett 1973). Moreover, intentionality enables learning from mistakes, regret of wrongs and wish to do right – all of which are typically human abilities. According to this view, both causal responsibility and intentionality are necessary for someone to be considered morally responsible for an act.

1.4.2 Pragmatic (Functional) Approach to Moral Responsibility

Questions of intentionality (Dennett 1994) and free will of an agent are difficult to address in practical engineering circumstances, such as development and use of intelligent adaptive robots/softboats. Consequently, Dennett and Strawson suggest that we should understand moral responsibility not as an individual duty but instead as *a role defined by externalist pragmatic norms of a group* (Dennett 1973; Strawson 1974). We will also adopt a pragmatic approach, closer to actual robotic applications, where the question of free will is not the main concern. Moral responsibility can best be seen as a social regulatory mechanism which aims at enhancing actions considered to be good, simultaneously minimizing what is considered to be bad. "Responsibility" can thus be ascribed an intelligent artifact in much the same way as "intelligence". Dodig-Crnkovic and Persson (2008), and Adam (2008) all emphasize the parallel between artificial intelligence and artificial morality.

Artificial/artifactual intelligence is an ability of artificial agents to accomplish tasks that are traditionally thought to require human intelligence.

In the same way, we define artificial/artifactual morality as an ability of an artificial agent to behave in a way that is traditionally thought to require human morality.

Does it mean that artifactual intelligence is the same thing as human intelligence? No. It just produces the same behavior and solves the same problems. And that is why we build intelligent systems. We want them to solve problems for us. As they become more and more intelligent and autonomous, we want them to behave in accordance with our value systems and ethical norms.

> *We take the instrumental approach that while full-blown moral agency may be beyond the current or future technology, there is nevertheless much space between operational morality and "genuine" moral agency. This is the niche we identified as functional morality.* (Wallach and Allen 2009)

1.5 Moral Responsibility[7] of Artificial Intelligent Systems

Responsibility in a complex socio-technological system is usually a distribution of duties in a hierarchical manner, such as found in military or government organizations and even in corporations.[8] Dennett views moral responsibility as rational and socially efficient policy and as the result of natural selection within cooperative systems (Dennett 1973; Järvik 2003). Moral responsibility as a regulative mechanism shall not only locate the blame, but more importantly assure future appropriate behavior of a system.

---

[7] Floridi (2008b) does not talk about responsibility but instead accountability of artificial agents.

[8] Coleman, K. G., Computing and Moral Responsibility, *The Stanford Encyclopedia of Philosophy (Spring 2005 Edition)*, Edward N. Zalta (ed.), Available: http://plato.stanford.edu/archives/spr2005/entries/computing-responsibility/

In a pragmatic spirit, moral responsibility is considered to be the obligation to behave in accordance with an accepted ethical code (Sommerville 2007). It is relevant as long as it influences the behavior of individuals who have been assigned responsibilities (Dodig-Crnkovic 2005). In Software Engineering practice moral responsibility is a subfield of system dependability. The practical questions of allocation, acceptance, recording and discharge of responsibilities and their reflection in that context are addressed in DIRC project. When attributing moral responsibility, the focus is usually on individual moral agents. However, Coleman and Silver argue that even corporations and similar socio-technological systems also have a collective moral responsibility (Coleman 2005; Silver 2005).

A common argument against the ascription of moral responsibility to artificial intelligent systems is that they do not have the capacity for mental states like intentionality, and thus cannot fulfill all requirements for being morally responsible (Johnson 2006; Johnson and Miller 2006). The weakness of this argument is that it is actually nearly impossible to know what such a mental state entails (Floridi and Sanders 2004a). In fact, even for humans, intentionality is ascribed based on the *observed behavior* as we have no access to the inner workings of human minds – much less than we have access to the inner workings of a computing system; see even Coeckelbergh (2010).

Another argument against ascribing moral responsibility to artificial intelligent systems holds that it is pointless to assign praise or blame when it has no meaning for an agent, see Floridi and Sanders (2004b). In that case, we should ponder over the meaning of a meaning. An agent may be programmed such that it has a meaning for praise and blame in the same way as it has a meaning for obstacles and goals. The question of building in emotions/synthetic emotions into artifacts is addressed in Coeckelbergh (2010), Becker (2009), Arkin (1998), Fellous and Arbib (2005), and Minsky (2006). Emotions appear to be a very suitable regulatory mechanism applicable in this case.

In addition, both above arguments against ascribing moral responsibility to artificial intelligent agents stem from a view in which artificial intelligent systems are seen primarily as isolated entities. However, in order to address the question of moral responsibility in intelligent systems, we must see them as parts of larger socio-technological organization. From such a standpoint, ascribing responsibility to an intelligent system has primarily a *regulatory role*.

Investigation of moral responsibility for systems involving technological artifacts must take into account the actions of the users and producers of the artifacts, besides the technological artifacts themselves (Johnson and Powers 2005). It is not only human agents that by engineering and operating instructions can influence the morality of artificial agents. Artifacts, as actors in socio-technological systems, can impose limits on and influence the morality of human actors too (Adam 2005; Latour 1992). Despite this, the existence and influence of artifacts up to now have always originated in a human designer and producer (Johnson 2006) and their role is undoubtedly central.

Delegation of tasks is followed by distribution of responsibilities in the socio-technological system, and it is important to be aware of the balance of responsibilities between different actors in the system (Adam 2005). Commonly, the distribution of responsibility for the production and use of a system can be seen as a kind of contract, a manual of operations, which specifies how and under what circumstances the artifact or system should be used (Matthias 2004). This clear distinction between the responsibilities of the producer and user was historically useful, but with increased distribution of responsibilities throughout a socio-technological system, the distinction becomes less clear-cut. Production and use of intelligent systems has increased the difficulty, as the intelligent artifacts themselves display autonomous morally significant behavior, which has lead to a discussion about the possibility of ascribing moral responsibility to machines, see Matthias (2004), Johnson (2006), Floridi and Sanders (2004a, b) and Stahl (2004). Many of the practical issues in determining responsibility for decisions and actions made by intelligent systems will probably follow already existing models that are now regulated by product liability laws (Stahl 2004). There is a doubt that this approach may not be enough, and that alternative ways of looking at responsibility for the production and use of intelligent systems may be needed (Stahl 2006).

In sum, having a system which "takes care" of certain tasks intelligently, learns from experience and makes autonomous decisions gives us good reasons to talk about a system as being "responsible for a task". Technology is morally significant for humans, so the responsibility for a task with moral consequences could be seen as moral responsibility. The consequential responsibility which presupposes moral autonomy will, however, be distributed through the system.

Numerous interesting questions arise when the issue of artificial agents capable of moral responsibility in the classical sense is addressed by defining autonomous ethical rules of their behavior. Those are issues addressed within the field of Machine Ethics (Moor 2006) which includes developing ethical rules of behavior for e.g. softbots which seems to be both useful and practical.

1.6 Distribution of Responsibilities and Handling of Risks in Technical Systems

When it comes to the practical applications, based on the experiences with safety critical systems such as aerospace, transportation systems and nuclear power, one can say that the socio-technological structure which supports their functioning consists of safety barriers preventing and mitigating their malfunction. The central and most important part is to assure the safe functioning of the system under normal conditions, which is complemented by the preparedness for abnormal/accidental condition mitigation. There are several levels of organizational and physical barriers ready to cope with different levels of severity of malfunctions (Dodig-Crnkovic 1999).

Handling risk and uncertainty in the production of a safety critical technical system is done on several levels. Producers must take into account everything from technical issues, through issues of management and of anticipating use and effects, to larger issues on the level of societal impact (Huff 2004; Asaro 2007). The central ethical concerns for engineers are: "How to evaluate technologies in the face of uncertainty" and "How safe is safe enough" (Shrader-Frechette 2003; Stamatelatos 2000; Larsson 2004).

Any technology subject to uncertainty and with a potentially high impact on human society is expected to be handled cautiously, and intelligent systems surely fall into this category, where the precautionary principle (Montague 1998) applies. Thus, preventing harm and having the burden of proof of harmlessness is something that producers of intelligent systems are responsible for. The analogy might be with a state sending an army to a battlefield, where responsibility is organized hierarchically, with the highest responsibility on the top of the hierarchy, but which includes responsibilities of each and every soldier, be they human or artifacts.

1.7 Computational Modeling and Information Ethics

There are numerous examples of info-computational agent-based models and their applications in social systems (Gilbert 2008; Epstein 2004). Computational modeling and especially Agent-Based Models present potential as a method for elucidating ethical issues within the framework of Floridi's Information Ethics. An interesting development is found in the research of the behavior of artificial agents and simulation of intelligent agent networks (Danielson 1992; Floridi and Sanders 2004a, b). Among studies based on explicit modeling, those of Ramchurn et al. (2004) and Lomi and Larsen (2000) present analysis of trust in multi-agent systems and simulated organizational societies, while Prietula (2000) addresses advice, agreement and trust among artificial agents. Dodig-Crnkovic and Anokhina (2008) give an IE analysis of typical information-communication phenomena of workplace gossip and rumor.

Building on the results from the social sciences, Lik Mui (2002) proposes a formal framework for modeling trust and reputation. This model makes explicit the importance of social information (indirect channels of inference) in helping members of a social network to choose whom they want to interact with. This framework is subsequently extended to address the evolution of cooperation, which is a fundamental problem of social science and biology. Lik Mui's research results show that provided an indirect inference mechanism for propagation of trust and reputation, cooperation among selfish agents can be explained for a set of game theoretic simulations.

Based on similar insights gained from explicit modeling, Information Ethics may be developed into a powerful tool of analysis of a wide range of phenomena in different informational environments. Business ethics, research ethics, publication ethics, computer ethics and robotic ethics are examples of fields where Floridi's approach gives a very suitable framework for study. In the case of moral responsibility, IE provides a vital support, especially in the case of info-computational modeling of processes and patterns. There are plenty of additional examples of computational models useful for the elucidation of basic informational mechanisms in ethical analysis. Recent criticism of Computational Modelers approach to IE (and related AI) (Johnson and Miller 2008) deserves a comment.

Computational Models are widely used cognitive tools. Computational Modelers who primarily focus on models also design computerized tomographs which are used for diagnostics in medicine. They do not believe that human being is "just" that aspect of the physical body which is modeled. But when they help to solve a health problem on a macroscopic level of abstraction, those Computational Modelers are extremely valuable even though they actually care mostly for physical mechanisms and completely neglect all other patient's characteristics but go down to tissue or molecular level. Looking on that level, they may find that something is wrong in the body of a patient and by connecting those fundamental levels with the knowledge of macroscopic body, the doctor will hopefully be able to devise a treatment. The problem on a microscopic level might be something we lack an idea of in our everyday life. And yet, this kind of fundamental-level modeling may improve our medical methods. Similar may be said of Information Ethics. Its advice is on a fundamental informational level of reality. On the more "macroscopic" level, we may find a classical variety of ethical theories useful, depending on circumstances.

Being manifestly macro-ethics, IE will certainly support the building of a new global ethics along with a host of traditional ethical theories and practices which will be enriched and further developed through global interactions between intelligent agents, biological as well as non-biological.

The claim of IE that being as informational structure possesses intrinsic value is natural for the fundamental level of abstraction. That would in our medical analogy be equivalent to the claim that tissues (or molecules or atoms) of human body possess an a priori intrinsic value. However, from some higher level of organization of human body one may know that human organs can malfunction, and sometimes destruction of tissue is necessary.

The starting point is a respect for the world or existence as such. This is a good start. On the most fundamental level, we really do not see those structures/relationships/processes that constitute knowledge or its more elaborate constructs that knowledge in its turn constitutes such as theories, paradigms or cultural horizons. All those higher forms of organization of information are invisible for an analysis on a fundamental informational level.

In analogy with analytical methodological tools in medicine, IE is not expected to deliver answers to all questions on all levels of abstraction. Instead, it supports those abstract informational ones.

Interesting to notice is that there is a self-reflective loop in the Info-Computational picture. As reality appears to be informational (Floridi 2008b) with information the whole way up and the whole way down, we have information about information and information about information about information, etc. – complex informational architectures which undergo continuous changes through computational processes. From the proto-information that constitutes the world in itself, new information emerges when existing information interacts with itself. Informational structures within agents interact with informational structures that agents identify as physical objects, changing agents own structures, saving memories of past interactions and so internalizing the world, adding intentionality to a cognizing agent.

1.8 Conclusions

There are many parallels between IE and environmental ethics, of which IE is a generalization and the Infosphere may be understood as our new cognitive environment. However, there is an important sense in which they differ. Environmental ethics is placed on the same "macroscopic" and detail-rich everyday-life level of description, while IE is on a more abstract fundamental level of information structures and their processes.

By generalizing the idea of an agent to include a wide spectrum of actors from cellular automata to humans, Info-Computational reading of Information Ethics leads to a conclusion that artificial agents in addition to (artifactual) intelligence can be ascribed (given) a certain degree of artifactual morality. In this approach, networks of agents can be analyzed with respect to type of agency and kind of communication. One of the properties of an artificial agent may be artifactual responsibility/accountability. According to the classical approach, free will is essential for an agent to be assigned moral responsibility. Pragmatic approaches (Dennett, Strawson), on the other hand, focus on social, organizational and role-assignment aspects of responsibility, which is directly applicable to Agent-Based Models.

This analysis argues that moral responsibility in intelligent systems is best viewed as a regulatory mechanism, and follows an essentially pragmatic (instrumental, functionalist) line of thought. For all practical purposes, the question of responsibility in learning intelligent systems may be addressed in the same way as safety in traditional safety critical systems. Long-term, wide-range consequences of the deployment of intelligent systems must be discussed on a wide democratic basis as the intelligent systems have a potential of radically transforming the future of human society globally.

IE is one of the tools of investigation which will help improve the understanding of ethical aspects of our life in increasingly densely populated Infosphere. We are far from being able to reconstruct/generate/simulate the structure and behavior of an intelligent agent from information as the stuff of the universe, and we are even less capable of understanding their ethical behavior based on a few basic informational principles. IE is not a machine for the production of the ultimate ethical advice, but a powerful complementary instrument of ethical analysis.

Acknowledgements The author wants to thank Mark Coeckelbergh for insightful comments on earlier versions of this paper.

References

Adam, Alison. 2005. Delegating and distributing morality: Can we inscribe privacy protection in a machine? *Ethics and Information Technology* 7: 233–242.

Adam, Alison. 2008. Ethics for things. *Ethics and Information Technology* 10(2–3): 149–154.

Arkin, Ronald C. 1998. *Behavior-based robotics*. Cambridge: MIT Press.

Asaro, Peter M. 2007. Robots and responsibility from a legal perspective. In Proceedings of the IEEE 2007 international conference on robotics and automation, Workshop on RoboEthics, Rome.

Becker, Barbara. 2006. Social robots – emotional agents: Some remarks on naturalizing man-machine interaction. *International Review of Information Ethics* 6: 37–45.

Becker, Barbara. 2009. Social Robots – Emotional Agents: Some Remarks on Naturalizing Man-machine Interaction. In *Ethics and robotics*, ed. R. Capurro and M. Nagenborg. Amsterdam: IOS Press.

Brey, Philip. 2008. Do we have moral duties towards information objects? *Ethics and Information Technology* 10(2–3): 109–114.

Capurro, Rafael. 2008. On Floridi's metaphysical foundation of information ecology. *Ethics and Information Technology* 10(2–3): 167–173.

Coeckelbergh, Mark. 2010. Moral appearances: Emotions, robots, and human morality. *Ethics and Information Technology* 12(3): 235–241. ISSN 1388-1957.

Coleman, K.G. 2005. Computing and moral responsibility. In *The Stanford encyclopedia of philosophy*, Spring edn, ed. Edward N. Zalta. Stanford: Standford University. Available: http://plato.stanford.edu/archives/spr2005/entries/computing-responsibility/

Crutzen, C.K.M. 2006. Invisibility and the meaning of ambient intelligence. *International Review of Information Ethics* 6: 52–62.

Danielson, Peter. 1992. *Artificial morality virtuous robots for virtual games*. London: Routledge.

Dennett, Daniel C. 1973. Mechanism and responsibility. In *Essays on freedom of action*, ed. T. Honderich. Boston: Routledge/Keegan Paul.

Dennett, Daniel C. 1994. The myth of original intentionality. In *Thinking computers and virtual persons: Essays on the intentionality of machines*, ed. E. Dietrich, 91–107. San Diego/London: Academic.

DIRC project. http://www.comp.lancs.ac.uk/computing/research/cseg/projects/dirc/projectthemes.htm (accessed October 26, 2010).

Dodig-Crnkovic, Gordana. 1999. *ABB atom's criticality safety handbook, ICNC'99 sixth international conference on nuclear criticality safety, Versailles, France*. http://www.idt.mdh.se/personal/gdc/work/csh.pdf (accessed October 26, 2010).

Dodig-Crnkovic, Gordana. 2005. On the importance of teaching professional ethics to computer science students. In *Computing and philosophy,* Computing and philosophy conference, E-CAP 2004, Pavia, Italy, ed. L. Magnani. Pavia: Associated International Academic Publishers.

Dodig-Crnkovic, Gordana. 2006a. *Investigations into information semantics and ethics of computing*. Västerås: Mälardalen University Press. http://mdh.divaportal.org/smash/get/diva2:120541/FULLTEXT01 (accessed October 26, 2010).

Dodig-Crnkovic, Gordana. 2006b. Professional ethics in computing and intelligent systems. In *Proceedings of the ninth Scandinavian Conference on Artificial Intelligence* (SCAI 2006), Espoo, Finland, October 25–27.

Dodig-Crnkovic, Gordana. 2008. Knowledge generation as natural computation. *Journal of Systemics, Cybernetics and Informatics* 6: 12–16.

Dodig-Crnkovic, Gordana. 2009. *Information and computation nets*. Saarbrücken: VDM Verlag.

Dodig-Crnkovic, Gordana. 2010. The cybersemiotics and info-computationalist research programmes as platforms for knowledge production in organisms and machines. *Entropy* 12: 878–901. http://www.mdpi.com/1099-4300/12/4/878 (accessed October 26, 2010).

Dodig-Crnkovic, Gordana, and Margaryta Anokhina. 2008. Workplace gossip and rumor. The information ethics perspective. In *ETHICOMP-2008*, Mantova, Italy.

Dodig-Crnkovic, Gordana, and Vincent Müller. 2010. A dialogue concerning two world systems: Info-computational vs. mechanistic. In *Information and computation*, ed. G. Dodig-Crnkovic and M. Burgin. Singapore: World Scientific Publishing Co.

Dodig-Crnkovic, Gordana, and Persson Daniel. 2008. Sharing moral responsibility with robots: A pragmatic approach. In *Tenth Scandinavian Conference on Artificial Intelligence SCAI 2008*, Frontiers in artificial intelligence and applications, vol. 173, ed. A. Holst, P. Kreuger, and P. Funk. Amsterdam: IOS Press.

Epstein, Joshua M. 2004. *Generative social science: Studies in agent-based computational modeling*, Princeton studies in complexity. Princeton/Oxford: Princeton University Press.

Eshleman, Andrew. 2004. Moral responsibility. In *The Stanford encyclopedia of philosophy*, Fall ed, ed. Edward N. Zalta. Stanford: Stanford University. http://plato.stanford.edu/archives/fall2004/entries/moral-responsibility (accessed October 26, 2010).
Fellous, Jean-Marc, and Michael A. Arbib (eds.). 2005. *Who needs emotions?: The brain meets the robot*. Oxford: Oxford University Press.
Floridi, Luciano. 1999. Information ethics: On the theoretical foundations of computer ethics. *Ethics and Information Technology* 1(1): 37–56.
Floridi, Luciano. 2002. What is the philosophy of information? *Metaphilosophy* 33(1/2): 123–145.
Floridi, Luciano. 2008a. A defence of informational structural realism. *Synthese* 161(2): 219–253.
Floridi, Luciano. 2008b. Information ethics: Its nature and scope. In *Moral philosophy and information technology*, ed. Jeroen van den Hoven and John Weckert, 40–65. Cambridge: Cambridge University Press.
Floridi, Luciano. 2008c. The method of levels of abstraction. *Minds and Machines* 18(3): 303–329.
Floridi, Luciano. 2008d. Ethics Information ethics: A reappraisal. *Ethics and Information Technology* 10: 189–204.
Floridi, Luciano, and J.W. Sanders. 2004a. On the morality of artificial agents. *Minds and Machines* 14(3): 349–379.
Floridi, Luciano, and J.W. Sanders. 2004b. On the morality of artificial agents. In *Minds and machines*, vol. 14, 349–379. Dordrecht: Kluwer Academic Publishers.
Gilbert, Nigel. 2008. *Agent-based models*, Quantitative applications in the social sciences. Los Angeles: Sage Publications.
Grodzinsky, Frances S., Keith W. Miller, and Marty J. Wolf. 2008. The ethics of designing artificial agents. *Ethics and Information Technology* 11(1): 115–121.
Hansson, Sven Ove. 1997. The limits of precaution. *Foundations of Science* 2: 293–306.
Hansson, Sven Ove. 1999. Adjusting scientific practices to the precautionary principle. *Human and Ecological Risk Assessment* 5: 909–921.
Himma, Kenneth E. 2009. Artificial agency, consciousness, and the criteria for moral agency: What properties must an artificial agent have to be a moral agent? *Ethics and Information Technology* 11(1): 19–29.
Hongladarom, Soraj. 2008. Floridi and Spinoza on global information ethics. *Ethics and Information Technology* 10: 175–187.
Huff, Chuck. 2004. Unintentional power in the design of computing systems. In *Computer ethics and professional responsibility*, ed. T.W. Bynum and S. Rogerson, 98–106. Kundli: Blackwell Publishing.
Järvik, Marek. 2003. How to understand moral responsibility?, *Trames*, 7(3), 147–163. Tallinn: Teaduste Akadeemia Kirjastus.
Johnson, Deborah G. 2006. Computer systems: Moral entities but not moral agents. In *Ethics and information technology*, vol. 8, 195–204. Dordrecht: Springer.
Johnson, Deborah G., and Keith W. Miller. 2006. A dialogue on responsibility, moral agency, and IT systems. In *Proceedings of the 2006 ACM symposium on Applied computing table of content*, Dijon, France, 272–276.
Johnson, Deborah G., and Keith W. Miller. 2008. Un-making artificial moral agents. *Ethics and Information Technology* 10(2–3): 123–133.
Johnson, Deborah G., and T.M. Powers. 2005. Computer systems and responsibility: A normative look at technological complexity. In *Ethics and information technology*, vol. 7, 99–107. Dordrecht: Springer.
Larsson, Magnus. 2004. Predicting quality attributes in component-based software systems. PhD thesis, Mälardalen University Press, Sweden. ISBN: 91-88834-33-6.
Latour, Bruno. 1992. Where are the missing masses, sociology of a few mundane artefacts, originally. In *Shaping technology-building society. Studies in sociotechnical change*, ed. Wiebe Bijker and John Law, 225–259. Cambridge, MA: MIT Press. http://www.bruno-latour.fr/articles/1992.html (accessed October 26, 2010).
Lik Mui. 2002. Computational models of trust and reputation: Agents, evolutionary games, and social networks. PhD thesis, MIT. http://groups.csail.mit.edu/medg/ftp/lmui/computational%20models%20of%20trust%20and%20reputation.pdf (accessed October 26, 2010).

Lomi, Alessandro, and Erik Larsen (eds.). 2000. *Simulating organizational societies: Theories, models and ideas*. Cambridge, MA: MIT Press.

Magnani, Lorenzo. 2007. Distributed morality and technological artifacts. In *4th international conference on human being in contemporary philosophy, Volgograd*. http://volgograd2007.goldenideashome.com/2%20Papers/Magnani%20Lorenzo%20p.pdf (accessed October 26 2010).

Marino, Dante, and Guglielmo Tamburrini. 2006. Learning robots and human responsibility. *International Review of Information Ethics* 6: 46–51.

Martin, Mike W., and Ronald Schinzinger. 1996. *Ethics in engineering*. New York: McGraw-Hill.

Matthias, Andreas. 2004. The responsibility gap: Ascribing responsibility for the actions of learning automata. In *Ethics and information technology*, vol. 6, 175–183. Dordrecht: Kluwer Academic Publishers.

Minsky, Marvin. 2006. *The emotion machine: Commonsense thinking, artificial intelligence, and the future of the human mind*. New York: Simon and Shuster.

Montague, Peter. 1998. The precautionary principle. *Rachel's Environment and Health Weekly*, No. 586. http://www.biotech-info.net/rachels_586.html (accessed October 26, 2010).

Moor, James H. 2006. The nature, importance, and difficulty of machine ethics. *IEEE Intelligent Systems* 21(4): 18–21.

Nissenbaum, Helen. 1994. Computing and accountability. In *Communications of the ACM*, vol. 37, 73–80. New York: ACM.

Pancomputationalism. 2009. http://www.idt.mdh.se/personal/gdc/work/Pancomputationalism.mht (accessed October 26, 2010).

Prietula, Michael. 2000. Advice, trust, and gossip among artificial agents, chapter. In *Simulating organizational societies: Theories, models and ideas*, ed. A. Lomi and E. Larsen. Cambridge, MA: MIT Press.

Ramchurn Sarvapali, D., Dong, Huynh, and Nicholas, R. Jennings. 2004. Trust in multi-agent systems. *The Knowledge Engineering Review* 19:1–25. Cambridge: Cambridge University Press.

Roboethics links. http://www.roboethics.org, http://www.scuoladirobotica.it, http://roboethics.stanford.edu, http://ethicalife.dynalias.org/schedule.html, http://www-arts.sssup.it/IEEE_TC_RoboEthics,http://ethicbots.na.infn.it,http://www.capurro.de/lehre_ethicbots.htm.ETHICBOTS seminar by Rafael Capurro http://www.roboethics.org/icra2009/index.php?cmd=program *ICRA2009 Roboethics workshop on IEEE Conference on robotics and automation* (accessed October 26, 2010).

Russell, Stuart, and Peter Norvig. 2003. *Artificial intelligence – a modern approach*. Upper Saddle River: Pearson Education.

Shrader-Frechette, Kristen. 2003. Technology and ethics. In *Philosophy of technology – the technological condition*, ed. R.C. Scharff and V. Dusek, 187–190. Padstow: Blackwell Publishing.

Silver, David A. 2005. Strawsonian defense of corporate moral responsibility. *American Philosophical Quarterly* 42: 279–295.

Siponen, Mikko. 2004. A pragmatic evaluation of the theory of information ethics. *Ethics and Information Technology* 6(4): 279–290.

Sommerville, Ian. 2007. Models for responsibility assignment. In *Responsibility and dependable systems*, ed. G. Dewsbury and J. Dobson. London: Springer. ISBN 1846286255.

Søraker, Johnny H. 2007. The moral status of information and information technologies: A relational theory of moral status. In *Information technology ethics: Cultural perspectives*, ed. S. Hongladarom and C. Ess, 1–19. Hershey: IGI Global.

Stahl, Bernd C. 2004. Information, ethics, and computers: The problem of autonomous moral agents. In *Minds and machines*, vol. 14, 67–83. Dordrecht: Kluwer Academic Publishers.

Stahl, Bernd C. 2006. Responsible computers? A case for ascribing quasi-responsibility to computers independent of personhood or agency. In *Ethics and information technology*, vol. 8, 205–213. Dordrecht: Springer.

Stamatelatos, Michael. 2000. Probabilistic risk assessment: What is it and why is it worth performing it? NASA Office of Safety and Mission Assurance. http://www.hq.nasa.gov/office/codeq/qnews/pra.pdf (accessed October 26, 2010).

Strawson, Peter F. 1974. Freedom and resentment. In *Freedom and resentment and other essays.* London: Methuen.
Veruggio, Gianmarco, and Fiorella Operto. 2008. Roboethics, Ch. 64 in Springer. In *Handbook of robotics*. Berlin/Heidelberg: Springer.
Wallach, Wendell, and Colin Allen. 2009. *Moral machines: Teaching robots right from wrong*. Oxford: Oxford University Press.

Chapter 2
Artificial Agents, Cloud Computing, and Quantum Computing: Applying Floridi's Method of Levels of Abstraction

M.J. Wolf, F.S. Grodzinsky, and K.W. Miller

2.1 Introduction

In his paper "On the Intrinsic Value of Information Objects and the Infosphere," Luciano Floridi asserts that the goal of Information Ethics (IE) "is to fill an 'ethical vacuum' brought to light by the ICT revolution, to paraphrase Moor" (1985). He claims "IE will prove its value only if its applications bear fruit. This is the work that needs to be done in the near future" (Floridi 2002). Our chapter proposes to do part of that work. Initially we focus on Floridi's Method of Levels of Abstraction (LoA). We begin by examining his methodology as it was first developed with J. W. Sanders in "The Method of Abstraction" (Floridi and Sanders 2004) and expanded in "The Method of Levels of Abstraction" (Floridi 2008b). Then we will demonstrate the general applicability and ethical utility of the method of levels of abstraction by considering three different computational paradigms: artificial agents, cloud computing, and quantum computing. In particular, we examine artificial agents as systems that embody the traditional digital computer (modeled as a single Turing machine). This builds on previous work by Floridi and Sanders (2004) and Grodzinsky et al. (2008). New contributions of this chapter include the application

M.J. Wolf (✉)
Department of Mathematics and Computer Science,
Bemidji State University, Bemidji, MN, USA
e-mail: mjwolf@bemidjistate.edu

F.S. Grodzinsky
Department of Computer Science and Information Technology,
Sacred Heart University, Fairfield, CT, USA
e-mail: grodzinskyf@sacredheart.edu

K.W. Miller
Department of Computer Science,
University of Illinois Springfield, Springfield, IL, USA
e-mail: miller.keith@uis.edu

H. Demir (ed.), *Luciano Floridi's Philosophy of Technology: Critical Reflections*, Philosophy of Engineering and Technology 8, DOI 10.1007/978-94-007-4292-5_2,

of the method of levels of abstraction to the developing paradigm of cloud computing and to the nascent paradigm of quantum computing. In all three paradigms, we emphasize aspects that highlight ethical issues.

Our focus throughout is on the levels of abstraction that are most relevant to computing professionals. What are the consequences of each paradigm, and how should computing professionals approach that paradigm to maximize benefits and minimize risks to the public? What virtues of computing professionals are most relevant to each paradigm? And do these paradigms significantly affect the responsibilities associated with the design, implementation and deployment of computing artifacts? As we consider each – artificial agents, cloud computing and quantum computing – we develop multiple LoAs to form a gradient of abstraction (GoA) for each of the systems under consideration. In our final analysis, we tie together the GoAs, observing their similarities and differences.

2.2 Floridi's Theory

The notion of observables is central to the application of Floridi's Method of Levels of Abstraction (Floridi 2008b). Observables are interpreted, typed-variables. A collection of observables forms a level of abstraction. Different collections of observables give rise to different LoAs. A collection of different LoAs that focus on a particular system or feature forms a gradient of abstraction (GoA).

In each of the systems we will apply Floridi's theory by identifying observables and determining the relationships that hold among the observables. We will identify multiple LoAs and compare and assess the corresponding systems. Our assessment focuses on the relevance of Floridi's theory to computing professionals as they consider questions such as: What are the consequences of the type of system? How can computing professionals approach the system to maximize benefits and minimize risks to the public? What virtues of computing professionals are most relevant to this particular system? How does the system affect the responsibilities associated with the design, implementation and deployment of these computing artifacts?

2.2.1 Levels of Abstraction

For Floridi, a LoA qualifies the level at which a system is considered and informs the discussion of such a system. When we analyze a system, we do so from a particular perspective or level of abstraction. This often results in a model or prototype that identifies the system at the "given LoA". Floridi refers to this as the system-level-model-structure scheme: "Thus, introducing an explicit reference to the LoA makes it clear that the model of a system is a function of the available observables, and that it is reasonable to rank different LoAs and to compare and assess the corresponding models" (Floridi 2008b).

When developers understand the particular LoA under which a system is being built, the discussion of the analysis and design of the system and eventually its realization can be more productive. Floridi (2008b) asserts that "[t]he definition of observables is only the first step in studying a system at a given LoA. The second step consists in deciding what relationships hold between the observables." He defines this as the concept of system "behaviour." A *behaviour* of a system, at a given LoA, is defined to consist of a predicate whose free variables are observables at that LoA. The substitutions of values for observables that make the predicate true are called the *system behaviours*. A *moderated LoA* is defined as a LoA together with a behaviour at that LoA.

There can be many LoAs applied to the same system; a helpful distinction is that of a Gradient of Abstractions. "A *Gradient of Abstractions* is a formalism defined to facilitate discussion of discrete systems over a range of LoAs. Whilst a single LoA formalizes the scope or granularity of a single model, a GoA provides a way of varying the LoA in order to make observations at differing levels of abstraction" (Floridi 2008b).

To effectively work with LoA's and GoA's, Floridi has created a Method of Abstraction. The steps of the method consist of: (Floridi 2008b)

- First, specifying the LoA means clarifying, from the outset, the range of questions that (a) can be meaningfully asked and (b) are answerable in principle. Knowing at which LoA the system is being analyzed is indispensable, for it means knowing the scope and limits of the model being developed.
- Second, being explicit about the LoA adopted provides a healthy antidote to ambiguities, equivocations and other fallacies or errors due to level-shifting.
- Third, by stating its LoA, a theory is forced to make explicit and clarify its ontological commitment. The ontological commitment of a theory is best understood by distinguishing between a committing and a committed component. A theory commits itself ontologically by opting for a specific LoA. A theory becomes ontologically committed in full through its model, which is therefore the bearer of the specific commitment.

We have seen that a model is the output of the analysis of a system, developed at some LoA(s), for some purpose. So a theory of a system comprises at least three components:

(i) an LoA, which determines the range of available observables and allows the theory to investigate the system under analysis;
(ii) an elaboration of the ensuing model of that system
(iii) the identification of a structure of the system at the given LoA.

2.3 Artificial Agents

In an earlier paper, we used the Method of Abstraction to analyze the ethics of designing artificial agents (Grodzinsky et al. 2008). In that paper we identified two different levels of abstraction, LoA1, which refers to a user's view of what is often called an "autonomous system," and LoA2, which refers to the designer's view of that same system. We extend those notions in this paper to refer generally to the user's view and to the designer's view of each system under consideration. That is, LoA1 is a set of observables available to a user of a system and LoA2 is a set of observables available to the designer of a system.

In that paper, we focused on LoA2 and described a model of computation whereby artificial agents could exhibit traits that at LoA1 appeared similar to, if not indistinguishable from, human traits we call learning and intentionality. This exploration of the interaction between these two LoAs demonstrated that if the designer failed to consider an expansive enough set of observables at LoA1 to be given consideration at LoA2, the designer might miss certain ethical responsibilities that arise at LoA1. If the designer is focused on low-level observables (LoA2) such as the changing of the value of a variable or the changing of the sequence of operations carried out by the artificial agent, the designer may well get the code for the agent "right." However, the observables properly associated with LoA1 take on new importance when the designer is producing an artificial agent that appears to be learning or demonstrating intentionality. We demonstrated that these sorts of agents are more prone to unpredictable future behaviors and are capable of emergent behaviors not initially programmed by the developer. Thus, we concluded that a designer of artificial agents is under an increased burden of care. That burden requires a thorough examination of observables at LoA1 and their implications. Once those are understood, the designer must consider the GoA, the interface between LoA2 and LoA1 and design the system (an LoA2 endeavor) in such a way as to minimize the risk of undesirable behaviors at LoA1.

In this paper, we are still interested in LoA1 and LoA2, but we also introduce a third Level of Abstraction: LoAS, where the "S" stands for "society." LoAS is the set of observables available to an observer of society. This set of observables consists of those social structures and relationships that are prevalent in the functioning of an information society. At LoAS, observables include a set of variables that describes the characteristics of entities that are or could be affected by a piece of software: descriptive observables concerning individuals, businesses, and governments are all possible members of the set. Questions that might be addressed could be, e.g., If individuals are among the buyers, is there a particular demographic that dominates the buyers? The set of observables at LoAS might be available to a user of a software system. It might be, however, that some observers at LoAS will have access to certain research that is not typically available to a user or even a designer of software systems. It might be that LoAS observables are largely disjoint from the observables typically considered at LoA1 and LoA2.

Our ethical analysis at LoAS focuses on the changes in the users from using the software, and on the changes in others because of the existence of the software in society. Our observations at LoAS are concerned with identifying not only the changes in individuals, but also the cumulative effect of these changes to larger groups and organizations, effects that may be attributable to the software, or to the software combined with other sociotechnical factors (Johnson and Miller 2009).

One particular GoA of interest is the combination of LoA2 and LoAS. The designer might be looking at the demographic in deciding who the users of the system are and their values (see work on Value Sensitive Design, such as Friedman 1996). For example, in designing e-voting software, the developers had to consider the user interface for able-bodied users, and for those with disabilities due to infirmities and age. In the state of Connecticut in the United States, for example, several interfaces were tested at several sites, to see what potential users actually preferred. The secretary of state contracted with a team of University of Connecticut engineering faculty "to provide advice to the state regarding new voting technology and to assist in the certification and acceptance testing of the AccuVote Optical Scan voting machines…" (UCONN 2010). This team conducted pre-election and post-election audits of the memory cards used in the machines. Once these cards are programmed the integrity of the vote falls upon the precinct polling personnel (LoAS). Misinterpretation of instructions, failure to conduct pre-election tests, inadequate training of precinct personnel all led to problems that were unlikely to have been anticipated at LoA1 or LoA2. Concerned with fair voting practices, Connecticut is using several safeguards to verify the accuracy of the election outcomes. In another example, developers of social networking sites like Facebook and Twitter did not accurately predict the impact of these products on the communication habits of the users when the products were launched.

LoAS can frame earlier work by social psychologists, sociologists, and society and technology scholars. In the early 1990s Chuck Huff developed a social impact statement for software developers based on an idea of Ben Schneiderman's. Huff encouraged software designers to "find out the social impact of the systems they design in time to incorporate changes in those systems as they are built" (1996). Cast in our terms, Huff was encouraging developers to consider LoAS as they manipulated a program at LoA2.

The Embedded Values approach of Friedman and Nissenbaum concerned itself with the ways in which biases emerge in computer systems. These authors examined preexisting biases of the individual or organization, technical biases and emergent biases which arise when "the social contexts in which the system is used is not the one intended by its designers." For example, an ATM that relies heavily on written instructions may be deployed in a neighborhood with an illiterate population (Friedman and Nissenbaum 1996). If designers are aware of biases (at LoA2) that have significant impacts at LoA1 and LoAS, they can use that awareness to design systems that avoid problems. An analysis that incorporates LoAS could be an effective method for managing emergent biases.

In his piece Moral Methodology and Information Technology, Jeroen van den Hoven states, "We need to give computers and software their place in our moral world. We need to look at the effects they have on people, how they constrain and enable us, how they change our experiences, and how they shape our thinking" (2008:50). He asserts that,

> We are now entering a third phase in the development of IT, where the needs of human users, the values of citizens and patients and some of our social questions are considered in their own right and are driving IT, and are no longer seen as mere constraints on the successful implementation of technology (van den Hoven 2008:60).

One theorist who has embraced this concept is Phillip Brey. Brey's Disclosive Ethics reveals embedded values in IT systems (2010). His theory concerns itself with the question: "Is it possible to do an ethical study of computer systems themselves independently of their use by human beings?" (Brey 2010). His answer is basically no. He espouses Disclosive Ethics as a method in which different parties responsible for the design, adoption, use and regulation of computer technology share responsibility for the moral consequences of using it, and in which the technology itself is made part of the equation (Brey 2010:53). The GoA of LoA1, LoA2 and LoAS suggests a formalism that could address Brey's concerns.

We contend that the addition of LoAS to the method of levels of abstraction is consistent with Floridi's desire to formulate "an ethical framework that can treat the Infosphere as a new environment worth the moral attention and care of the human inforgs inhabiting it" (Floridi 2010:19). LoAS consolidates the concerns of those working on embedding values in design and those concerned with the effect of technology on society. It expands Floridi's method beyond the levels of designer and user and includes society in the mix.

A plausible criticism of using LoAS is that LoAS adds nothing to the existing work described above, and merely muddies the water with new (superfluous) terminology. We disagree. Our contention is that the idea of LoAS is a concept that unifies, rather than obscures, the underlying commonality in the existing work of Huff, Friedman, Nissenbaum, van den Hoven, Brey, and others. The similarities in their work derive, at least in part, from the high level of abstraction (as compared to LoA1 and LoA2) at which they work. Their different emphases can be seen as a consequence of their different choices of observables at LoAS.

In addition to providing a framework for better understanding existing work at the sociotechnical level, LoAS also helps integrate work on technology and society at the different levels LoAS, LoA1 and LoA2. When ethical analysis at these three different levels is perceived as being in competition, or at odds with each other, unnecessary conflicts can arise. If work at these different levels is seen as similar analyses, recognizably using the same fundamental concepts, but using different observables, we are convinced that a more effective coherence can be perceived and refined. This theoretical coherence could, and we hope will, lead to practical negotiations and agreements between academics and practitioners who will be better able to understand, together, the important differences and similarities at LoAS, LoA1, and LoA2. In the next sections, we explore how these levels of abstraction can be used in concert to examine carefully the ethical significance of three computing paradigms.

2.4 Artificial Agents and Mapping Table Processing

Floridi and Sanders originally presented a notion of a transition system to describe the internal actions of an agent (2004). In Grodzinsky et al. we developed a more detailed description of the transition system and made important distinctions regarding the burden of care borne by those who design artificial agents (2008). We include a brief description of it here to give readers a sense of how the concept of theoretical computational machines complements Floridi's notions of LoA, and how technical details of an artificial agent's implementation can have significant impacts for LoAS. Readers are referred to the original work for a more detailed presentation.

Our model closely follows the Turing Machine model of computation and includes a large mapping table with a mechanism for mapping inputs and the current state to a next state and output values. In any practical situation, the mapping table is prohibitively large, though finite. The table is a model for the programming (and therefore the design) of the agent. We explored two variations of the model in which the agent had the ability to modify part of its mapping table. In the first, the agent can modify any part of the table that defines the intelligent agent's behavior during its execution; in other words, the agent can self-modify. In this variation, the agent can add new entries to the table, delete entries from the table and modify entries that exist in the table. Its execution proceeds as in the original case, except when the table fails to contain a valid mapping. In this case, the agent is forced to stop. An agent with a table with this variation (called "fully modifiable") has enough power to render itself useless by introducing changes that force it into a state for which the table contains no mapping. Note that it is also possible for an agent with such a table to add an entry to the table that would duplicate an existing entry except with different outputs or a different next state. A table with multiple identical entries except for the next state seemingly exhibits nondeterminism,[1] since the same input/state would have two different output/state mappings. Although the steps outlined above are deterministic, the choice of which of the two mappings might indeed be arbitrary since the possibility of multiple mappings are not explicitly dealt with in the fundamental behavior of the agent.

In the second variation (called "modifiable"), the mapping table is divided into two parts: in one part, the mappings can be modified; in the other part, the mappings cannot be modified. In other words, some parts of the mapping table are protected from self-modification by the agent. Since the mapping table governs the entire operation of the agent, the designer may wish to prevent the artificial agent from carrying out certain modifications like the one mentioned above. Thus, the designer may opt to protect the entries that govern self-modification from self-modification. While this idea has a certain appeal, especially from the perspective of designing the mapping table in such a way that a modifiable agent always behaves properly, we showed that the modifiable variation can readily promote itself to a fully modifiable machine.

This argument suggests that there is no absolute distinction between a modifiable agent and a fully modifiable agent. However, the two models give different perspectives on the ethical intentions of the designer, even if the designer's intentions may eventually be thwarted.

[1] Note that we are referring to computational nondeterminism. Computational nondeterminism is a theoretical construct that allows a device to be in two or more states simultaneously, with each state experiencing independent sets of inputs and producing independent sets of outputs. This notion is not to be confused with the philosophical notion of nondeterminism.

We can expand on our earlier work, which emphasized LoA1 and LoA2 for artificial agents, by examining LoAS for artificial agents. A natural question at LoAS is about the consequences on society as artificial agents become increasingly common; one possible and ethically significant consequence is that many jobs previously held by people will be done by artificial agents. The use of machines to replace human employees is nothing new, but the sophistication of modern artificial agents may result in people being displaced in jobs that used to be considered immune from automation. At LoAS, we could examine observables such as the number of artificial agents deployed in, for example, care of the elderly or as receptionists; then we could examine the number of humans in jobs in this same area. Next, we could see if people displaced from jobs in these areas found work elsewhere. Finally, we could examine what groups of people had gained from the increased use of artificial agents (perhaps corporations and employers), and what groups of people had lost from that same use (perhaps former employees) (Grodzinsky et al. 2009). Another LoAS consideration might include benefits and risks from the increased use of artificial agents; in health care, for example, an observable might be accidental deaths among the elderly. Perhaps the use of artificial agents to care for the elderly would, overall, reduce such deaths; perhaps not.

Paying attention to LoAS observables during and after artificial agents are designed, developed and deployed should help computer professionals build artificial agents that are more likely to benefit people, and less likely to harm them. At LoA1, the people who directly use a computing artifact are directly in focus; at LoAS, people who are affected by, but do not directly use, an artifact are also in focus.

2.5 Cloud Computing

Cloud computing is a computational paradigm that has been of concern in recent years because of its apparent differences with a more traditional desktop paradigm. In cloud computing, many of the complexities of computing applications and infrastructure are hidden behind abstractions afforded by the Internet. Information system developers already use this level of abstraction to create virtual local area networks and virtual servers which abstract the complexities of the underlying computer network. In cloud computing, we take this a step further and use the cloud to make "an entire data-center's worth of servers, networking devices, systems management, security, storage and other infrastructure, look like a single computer, or even a single screen" (Fogarty 2009). One concept that has arisen with the advent of cloud computing is "software as service" (SaS). Rather than run software that is present on the computer under the direct local control of the user, SaS requires the user to submit his/her data to the owner of the SaS system; the SaS system carries out the computation using hardware and software that is completely owned by the vendor and then returns the result to the user. Issues of legacy applications, interoperability and accessibility all affect the user of SaS who may or may not be aware of what is happening to his/her data at all times.

The paradigm for computation that most users have experienced since the advent of the personal computer includes the user as owner of the hardware and software that is holding and manipulating the user's data. Typically at this level, the software itself is opaque to users, except for software where the source code is freely available, e.g. free software and open source software. The user is familiar with his or her data and its meaning, and by the locality of the media, controls access to the data – at least to a first approximation. The user decides which software to install on the computer, which programs get access to which data files and how long they get that access.

The cloud computing paradigm brings a different set of access and control features. For example, it is quite possible that the data is no longer stored on hardware owned by the user, but stored "in the cloud." Both Facebook and Google docs are early examples of this kind of service, and now many other providers have entered, or are planning to enter, the market. Another distinction with cloud computing is that the software that manipulates the data is not necessarily present on the same device that is used to access or compute the data. Instead, the user submits data to a software service, the service carries out the computation on its hardware with its software and returns the result to the user. A search executed by a commercial search engine is a common example of this protocol. Only the search query and the search results are ever local to the user; the algorithms and data necessary to carry out the search are owned by the search engine company, and are located on its servers. Google is a prominent example of this kind of "in the cloud" service. To the user, the observables are the same: click on an icon, the program runs and a result is produced. Thus at LoA1, the user may not even be in a position to distinguish SaS from a traditional program. At LoA2, there are significantly more observables in the cloud computing paradigm, visible to a developer but not to a user. Everything from web addresses to the type of compression make a difference at the designer's level of abstraction.

There are also regulatory issues and control issues that impact the user who may or may not have the means of supplying the supporting data if it is stored in the cloud. An analysis of cloud computing at the LoAS level would include issues of trust between cloud providers and customers; issues of control, security and confidentiality, standardization attempts, and consequences of the outcomes. In each of these issues, knowledge of hardware and software are not sufficient; instead, people, institutions and events would have to be taken into account. While these issues do involve technical details at LoA2, they are driven by human values that may be reflected in observables at the LoAS level; thus, forming a GoA is helpful.

Observables at LoAS can be used to gather empirical data useful in making an ethical analysis. For example, it was initially thought that cloud computing could be used to reduce overall energy consumption, but some scholars now dispute that claim (Berl et al. 2010). The data necessary to test claims about energy consumption and the cloud would be available in LoAS observables.

We briefly consider Facebook's role as a cloud-based document storage service provider as an example. Among other things, Facebook stores users' pictures. In the typical flow of operations, the user has complete control over who gets to see the photos and how long the photos remain with Facebook. Facebook has a fiduciary relationship with the user in which it agrees to show the photos to only those people the user has identified and to delete the photos when the user asks for them to be deleted. Of course, there are no assurances that Facebook complies with these sorts of requests. This is especially true for any backup copies of the photos that Facebook may have made to maintain a high quality of service level.

Facebook's recent difficulties with users unhappy about its policies, illustrates that social forces can influence technical decisions (either proactively or retroactively). Issues of privacy and confidentiality will be played out on the LoAS level as cloud computing becomes an increasingly competitive marketplace. It may be that ethical behavior and good business will coincide when users gravitate to vendors that treat their users with respect. People's trust for cloud computing (LoA1 and LoAS) will be affected by whether cloud computing providers are trustworthy stewards of users' data. Users will have to trust cloud providers in order to be comfortable giving up a large measure of control over their data and processes and should choose vendors wisely. Therefore, as with artificial agents, we contend that computing professionals (at LoA2) should pay careful attention to LoAS observables as part of the development process.

Cloud computing, more specifically SaS, presents a potential ethical impact on the Free and Open Source Software (FOSS) communities. When the Free Software Foundation developed version three of the GNU General Public License (GPLv3) there was controversy surrounding provisions dealing with SaS. As a result of those controversies the provisions addressing SaS were removed from GPLv3 and included in a second, companion license, the Affero General Public License (AGPL). Our analysis of GPLv3 and the AGPL identified a piece of software where the observables at LoA2 were the same, yet the observables at LoAS were very different and had a different social impact (Wolf et al. 2009). Depending on how the software was deployed, the developer was under different legal obligations regarding the release of modified source code. That is, in one scenario, the developer was required by the AGPL either to share the modifications with the community, or in another, seemingly ethically equivalent scenario, the developer was under no legal obligation to share. Our interest here, however, is in showing how analysis of LoAS raises the question of the impact that SaS will have on the sharing ethic that is prevalent in FOSS communities.

2.6 Quantum Computing

When considering quantum information, there are two different aspects that are of importance. One is the notion of quantum computation and the other is the notion of quantum information transfer. As a practical matter, both are currently feasible. A quantum computer that factors an integer has been built. That integer is 15 (Blatt 2005:244). ID Quantique offers a quantum computer that uses quantum principles to generate truly random numbers (rather than pseudo-random numbers common in classical computers) for 1,000–2,500€. The same company offers a quantum computer embedded in a quantum networking device – a product that implements secure classical information transfer using both quantum and classical means (see: http://www.idquantique.com).

The quantum network device is an important example, since it uses a combination of quantum techniques and classical (non-quantum) encryption algorithms to transmit secret data. Researchers and others have made claims such as: "Quantum cryptography makes an absolutely safe communication possible for the first time" (Weinfurter 2005:166). The physics of the quantum mechanics ensure that should an eavesdropper "listen in" on the communication, both the sender and the receiver will know that the communication has been intercepted. Yet, European researchers have recently demonstrated that with easily obtainable components they can remotely control a key component of the system and obtain the data in the communication and remain undetected (Lyderson et al. 2010). Clearly, there are ethical problems lurking in the development of, and understanding of, quantum computing.

General quantum computing and quantum teleportation are in the research stage and barring an unexpected breakthrough will not be used or available in a general sense for quite some time. However, much is known about the nature of quantum information and fundamental quantum computation techniques, giving us the opportunity to begin exploration of ethical issues that are emerging along with this model of computation. As in the previous sections we will draw attention to the three different levels of abstraction. However, our main focus will be at LoA2 and how, as in the artificial agent case, quantum developers will carry an increased burden of care. We will find that due to the nature of quantum computation, "quantum developers" seems to include a broader range of people than we normally consider in the development of traditional computing applications.

Next we give an overview of some of the fundamentals of quantum-information processing and transfer. We are especially concerned with two distinctions that make the quantum case different from the classical case: superposition and entanglement. We will then look at three applications of quantum techniques: factoring, searching and cryptography. Once these ideas are presented we will consider the impact these distinctions have on LoA2, and in particular quantum developers. We will conclude this section with an analysis of quantum computation's impact at LoAS. We anticipate that fundamental differences between quantum and classical computation will raise significant ethical issues when users routinely access machines based on quantum computing.

2.6.1 Distinguishing Quantum and Classical Approaches to Computation

Perhaps the most striking difference between classical computation and quantum computation is the way that information is conceived. In classical computation, the smallest piece of information is the bit – either a 0 or a 1 – and it is given a physical realization. Once the bit is given a physical realization, it can be read again and again and it should always yield the same information. Quantum information on the other hand, is stored in a superposition of classical states. That is, to a first approximation, a single quantum bit (qubit) in a given physical realization is in a probabilistic state where with probability p it is 0 and probability of $1 - p$ it is 1. Quantum computation typically proceeds by repeatedly refining the probabilities of qubits until the probability that they contain the correct answer to a given computation crosses a given threshold, a threshold that is arbitrarily close to 1, but never exactly 1.

Qubits possess another property that distinguishes them from classical bits. Once a qubit is read, the superposition is destroyed and the qubit reverts to being a classical bit. "Read" needs to be broadly interpreted here. It can mean the usual sense of reading a bit. But it can also mean the qubit interacts with any particle or photon that is not part of the intended computation. In such a case, a qubit exhibits the same behavior of having its superposition destroyed.

This destruction of superposition has an important consequence: it is impossible to clone or make a copy of a qubit. The "copy" operation is an important part of classical computation. Yet, it is impossible to copy a qubit (Werner 2005:176). The proof of this statement goes even further. Not only is impossible to copy a qubit, it is impossible to translate all of the quantum information contained in a qubit into classical information. It is impossible to classically describe the state of a qubit in a complete and accurate way. Thus, developers of software that require multiple copies of a particular qubit must anticipate that need, so that multiple qubits can be set up in the same way and have the same computation applied to all of them to prepare them. They can then be used as if they are true copies of each other. However, their quantum nature ensures that they are clearly not.

Closely related to the inability to copy quantum systems, is the inability of a quantum system to carry its identity with it. "We cannot mark a quantum system and then recognize it again" (Esfeld 2005:278). This inability provides challenges to those who develop algorithms for quantum computation systems.

Another important distinction between qubits and classical bits is that qubits can be entangled. More correctly, two quantum systems carrying quantum information can be entangled. That is, they form a single quantum system where all that can be described externally is the state of the relationship between the two quantum systems. The states of the individual subsystems are not knowable without destroying the entanglement. Only the relationship between the two entangled systems is knowable. Thus, in an entangled system, there is no information that is intrinsic to the subsystems therein. When a subsystem in such an entangled quantum system is measured (or read), the entanglement is lost and both the subsystems contained therein lose their superposition and revert to classical information.

We use some standard notation to form an example:

$$(|00\rangle + |11\rangle) / \sqrt{2}$$

describes an entangled quantum system in the following state:

> There are two entangled quantum systems such that should the system decohere, both of the bits will be identical and with equal probability, they will be 0 or 1.

Note that the state implicitly includes the description that under no decoherence scenario will the two subsystems register different bits. Entanglement introduces two additional properties that are important for quantum computation and quantum information transfer systems and challenge usual assumptions about information and computation. The first is that locality of information is no longer required. Once two quantum systems are entangled to form a single quantum system, there is no requirement that they be kept in close physical proximity. Thus, from the notation example, the two subsystems can be separated, one of the two subsystems can then be measured, and without measuring the other, its state can be known with certainty.

Researchers on quantum teleportation systems have recently separated entangled photon pairs 16 km (Jin et al. 2010). These photons were sent through free space, rather than a fiber optics cable. While the current research is obviously experimental, our point is to demonstrate that locality of quantum systems should not be assumed in consideration of ethical concerns.

The final property to consider is that is possible to entangle multiple quantum systems into a single quantum system. Under certain conditions, measurement of a single subsystem can result in either complete decoherence or partial decoherence. Roos et al. entangle three quantum systems in two different ways (2004). Using the notation above, they are:

$$(|000\rangle + |111\rangle)/\sqrt{2}$$

and

$$(|110\rangle + |101\rangle + |011\rangle)/\sqrt{3}.$$

The first entangled system can be described as:

> There are three entangled quantum systems such that if the system decoheres, all of the bits will be identical and with equal probability, they will be 0 or 1.

Note that all of the entanglements and superpositions are lost when any one of the bits is read. The second one is different. Say that the first bit is read, if it is a 1, then the second and third bits retain their coherence in the state $(|01\rangle + |10\rangle)/\sqrt{2}$. If the first bit is a 0, the second and third bits still retain their coherence, but in the state $|11\rangle$. Note that these experiments demonstrate similar behavior when reading any of the three bits in the second entangled system.

2.6.2 *Quantum Approaches*

Two of the more well-known quantum algorithms are for the factoring problem (given an integer, find its prime factors) and the database searching problem. In addition to their interesting technical attributes, these algorithms demonstrate that practical implementations of "quantum computation" are really a combination of classical computation and quantum computation. Quantum algorithms typically involve some initial classical computation, preparation of the quantum register where the quantum information is stored, quantum computation, decoherence (reading) of the quantum register, evaluation of the results. Notice that it is trivial to see if a factoring process has yielded the correct answer: multiply the proposed factors and see if the product matches the integer to be factored. If the results are incorrect (say the quantum register reports that 12 and 13 are the prime factors of 143), the process repeats itself. If the results are correct (11 and 13, 11*13=143), the results are reported and the computation ends. When quantum algorithms are implemented, the system consists of a classical computer with a quantum subprocessor. The quantum part of the computation requires complete coherence of the quantum subprocessor. During that time, the computation is completely reversible and none of the information in the quantum register is or can be revealed to any system outside of the quantum subprocessor (Blatt 2005:239). It is only when the quantum subprocessor has completed its run that reading can take place.

There is currently no known way to factor integers using a classical computer that works efficiently for large integers, say fifty or more digits. For each additional digit in the integer, the amount of time it takes to determine its factors on a classical computer doubles. This problem is of particular importance in cryptographic systems as their efficacy relies on the assumption that integers cannot be factored efficiently. The factoring problem has been studied by mathematicians and computer scientists for many years using myriad techniques and relatively little progress has been made on producing an efficient algorithm for factoring (or on showing that no such algorithm exists, for that matter).

Perhaps the biggest breakthrough on the problem occurred in 1994 when Shor developed a quantum algorithm for the factoring problem (1994). Shor used a variety of mathematical techniques to transform the factoring problem into a different problem for which quantum computers are well-suited. The quantum computation can actually fail and produce wildly wrong results for the problem, yet Shor's algorithm provably detects these situations and correctly resolves them. With Shor's quantum algorithm, it is possible to factor integers efficiently.

Grover's quantum algorithm for the database searching problem is another potentially significant development (1997). Given an unsorted database of items with unique keys and a search key, Grover's algorithm retrieves the item from the database whose key matches the search key. During this algorithm's quantum computation, it too reaches a correct answer, but again, probabilistically. The algorithm accommodates the possibility of an incorrect response and easily verifies any result that it gives is indeed correct.

As suggested earlier, both of these algorithms are of little practical use today outside of experimental research set ups. For the algorithms to be useful in a general sense, we need the physical realization of quantum registers consisting of more than a few bits. However, quantum registers with just a few bits are quite useful in the world today as part of a secure information transfer systems. Secure information transfer today relies on cryptography: any message to be sent from Alice to Bob is encrypted by Alice, transmitted, and then decrypted by Bob. In order for this process to work, Bob needs to have the decryption key. Thus, Alice needs to send a key to Bob via a secure medium. Obviously, the key cannot be encrypted, otherwise Alice would need to send Bob a key to decrypt the encrypted key.

Quantum cryptography, or more properly, Quantum Key Distribution, solves this problem in a secure way. Using superposition of photons, Alice transmits the key to Bob. Through unsecured communication Bob and Alice agree on a key based on the photons Alice sent. Quantum properties of the photons ensure that if the photons are intercepted, both Alice and Bob know that the key has been compromised. Once a key is agreed upon, Alice uses the key to encrypt the data and then transmits the encrypted data to Bob. Bob can then decrypt the data. The actual transmission is secure. But, as we will note in the next section, this does not mean that it is impossible for an eavesdropper to intercept the message without being detected.

Quantum approaches to computation and information transmission can take advantage of both superposition and entanglement. Superposition and entanglement are the resources that leads to new possibilities for data transmission and the speed-ups found in quantum algorithms (Werner 2005:183).

2.6.3 Ethical Concerns

During its research stage, quantum computing has already begun to bring to light some ethical concerns. If quantum computing becomes a common, practical technology, we expect significant ethical issues will arise. Although there are important practical speed and efficiency advantages to quantum computing, qubits by their very nature do not register information in the same way that conventional digital memories do, thus challenging some of the most fundamental assumptions of Information Ethics. If quantum computing is to become practical for most users (many researchers believe that it will eventually), it seems likely that the probabilistic nature of quantum memory and computation will be hidden from users. Further, users will likely not even know when the quantum subprocessor has been used to determine a result. Seen another way, when most users enter input, they are not going to include a probability threshold to be used to determine whether an output is correct. Users will want to assume confidently that the output is correct; it will be left to those developing and implementing quantum algorithms to determine the level to set the threshold for correctness. Only someone with an LoA that includes at least some knowledge about the intricacies of quantum computing can make an informed ethical choice about picking the threshold (LoA2). There is power and responsibility in that choice. The autonomy of users at LoA1 is impinged upon when they receive output without knowing about the inherently probabilistic nature of quantum computation.

For some algorithms (such as factoring, described above), a conventional program can easily and efficiently check to see if the quantum algorithm has delivered a correct answer. However, many useful applications of quantum computing in which such "post quantum checking" will not be practical. These applications include functional versions of NP-complete problems such as the Traveling Salesman Problem. In these cases, the setting of a threshold will carry important ethical weight.

Another concern at LoA2 has to do with claims made by those developing quantum-based devices. Statements such as, "Quantum cryptography makes an absolutely safe communication possible for the first time" (Weinfurter 2005) have the potential to be misleading. While it is true that the laws of quantum physics ensure that the communication of the quantum key cannot be eavesdropped, "absolutely safe communication" has at least two additional requirements. First, the implementation of the system must be done correctly. As Lyderson's team has shown, currently available quantum cryptography has an exploitable implementation flaw (2010). With these systems, it is possible for an eavesdropper to take control of detectors in two different commercial quantum cryptography systems. Second, absolutely safe communication requires the nature of computation to remain unchanged. We argue this point more thoroughly later in this section.

In addition to the concerns about the user's ability to retain autonomy at LoA1, we have concerns about the impact that quantum computation will have on LoAS. Right now, quantum computation is in its infancy, just as classical computation is maturing. An attribute of that maturation process is the relatively stable role computers play in the lives of many people – at home, work and school. People have begun to expect, perhaps unconsciously, that the computer behaves in a particular way. Part of that expectation is based on the way that software is developed. That is, the enterprise of developing software has matured to a point where there are well known guidelines and techniques that help ensure better-developed software. There is much controversy about which techniques are the best and about whether sufficient reliability is typically achieved. However, many scholars agree that significantly effective techniques are available, even if many developers choose to ignore them (For example, see Parnas 2009). Regardless of the particular software engineering technique used, debugging software is part of the software development process.

Quantum computation is fundamentally different from traditional computing in at least two ways: the impossibility of making copies of quantum objects, and the inability to inspect quantum objects part way through a computation. Because of these two important differences, debugging quantum software will be profoundly affected. This calls for special ethical care for those implementing quantum software.

A commonly used debugging techniques is one in which software is stopped mid-execution and the values stored in registers are inspected and, if need be, changed, to better understand the reason the software is not producing the expected results. This technique cannot be used to debug quantum software because any inspection of a quantum register causes it to decohere. A simple variant of this technique that entails making a copy of the quantum register is similarly foiled due to the physical impossibility of making copies of quantum objects. The aforementioned possibility of using multiple quantum registers that are prepared in the same way may prove to be effective. However, the probabilistic nature of quantum computation should serve as a constant reminder to the debugger of quantum software that no two of the quantum registers can be assumed to be identical as is so often done (and rightly so) with the information stored in classical registers.

The message here is clear. In order to avoid the sorts of ethical concerns that arise from the release of poorly designed and tested classical software, we need to develop methods of designing, implementing, testing and debugging that are appropriate for quantum software. The old ways are insufficient to meet high ethical standards. In a rush to exploit the considerable advantages of quantum computing, it is vital that computing professionals insist on sufficiently mature quality control of quantum applications before they are deployed in situations that might be dangerous for the public.

Looking at quantum computing from LoAS, it seems reasonable to assume that once quantum computers are widely available, the existing classical computers will not all be replaced at the same moment. It is during this transition time that the development of quantum computers has the potential to have a profound impact on all of society. For example, almost all security on the Internet today is provided by some cryptographic means. Today's cryptographic methods rely on the assumption that factoring is computationally difficult (the amount of time it takes to factor an integer doubles with the addition of each digit). As Shor's algorithm demonstrates, a quantum computer can efficiently factor large integers. Anyone who possesses a quantum computer with a sufficiently large quantum register will have the means to break any cryptographic code that relies on today's most popular cryptographic techniques. Thus, even though two parties may engage in a quantum key exchange, an eavesdropper who has access to a quantum computer and intercepts an encrypted message will be in a position to decrypt that message quickly without the having knowledge of the decryption key.

While it may be possible that new secure communication techniques will be developed for communication between parties who possess quantum computers, we are concerned about a "have's" and "have not's" situation developing. In a communication scenario where quantum computers are available to some but not to others, those with sufficient funding to obtain a quantum computer will be able to ensure their secure communication, while the rest will not. Not only will these entities be in a position to ensure their secure communication, they will also have the computational ability to decode any communication they intercept. This clearly creates a power imbalance. Furthermore, this threatens to disrupt much of the e-commerce system currently in place.

Thus, should large quantum computers be developed, the first to develop them will be in a position of power over those who do not. Since it is impossible to determine the virtues of the first developer *a priori*, a prudent practical and ethical argument can be made for the development of systems that will ensure secure private communication between parties that do not possess quantum computers, when there are those in the world who do possess them.

In addition to the ethical challenges inherent in quantum computing, we also contend that quantum computing presents a fundamental theoretical challenge to

Floridi's Information Ethics. Floridi and Sanders discuss the nature of an act by an actor *a* to a patient *p*:

> Evil action = one or more negative messages, initiated by *a*, that brings about a transformation of states that (can) damage *p*'s welfare severely and unnecessarily; or more briefly, any patient unfriendly message (Floridi and Sanders 2001:57).

It is important for our purposes at LoAS to note that the patient *p* in Floridi and Sanders' formulation may be human, biological but not human, or artificial.

We contend that this definition of evil means that the probabilistic nature of quantum computing may be considered fundamentally evil, or at least not entirely commendable. Quantum computing introduces an inherent uncertainty. Such uncertainty can sometimes be managed (as in Shor's quantum factoring algorithm), but that does not remove the objection that quantum computing is, at its core, less certain that traditional computing. If less certain, then it can be argued, it is less good in Information Ethics (Floridi 2005).

2.7 Conclusions

In this chapter we have approached three cases using Floridi's Method of Levels of Abstraction. It is clear to us that this method offers a usable framework in the analysis and development of software applications. The addition of LoAS provides us with an added dimension which addresses the direct and indirect effects of software on society. The three levels that we have chosen to define, LoA1, LoA2 and LoAS, are clearly applicable to Artificial Agents and the emerging paradigm of Cloud Computing. The use of the method with Quantum Computing demonstrates its effectiveness even with nascent notions of computing. The challenge for Quantum Computing developers is to find a way to address the ethical concerns that the intrinsic nature of quantum computing presents at all three levels of abstraction, and the challenge to IE theorists is to address how or if quantum applications fit into their conception of the Infosphere and IE. We have begun part of that work; there is much more to be done.

References

Berl, A., E. Gelenbe, M. Di Girolamo, G. Giuliani, H. De Meer, M. Dang, and K. Pentikousis. 2010. Energy-efficient cloud computing. *The Computer Journal* 53(7): 1045–1051.

Blatt, R. 2005. Quantum information processing: Dream and realization. In *Entangled world: The fascination of quantum information and computation*, ed. J. Audretsch, 235–270. Weinheim: Wiley-VCH.

Brey, P. 2010. Values in technology and disclosive computer ethics. In *The Cambridge handbook of information and computer ethics*, ed. L. Floridi, 41–58. Cambridge: Cambridge University Press.

Esfeld, M. 2005. Quantum theory: A challenge for philosophy! In *Entangled world: The fascination of quantum information and computation*, ed. J. Audretsch, 271–296. Weinheim: Wiley-VCH.

Floridi, L. 2002. On the intrinsic value of information objects and the infosphere. *Ethics and Information Technology* 4(4): 287–304. doi:10.1023/A:1021342422699.
Floridi, L. 2005. Information ethics, its nature and scope. *Computers and Society* 35(2): 3. June 2005.
Floridi, L. 2008a. Foundations of information ethics. In *The handbook of information and computer ethics*, ed. K. Himma and H. Tavani, 3–23. Hoboken: Wiley.
Floridi, L. 2008b. The method of levels of abstraction. *Minds and Machines* 18: 303–329. doi:10.0007/s11023-008-9113-7.
Floridi, L. 2010. Ethics after the information revolution. In *The Cambridge handbook of information and computer ethics*, ed. L. Floridi, 3–19. Cambridge: Cambridge University Press.
Floridi, L., and J.W. Sanders. 2001. Artificial evil and the foundation of computer ethics. *Ethics and Information Technology* 3: 55–66.
Floridi, L., and J.W. Sanders. 2004. On the morality of artificial agents. *Minds and Machines* 14(3): 349–379.
Fogarty, K. 2009. Cloud computing definitions and solutions. http://www.cio.com/article/501814/Cloud_Computing_Definitions_and_Solutions?page=1&taxonomyId=3024. Accessed June, 2010.
Friedman, B. 1996. Value-sensitive design. *Interactions* 3(6): 16–23.
Friedman, B., and H. Nissenbaum. 1996. Bias in computer systems. *ACM Transactions on Computer Systems* 14(3): 335.
Grodzinsky, F.S., K.W. Miller, and M.J. Wolf. 2008. The ethics of designing artificial agents. *Journal of Ethics and Information Technology* 10(2–3): 115–121. doi:10.1007/s10676-008-9163-9.
Grodzinsky, F.S., K.W. Miller, and M.J. Wolf. 2009. *Why turing shouldn't have to guess*. Asia-Pacific Computing and Philosophy Conference, Tokyo, October 1–2, 2009.
Grover, L. 1997. Quantum mechanics helps in searching for a needle in a haystack. *Phys Rev Let* 79(2): 325–328.
Huff, Chuck. 1996. About social impact statements. http://www.stolaf.edu/people/huff/prose/SIS.html. Accessed September, 2010.
Jin, X., J. Ren, B. Yang, Z. Yi, F. Zhou, X. Xu, S. Wang, D. Yang, Y. Hu, S. Jiang, T. Yang, H. Yin, K. Chen, C. Peng, and J. Pan. 2010. Experimental free-space quantum teleportation. *Nature Photonics* 4: 376–381. doi:10.1038/nphoton.2010.87.
Johnson, D., and K. Miller. 2009. *Computer ethics: Analyzing information technology*, 4th ed. Upper Saddle River: Prentice-Hall.
Lyderson, L., C. Wiechers, C. Wittmann, D. Elser, J. Skaar, and V. Makarov. 2010. Hacking commercial quantum cryptography systems by tailored bright illumination. *Nature Photonics* 4: 686–689. doi:10.1038/nphoton.2010.214.
Parnas, D.L. 2009. Document based rational software development. *Know-Based Syst* 22(3): 132–141.
Roos, C.F., M. Riebe, H. Häffner, W. Hänsel, J. Benhelm, G. Lancaster, C. Becher, F. Schmidt-Kaler, and R. Blatt. 2004. Control and measurement of three-qubit entangled states. *Science* 304(5676): 1478–1480. doi:10.1126/science.1097522.
Shor, P. 1994. Algorithms for quantum computation: Discrete logarithms and factoring. In *Proceedings of the 35th annual symposium on foundations of computer science*, ed. S. Goldwasser, 124–134. Los Alamitos: IEEE Computer Society Press.
University of Connecticut (UCONN). 2010. http://www.engr.uconn.edu/votercentertechnology.php. Accessed October 25, 2010.
van den Jeroen, Hoven. 2008. Moral methodology and information technology. In *The handbook of information and computer ethics*, ed. K. Himma and H. Tavani, 49–67. Hoboken: Wiley.
Weinfurter, H. 2005. Quantum information. In *Entangled world: The fascination of quantum information and computation*, ed. J. Audretsch, 143–168. Weinheim: Wiley-VCH.
Werner, R.F. 2005. Quantum computers – The new generation of supercomputers? In *Entangled world: The fascination of quantum information and computation*, ed. J. Audretsch, 169–201. Weinheim: Wiley-VCH.
Wolf, M.J., K. Miller, and F.S. Grodzinsky. 2009. On the meaning of free software. *Ethics and Information Technology* 11(4): 279–286. doi:10.1007/s10676-009-9207-9.

Chapter 3
Levels of Abstraction and Morality

Richard Lucas

3.1 Introduction

Floridi and Sanders' work on Levels of Abstraction (*LoA*) is one of philosophical depth and innovation, one of significance for the field of philosophy of information generally and ethics and information particularly. However, with this significance comes a price. This price is that the concept of *LoA* contains a number of innovative and controversial concepts that require lengthy and careful examination to appreciate.

In many papers, Floridi (2004, 2008 and with Sanders in several papers (especially 2001)) has persuasively argued the case that systems (generally thereby artificial agents) can be conceived of as moral agents. To do this, he and Sanders introduced the notion of Levels of Abstraction and combined this with state-transition theory to produce what they call an effective characterisation of moral agents. I examine this claim in general and *LoA*s in particular from the point of view of systems as agents, ordered levels of abstraction, state transitions, moral agency, LoA_2, and interactivity, adaptability, autonomy, and cognition.

The structure of this chapter is as follows: first I will examine some basic terminology, such as their view of action, their take on agency, and their view of morality. I will then examine and critique some so-called natural *LoA* examples.

I critique their schema in two ways: their characterisation of morality as a threshold function and the conception of *LoA*s as systems. I claim that there are difficulties with *LoA*s as systems (especially *LoA*s as closed systems) and that most *LoA*s cannot

R. Lucas (✉)
Head of Discipline, Information Systems, Faculty of Information Sciences and Engineering, University of Canberra, Canberra, ACT, Australia

CAPPE, Australian National University, LPO Box 8079 ANU, Acton, ACT 2601, Australia
e-mail: richard.lucas@canberra.edu.au

H. Demir (ed.), *Luciano Floridi's Philosophy of Technology: Critical Reflections*, Philosophy of Engineering and Technology 8, DOI 10.1007/978-94-007-4292-5_3,

say anything more than that the system meets the criteria. I find difficulties with Floridi and Sanders then calling that meeting of criteria a kind of morality.

3.2 Preliminary Concepts

3.2.1 Action

For Floridi and Sanders, the evaluation of actions as moral actions is dependent upon two things: thresholds and humans.

The idea of a threshold and a threshold function is used to define an action as a moral action. So what is a threshold function?

> A threshold function … is a function which, given values for all the observables … returns another value (Floridi and Sanders 2004, p. 369).

This is a straightforward and usual definition of function. What is unusual is their claim that the types of all observables can, in principle, at least, be mathematically determined. This claim, in principle, is false. See Chaitin (1998, 1999) for a proof that not everything mathematical can be determined; however, presenting Chaitin's argument is beyond the purpose of this work.

In spite of this obvious flaw, they proceed as if their claim about observables holds and continue by saying that:

> In such cases, the threshold function is also given by a formula (Floridi and Sanders 2004, p. 369).

This must also be false.

They then say that an action is moral if the value that the threshold function produces exceeds a pre-agreed value. They call this pre-agreed value a tolerance. It seems that the morality of an action is merely a matter of arithmetic and that the arithmetic is beyond the agent's control.

As I see it, this leads to two problems. The first is the problem of determining the tolerance and the second is the problem of the determiner that determines.

For the first problem of determining the value of the tolerance, it seems simplistic to say that "the threshold [is] determined … by human agents" (Floridi and Sanders 2003a, p. 20). There is absolutely nothing to give the reader any insight into this process or any sense of what this value might be. Nothing is spelled out about this function, such as what kind of function it is, what sorts of inputs it might accept, or what sort of outputs (values) it might provide. All of these concerns would need to be addressed before the reader could reasonably be asked to evaluate the worth of these statements.

On the problem of determining the determiner, Floridi and Sanders say that such a tolerance is determined by "human agents exercising ethical judgments" (Floridi and Sanders 2004, p. 369). But why should human agents be in such a privileged position? What gives them the right to determine for other moral agents what this tolerance ought to be? While they give no direct answers to these questions, it seems reasonable to posit that what they had in mind is that we human beings simply know that we are moral agents and ought to be able to judge and recognise when other agents are moral agents. The first part of this is relatively unproblematic, but the second part is not. The problem with this, of course, is an old one. There is much dispute among humans about what counts as a moral agent and, under such uncertainty, which group is to be accorded the right to make the final determination of the moral agency of another.

3.2.2 Agency

Much of what Floridi and Sanders say about the moral agency of artificial agents hinges on their conception of agency. While they do spend some time on groups as moral agents, I will not pursue that line here. I will instead concentrate on their depiction of agents as individual units. I will take without argument that the term "agent" includes both human and artificial agents. I also accept Floridi and Sanders' assertion that both of these kinds of agents are legitimate sources of moral action (though perhaps not sources of moral agency).

Agents vs. patients: Floridi and Sanders' first move in defining agents is to suggest that agents can be both moral patients and moral agents; moral agents are originators of moral action and moral patients are receivers of moral action. This discrimination has two purposes: to allow them to separately focus on the possibility of agents being moral agents without having to consider whether they might also be moral patients, and to thus narrow the focus of their exploration to moral agents only.

Agents as systems: Taking their inspiration from classical information systems theory, Floridi and Sanders provide a new way of conceiving of agents. They begin with the idea that agents are systems and show that, indeed, most things can be systems. They further show that systems have necessary and sufficient conditions for determining whether any suggested entity is a particular kind of system.

Floridi and Sanders do this because they recognise that it is impossible to always be definite about a definition, or, in this case, to be definite about what an agent is. They see the use of agents as systems as a way around this problem. This is a way out of the vagueness because it now allows us to think of such notions as agents as sets. After all, systems are just sets of values and processes that go together to produce some kind of output. To further give this notion a concrete appearance, they defer to the mathematical/logical conception of a set, where a set is seen as a collection of members (parameters). They then go on to say that, for a particular set, it is possible to define a set of these parameters while still allowing any definition of that set to remain fuzzy. They call this defining of the set of parameters specifying a Level of Abstraction (*LoA*) and call the set of parameters a *LoA*. The idea of *LoA*s is explained in the next section.

Agents are, then, for Floridi and Sanders, simply systems that are examined using a particular but not necessarily unique *LoA*. All agents are systems, but not all systems are agents.

The important conclusion to take from this characterisation of agents is that moral agents are systems as viewed through a particular *LoA*.

3.2.3 On the Very Idea of Levels of Abstraction

Everything in their account of the morality of artificial agents hinges on the idea of Levels of Abstraction.

> A LoA consists of a collection of observables, each with a well-defined possible set of values or outcomes (Floridi and Sanders 2004, p. 354).

They also say that:

> A level of abstraction, LoA, is a finite but non-empty set of observables (Floridi and Sanders 2004, p. 355).

The observables referred to here amount to the information held by a system (what Floridi and Sanders call an entity) that some observer has available to them, information that the observer can observe. Using the characterisation of agents as sets, it is straightforward that observables are the values that the parameters (members) can have. It is important to realise that not all of the parameters a system has will be available for inspection to an observer when viewing the system at a given *LoA*. Also, a *LoA* is not a view of the system as a whole; only a complete set of *LoA*s for a system would provide such a view. A *LoA* is a partial view of the system.

This then allows that any system might, and in practice usually will, have multiple *LoA*s. Trying to determine the number of *LoA*s for a given system results in two interesting results. The first is that the number of *LoA*s is the set of all proper subsets of all of the variables of the system, $((n(n+1)/2))$ times the number of possible values of each variable. The second is that, unless a *LoA* is intended to admit to the model only certain values of a particular variable, then the number of *LoA*s tends towards infinity.

Also implied in the examples they give is the idea that a set of *LoA*s for a system can be ordered. Floridi and Sanders then go on to say that "an entity may be described at a range of LoAs and so can have a range of models" (Floridi and Sanders 2004, p. 354). Thus, each *LoA* can be seen as a model of the system. I agree that models are the outcome of the analysis of a system. However, much more can be said about the process of analysis and its relationship with both the resulting models and the *LoA*(s) chosen for analysis. This deserves additional treatment but is beyond the scope of this work.

This conception is both straightforward and deceptive at the same time. It is straightforward because it seems obviously true; it simply follows from the fact that any entity may have a number of differing ways of being described, each of which provides a different point of view, a different focus on the entity. It is deceptive because hidden inside it is an implication that there is something about the entity that remains constant, fundamental, while the selection of bits of information about the entity that are being abstracted from it varies.

This constancy might concede something like Plato's ideal forms or Kant's things-in-themselves. It at least implies that there might be a superset of information about an entity from which all models are extracted. If it does not, then the information about the entity must vary or change in some way. That the set of information about an entity varies is common sense. After all, if for no other reason than the passage of time, information about me is constantly being added. However, if the notion of a superset of information is construed broadly enough, any possible change could be claimed to be a part of the superset of information. But this, in turn, must complicate the abstraction process. Which variation of the entity is it being abstracted from? How much variation can take place before the entity becomes a different entity? Any amount, or some threshold amount? None of this is explored in any of the relevant papers.

The next step that Floridi and Sanders take is to say that some of these models can be seen to represent the morality of an agent. That is, some *LoA*s/sets/models can be seen as representing a moral agent.

Finally, depending upon the *LoA* chosen, a system may or may not appear to be an agent, particularly a moral agent. On one *LoA*, a system may be a moral agent, and the same system examined using a different *LoA* may appear to not be an agent at all, but simply a collection of molecules. An example used by Floridi and Sanders in their paper is one in which a human being called Henry is said to be merely an agent and not a moral agent. Floridi and Sanders say that this *LoA* for Jan is LoA_1. They go on to say that:

> At *LoA1*, there is no difference between Henry and an earthquake. There should not be. Earthquakes, however, can hardly count as moral agents, so *LoA1* is too high for our purposes: it abstracts too many properties (Floridi and Sanders 2004, p. 357).

To further reinforce the idea that systems can be agents, Floridi and Sanders say that they are particularly interested in those kinds of systems that can be seen as agents and that agents are agents of change. From this, they therefore see agents as systems that must be capable of change.

A natural way to conceive of a system, and hence an agent, that changes is to use the idea of states and state transitions.

State transitions: Standard computing and information science conceive of systems as having states. Floridi and Sanders use this idea to show how *LoA*s can be attributed to systems. These states are merely the set of values that the variables of a system have at some particular time, say, T_0. Systems are always in a state; that is, they always have a particular configuration of values and any system might change from the state it is in to another state. This is known as a state transition: systems are state-transition models.

By combining the idea of a *LoA* and state transitions, Floridi and Sanders claim to be able to achieve the level of precision that they think is necessary to be able to sufficiently characterise a system as a moral agent.

Floridi and Sanders (pace Allen et al. 2000) do this by saying that the right *LoA* necessary for moral agenthood (called LoA_2) is one that satisfies three criteria: interactivity, autonomy, and adaptability. Here I quote Floridi and Sanders' definitions of these three criteria:

> Interactivity means that the agent and its environment (can) act upon each other. Typical examples include input or output of a value, or simultaneous engagement of an action by both agent and patient – for example, gravitational force between bodies. … Autonomy means that the agent is able to change state without direct response to interaction: it can

> perform internal transitions to change its state. So an agent must have at least two states. This property imbues an agent with a certain degree of complexity and decoupled-ness from its environment. … Adaptability means that the agent's interactions (can) change the transition rules by which it changes state. This property ensures that an agent might be viewed, at the given LoA, as learning its own mode of operation in a way which depends critically on its experience. Note that if an agent's transition rules are stored as part of its internal state then adaptability follows from the other two conditions (Floridi and Sanders 2004, pp. 357–8).

Of importance here is to note that interactivity is the same as a state transition and that adaptability amounts to the rules that describe how systems can change from one state to another (called transition rules). This importance will become apparent in the discussion of morality that follows.

There is, then, an explanation of how these three conditions contribute to an understanding of how an ordinary agent can become a moral agent merely by determining that an entity has these three added conditions. In their examples, human beings answer affirmatively to being interactive, autonomous, and adaptable, while rocks answer negatively to all of the three. More examination of systems at various non-moral levels of abstraction is then pursued. These examples of systems are noughts and crosses, Webbots, futuristic thermostats, smart paint, and organisations. Whether these are appropriate examples for non-moral *LoA*s is not examined.

Of state transitions, they say that:

> A transition may be non-deterministic. Indeed, it will typically be the case that the LoA under consideration abstracts the observables to make the transition deterministic. As a result, the transition might lead from a given initial state to one of several possible subsequent states (Floridi and Sanders 2004, p. 356).

I have difficulty making sense of this use of the term "non-deterministic." If determinism is used in the finite-state machine sense, then determinism relates to the certainty of the set of state transitions and outputs (see Gill 1962, p. 7).

In a finite-state machine, all of the paths from, initially, the initial state, and, thereafter, from any state to any other state, can be determined simply by examining the inputs and the rules governing transitions. This will then determine exactly which state the system will transition to: there is no uncertainty and hence no non-determinism. Non-determinism usually means that even given the possible inputs and the finite number of states, the result is not able to be determined ahead of time. However, non-determinism is simply not possible for systems that can be characterised as finite-state machines. What Floridi and Sanders must mean then is that which particular state a machine will transition to is not able to be known ahead of time. Now, this is true only when it cannot be known which particular input will occur next, so it must be that they mean "non-deterministic" to be applied to the inputs of the system and do not mean for the system itself to be non-deterministic. This, however, is going to be true for practically all systems. In general, it is not and cannot be known which inputs are going to occur next.

Further adding to the complexity of their equating systems with finite-state machines is that state transitions can be influenced by more factors than only when the properties (i.e. information) of an entity change value. The idea of state transition is broader than merely changing the value of a variable. An example of a system changing states without changing any of the system variables would be that in which a system changes from one state to another simply by the passage of time. This way, a change in state does not imply a change in an entity. Normally, the passage of time is not considered to be a property of the system, but rather separate from the system. It is commonplace in the field of computer systems analysis to not include the system clock as a property of the essence of the system (Lucas 2009, p. 18). It is usually seen more as a property of the infrastructure used to implement the system.

Floridi and Sanders also write about the differences between internal and external transitions. They make the claim that internal transitions are those that give a system the appearance of choice. Where they write about the differences between internal and external transitions, they say that internal transitions are those that are not influenced by the environment and external ones are those that are influenced by the environment. Once again, those things that can influence a system and come from the environment are just inputs. As specified earlier, determining what gets recognised as an input by the systems is just what setting the boundary of the system does. Thus, it seems to be the case that determining the difference between internal and external transitions of a system is the same as specifying the *LoA* of a system, which is the same as determining the boundary of a system, which is the same as determining the inputs and outputs (I/O) of a system. Since boundaries and I/O are always contingent upon specifying the purpose of a system and the purposes of systems are constructed, boundaries and I/O are, in a sense, not natural, because there is nothing inherently natural about the purpose of a system. This, however, is not the place to further explore so complex of an idea or to examine in detail the notion of purpose. The classical consequence of this is that there is no such thing as a right or natural system; all systems are constructed. Any coherently determined boundary is acceptable so long as it correctly identifies the boundary such that the inputs are necessary and sufficient for the correct determination of the (pre-determined) outputs and only those outputs; that is, it satisfies the purpose of the system. In other words, all inputs are needed and no unnecessary inputs are a consequence of the chosen boundary. There are no purpose-built systems that do not pre-specify their inputs, outputs, processing, or some pairwise combination of the three. Determining the boundary, which necessarily means that the *LoA* is then known, is simply the process of making that choice clear. That Floridi and Sanders insufficiently consider this aspect of *LoA*s undermines the grounding for their use of it in specifying the morality of artificial agents.

However, the above goes against Floridi and Sanders' hope in making believable the idea that there are natural *LoA*s that we can subscribe to; in particular, it goes against their attempt at getting us to accept one such natural *LoA*, the one in which systems/agents can be seen as moral agents. This reference by Floridi and Sanders to the idea of natural *LoA*s is reminiscent of Putnam's natural-kinds argument, but the nature and extent of this similarity will not be examined here, as it is peripheral to my central question. See Putnam (1975) and Weckert (1986) for examples and explanations of arguments concerning natural kinds.

To further their effort to establish the idea of *LoA*s as legitimate constructs with which to characterise agents, in particular moral agents, they provide what they call an effective characterisation of agents. What they have in mind is a sufficient characterisation for determining if a system/agent is a moral agent.

The first thing to notice about this characterisation is that the idea of a *LoA* is given a subscript: for example, LoA_1. This implies that there are many more *LoA*s (LoA_1, LoA_2, …, LoA_n). They go on to describe *LoA*s and imply that there is some sort of hierarchy or organisation of *LoA*s.

Consider the following:

> Described at this *LoA1*, Henry is an agent if Henry is a system, situated within and a part of an environment, which initiates a transformation, produces an effect or exerts power on it, as contrasted with a system that is …. acted on or responds to it, called the patient. At *LoA1*, there is no difference between Henry and an earthquake (Floridi and Sanders 2004, p. 357).

The difficulty with their description is that it implies that all, or at least most, systems can be so organised. This is not so. While some *LoA*s may be hierarchically related, most are not. Many systems have common elements, but few are imbedded in such a way as to accommodate a hierarchy. This hearkens back to their earlier discussion of the relationship between moral agents and moral patients. It would seem much stronger if these (LoA_1, LoA_2, …, LoA_n) had some sort of ordering criteria against which to judge a particular selected set of characteristics. That is, pick some *LoA* and compare it to the scaling and see where it fits. This is not done, and the reader is left wondering what such an ordering might be like. The idea of multiple (and related in some strong sense) *LoA*s also leads us to ask whether some different *LoA*s might simply be a case of Wittgenstein's seeing-as (Wittgenstein 1997). The account of abstraction theory would need to have something to say about this. If there is a set of *LoA*s, then some might conclude that there is something that can be said about the set of information that captures all of the *LoA*s for a given entity. As described earlier, this can be either a fundamental subset/core/essence, a minimal set, or a superset of all of the characteristics that might be called on in creating all of the *LoA*s for a given entity.

3.2.4 Morality

Having accounted for the idea that *LoA*s can be ordered and that there is something natural about that ordering, Floridi and Sanders move to the idea that morality is a part of this natural ordering and offer a definition of the *LoA* that fits with their idea of morality. Floridi and Sanders characterise morality as *interactivity*, *autonomy*, and *adaptability*. They then match this definition with the conception of *LoA* and say that a *LoA* that can be seen to naturally have moral characteristics (a moral *LoA*) is called LoA_2.

To support this claim, they describe two hypothetical systems, H and W, characterising them as having interactivity, autonomy, and adaptability, and then ask the question: are they moral? Answering this question, they say, requires expanding the criterion of identification to the following:

> An action is said to be morally qualifiable if and only if it can cause moral good or evil. An agent is said to be a moral agent if and only if it is capable of morally qualifiable action (Floridi and Sanders 2004, p. 364).

They continue with the example and add that, for us to be able to use this new definition, H and W must perform some action that qualifies as moral. For this, they say that H kills a patient and that W cures a patient. After some discussion, they conclude that both H and W are moral agents, and then reveal that H is a human and W is an artificial agent (AA).

Anticipating that some might object to them saying that W (the AA) is a moral agent, Floridi and Sanders (2003a, pp. 16–19) discuss the reasons why someone might object. Four of these objections centre on the idea of a responsible morality and are called the teleological objection, the intentional objection, the freedom objection, and the responsibility objection.

The teleological objection is that "an AA has no goals," and that this matters morally speaking. Here their characterisation of this objection seems incomplete. Some might argue that, usually, simply having goals is not what is meant. Crucial to a more complete teleological objection is that the goals are of the right kind and that they are not simply added simpliciter. Their claim that the *LoA* can be "readily … upgraded" so that both H and W have goals seems like merely changing the *LoA* so as to meet the objection. The notion of upgrading a *LoA* seems arbitrary and self-serving. Once a *LoA* is chosen, is one not obliged to stick with it? This analysis simply does not counter the claim that it matters that an AA has no goals.

The intentional objection is that "an AA has no intentional states," with the implication that having intentional states is crucial to being a moral agent. Floridi and Sanders' (On the morality of artificial agents [private communication]. (Received 24 December 2003), 2004) counter to this is that intentional states are nice but unnecessary for moral agency as they have conceived it and that intentional states require some form of privileged access (something like a God view), and that is not possible. But this is exactly what Floridi and Sanders rely on when describing the examples in which they seem to want to have access to internal states without interactivity. Thus, their argument against intentional states, because they require this privileged access, is the same one that they rely on to make their earlier case. They cannot have it both ways.

The freedom objection is that "an AA cannot be held responsible for its actions." That is, an AA is not free. Floridi and Sanders' counter to this is that AAs are "already free in the sense of being non-deterministic systems" (Floridi and Sanders 2004, p. 366), assuming the stance on determinism taken in their Sect. 2.4. I raise the same objection here as I did above, that the use of "non-determinism" is confused. Floridi and Sanders go on with the claim that the AAs "could have acted differently if they had chosen differently and they could have chosen differently because they are informed, autonomous and adaptive" (Floridi and Sanders 2004, p. 366). I contend that they have not shown that their definitions of autonomy and adaptability give what is needed.

The responsibility objection is that AAs simply lack responsibility. My objections to most of the proposals stated in Floridi and Sanders' Sect. 2.5 notwithstanding, this section is well argued, especially in the differentiation of identification and evaluation and the separation of accountability from responsibility. However, the following statement needs clarification: "This means that [parents] identify [children] as moral sources of moral action, although as moral agents [children] are not yet subject to the process of moral evaluation" (Floridi and Sanders 2004, p. 368). What parents do is treat children as future or potential moral sources, and hence as future or potential moral agents. All going well, children will become moral agents, not that they are moral agents presently. Parents act *as if* their children were moral agents, but do not actually believe that they are.

In the next example, of search-and-rescue dogs, they write:

> …but for the dogs it is a game and they cannot be considered morally *responsible* for their action. The point is that the dogs are involved in a moral game as main players and therefore we can rightly identify them as moral agents *accountable* for the good or evil they can cause (Floridi and Sanders 2003a, p. 19).

It seems that the dogs are not main players (agents), but rather tools used by the main players, that is, those organising and doing the searching, towards moral ends. Surely to count as a moral agent means that the agent must be aware of being such an agent. As Floridi and Sanders say, the dogs have no sense that this is anything other than a game, and there is no evidence that they are aware of themselves as sources of moral action. If this is true, then these dogs are not moral agents according to most other accounts of moral agency. This would seem to reinforce the prevailing view that Floridi and Sanders' characterisation is simply wrong. This sort of example does nothing to convince doubters of the veracity of their argument.

In their third example, citing the trials and tribulations of Oedipus, it should be noted that while Oedipus did not try to kill his *father*, he did try to kill the *king*. That the person was his father is of less importance; the greater importance is the intention to kill. The example of marrying his mother is more to the point; there is nothing inherently wrong with marriage. In this example, as he is ignorant of the fact that his bride is his mother, Oedipus is not morally responsible due to his ignorance. It is his ignorance that mitigates his responsibility. He is accountable in the sense that we can account for or attribute the source of the moral wrong without attaching responsibility. However, once his ignorance is addressed, then the responsibility adheres.

If my understanding of Floridi and Sanders' account of accountability is correct, then many would say that what they are doing is simply attaching the word "moral" as a field of study to the notion of "accountable." They would then go on to say that this, on its own, does not count as moral.

It does seem somewhat disingenuous to set up a *LoA* where the fact of his bride being his mother was permanently omitted when, in fact, in the story this information comes to light. Surely all of the morally relevant facts must be included in a *LoA* that is being used to make an assessment of the moral agency of one subject to a moral claim.

Interactivity:

Interactivity means that the agent and its environment (can) act upon each other (Floridi and Sanders 2004, p. 357).

This description of system interactivity seems too simplistic and needs to be accounted for more deeply. In classical information theory, there are four possible types of interaction a system can have with its environment.

Type	Name	Description
1	Closed	No inputs or outputs
2	Black hole	Inputs, but no outputs
3	God	No inputs, just outputs
4	Open	Inputs and outputs

This gives the following.

Type	Name	Inputs	Outputs
I_1	Closed	No	No
I_2	Black hole	Yes	No
I_3	God	No	Yes
I_4	Open	Yes	Yes

To reinforce their view of the interaction of a system with its environment, Floridi and Sanders refer to the example of "gravitational forces between bodies" (Floridi and Sanders 2004, p. 357), where there is simultaneous interaction. This hardly seems to be in tune with the shift from mere agency to moral agency. Common sense would reject gravity as even remotely analogous to a moral issue. It would seem that one of the hallmarks of moral interactivity is choice, and, indeed, Floridi and Sanders make this very point. The agent must be able to decide to be the source of moral action. Deciding that the laws of gravity are as morally troublesome as, say, local parking regulations flies in the face of sense. A much better example is needed here. Might there be a better argument than the one put forward by Floridi and Sanders?

It is true that *LoA*s that are moral agents must interact with their environment and so must be of either type I_3 or I_4. Now, it is possible that moral agents exist that know a priori all they need to know in order to be moral (that is, type I_3 agents), but it seems that neither humans nor artificial agents have such a priori knowledge. So, type I_3 agents can only be moral agents in the sense that humans are moral agents.

Autonomy:

Autonomy means that the agent is able to change state without direct response to interaction: it can perform internal transitions to change its state (Floridi and Sanders 2004, p. 357).

By "without direct response to interaction," I take it to mean that the system is not only simply responding to an input or creating an output, but, rather, it has some mechanism for determining when to change from one state to another, independent of inputs and outputs.

The problem with this is the following: All computer systems change state due to some stimuli. There are none that change state for no apparent reason. Now, this stimulus can be either external or internal. This is a problem for Floridi and Sanders because, while the idea of external stimuli equates to Floridi and Sanders' inputs from the environment, they have no corresponding notion for internal stimuli.

One account of internal stimuli might be the following.

Internal stimuli come from one of two sources. They can come from some background subsystem that is always running while the machine is active. That is to say, the subsystem checks for some particular internal state configuration, which, when detected, causes some action (state change) to take place. Conversely, they can come from the passing of time. Note that the passage of time can trigger a state change in two ways: either when a predetermined time has been reached, or when some previously determined time period has elapsed.

In the background subsystem case, the particular state can be reached by one of only three ways:

(i) External stimuli (*Se*), same as a straightforward direct response as a consequence to external stimuli;
(ii) Passage of time (S_t) such that eventually the required internal state may occur; or
(iii) Creation of new information (*Si*), and hence new states, based on an analysis of existing states.

From this, we can see that there are only two cases that we need to consider. The first possibility (S_e) is out, as it is just external stimuli. The only plausible candidates for non-direct response stimuli are thus time and creation. This means that autonomy, in the sense used here and discounting magic as stimuli, is equivalent to either time-based transition S_t or the analysis algorithms necessary for S_i.

The task is to now analyse S_t and S_i to have an account that will decide if S_t can ever occur, and to analyse the set of state transitions to determine whether the particular internal state can ever be reached and, if so, how.

Adaptability:

> Adaptability means that the agent's interactions (can) change the transition rules by which it changes state (Floridi and Sanders 2004, p. 358).

The next discussion is of Floridi and Sanders' definition of adaptability. This is described in terms of state changes. Internal states representing the transition-change rules can be viewed in two ways: either these transition-change rules are at the same conceptual level as all of the other transition rules, or they are at a higher conceptual level. If they are not at the same conceptual level as the normal transition rules, then they are meta-rules, rules for changing the rules. If they are at the same conceptual level as the normal transition rules, then they are just more rules for state change.

If the ways that a system can change its transition rules are to be treated as a special case, as meta-rules, then the infinite regression problem must be considered.

Are there to be meta-meta-rules, and, if so, when does this regression stop? For the kinds of moral agents that humans are, there is, in principle, no limit to this regression, and so, if Floridi and Sanders wish to continue relating the moral agency of artificial agents with human agency, there seems to be no reason to place a limit on this regression for artificial moral agents. This then turns into the halting problem that Turing highlighted (see Lucas 2009, p. 74).

If Floridi and Sanders do not mean for their rule changing to be of the meta-rule sort, then the fact that the rules change must simply mean that some (particular) states change, but the fact that particular rules change is no different to any other state that can change. This is saying that the particular pieces that allow for or cause these rules to be changed are treated as just another ordinary state, no different than any other state within the system. A state is a state is a state, whether it changes the rules for state transitions or it changes the values that variables in the system can hold.

But this ordinariness of rule changeability as a state change eliminates the specialness of adaptability and makes the stating of adaptability explicitly at least unnecessary and, probably, pointless. Adaptability under this is then an inherent character of all systems. This is what I take Floridi and Sanders to mean by their note that adaptability follows from interactivity and autonomy. If the rules are discernable at a *LoA*, then adaptability as a separate *LoA* characterisation is unnecessary. However, if they are not, then it seems that adaptability is not possible at that *LoA* unless there is return to transition-rule changes being meta-rules.

Where this leads to a problem is in the thermometer example, which is frequently invoked in arguments over agency. The case in which a thermometer becomes too hot and bursts, with the mercury leaking out, could be viewed as simply just another state change, the change from the state of being able to report the temperature to the state of not being able to report the temperature. To say anything sensible about this, we would then need to return to what the purpose of the thermometer is. If we change the rules about what counts as a state change, as in the example above, then it seems that we are changing its purpose from a device that reports the temperature to a device that does not. Now, normally, this sleight of hand is seen for what it is: redefinition for the sake of it, and it would be rejected as being not in the spirit of what was intended when the thermometer was specified. It is no longer a thermometer.

If they have something substantial and different to say about adaptability, then all of the above seems to imply that we cannot simply treat the rules for state change as ordinary states, but rather must take the meta-rule stance. Of course, adaptability might be a layered concept. In that event, it also implies that we must include this stance and its layering parametric value in the *LoA*. It seems straightforward to say that the *LoA* chosen will determine the depth of recursion.

Morality as a threshold function: Given the above, Floridi and Sanders also claim that morality can be seen as a threshold function that can "in principle at least be mathematically determined," and that this threshold function can be subject to "some pre-agreed value." This value is called a tolerance. Once this tolerance is reached, then the agent is considered to be a moral agent.

This idea of a "pre-agreed" tolerance seems to present Floridi and Sanders with a problem. It seems to conflict with the freedom objection and their claim that artificial agents are non-deterministic (Floridi and Sanders 2004, p. 354). The way out that Floridi and Sanders take is to claim that this tolerance is "identified by human agents exercising ethical judgments" (Floridi and Sanders 2004, p. 369), thus introducing a non-deterministic element.

Floridi and Sanders then consider the general and moral agency of two entities, H and W. In their description of the actions of H and W, they write: "They both acted autonomously: they could have taken different courses of actions…" (Floridi and Sanders 2004, p. 364). This is the first mention of the connection between autonomy and the ability to take different actions. Originally, their characterisation of autonomy was of the ability to change states. There is a difference between ability to change states and ability to take different actions. What might Floridi and Sanders mean? Well, there are two senses of taking action that might apply here: changing states, and acting in the world (i.e., the outputs). The first refers to Floridi and Sanders' original definition in their Sect. 2.2, but the second does not. It is this second one that is both the normal, default meaning of action and the only one that we have with which to judge autonomy.

With the notion of morality being a threshold concept, there seems to be a difficulty with the claim that "the types of all observables can in principle at least be mathematically determined" (Floridi and Sanders 2004, p. 369). Why does it matter that the types of observables can be so determined? Surely the only things that matter are the observables themselves. In the next paragraph, Floridi and Sanders write that it is not known if "all relevant observables" can be determined. It seems that this passage is comparing observables with types of observables. If it is not known that "all relevant observables can be mathematically determined," then how can it be known that all types of variables can be known? Surely there might be an unknown observable that is of an unknown type of variable. Even more, if not all relevant observables can be known, then how can it be known if the threshold has been reached? Floridi and Sanders do not say.

The idea that morality is a threshold function seems problematic, and Floridi and Sanders do not adequately account for the difficulties noted above to make the concept clear.

3.3 LoA_2 and Examples of Systems

As part of the argument that *LoA*s generally have some natural correspondence with natural entities in the physical world, Floridi and Sanders (2004, p. 359) offer a number of examples to support the idea that *LoA*s can be interpreted as indicating when a system is a moral agent. I reproduce the table here so that the reader can compare it with my expanded conception.

Examples satisfying the properties constituting agenthood:

Interaction	Autonomy	Adaptability	Examples
No	No	No	Rock
No	No	Yes	??
No	Yes	No	Pendulum
No	Yes	Yes	Closed ecosystem, solar system
Yes	No	No	Postbox, mill
Yes	No	Yes	Thermostat
Yes	Yes	No	Juggernaut
Yes	Yes	Yes	Human

There are, however, several problems with the examples in this table. I choose two:

Interaction	Autonomy	Adaptability	Examples
No	No	No	Rock
Yes	No	No	Postbox, mill

Floridi and Sanders go on to say that:

> For the sake of simplicity, all examples are taken at the same *LoA*, which consists of observations made through a typical video camera over a period of say 30 seconds (Floridi and Sanders 2004, p. 358).

Using the sensory apparatus of a video camera over 30 s as a way of specifying a *LoA* that Floridi and Sanders have chosen, it is difficult to see how a rock differs from a postbox. Indeed, in the early days of postal deliveries, it was commonplace to have rocks as postboxes. Just watching either for 30 s may not help to establish which is which, rock qua rock or rock qua postbox. Depending upon which 30 s were used, a postbox might have the same values as a rock. It seems to be fixing the results to say that the observation of the postbox (or rock) as postbox (or rock) occurs just when it is being used as a postbox. Furthermore, one might spend a lifetime at a postbox (or rock) waiting for the right 30 s.

These kinds of difficulties with *LoA*s indicates that most *LoA*s cannot say anything more than that the system meets the criteria, and then call that meeting of criteria "morality."

There is also a problem with the two closed-system examples:

Interaction	Autonomy	Adaptability	Examples
No	Yes	Yes	Closed ecosystem, solar system

First, it seems difficult to make any sense out of the idea of being able to use a video camera for 30 s to be able to comprehend a solar system qua solar system. Perhaps it is my limited imagination, but I cannot find any way of conceiving of that at all. I have asked others, but no systems analyst or philosopher I contacted was able to offer an explanation.

Second, there is difficulty in understanding the example of a closed ecosystem. The difficulty with closed systems is that if they are closed, then the existence of particular ones can be speculated about, and perhaps deduced (in, say, a Kantian sense), but not known. An analogy would be deducing the existence of a planet from the effect it has on the orbits of other planets and not by direct observation; another might be deducing the existence of a particular black hole.

Merely identifying a system means that we have some information about it, which must come from being (at least) able to discern its inputs/outputs; this means that that discrimination must be part of the *LoA*. There is no other way of knowing the states of a system without either being sufficiently superior or by the system making available its states by direct inspection. This availability is just making its states some form of output.

Being able to pass judgment on its adaptability and its autonomy means having access to its internal states. This level of accessibility must mean a different *LoA*. This, in turn, must imply that there are different *LoA*s for interaction, autonomy, and adaptability. It is either that, or all *LoA*s must have some kind of conception of internal states that allows for their inspection without that inspection taking the form of an output; this seems implausible.

With the example of a video camera over 30 s, we cannot know about a closed system's adaptability or its autonomy because we have no access to the internal states of a closed system. The camera cannot work. There is no unobserved observer.

Now, I look at those cases where interaction is given as NO.

Interaction	Autonomy	Adaptability	Examples
No	No	No	Rock
No	No	Yes	??
No	Yes	No	Pendulum
No	Yes	Yes	Closed ecosystem, solar system

The problematic nature of finding an example for the second of these is simply because of the above discussion. If there is no interaction, then we cannot know nor have any evidence for anything about either its autonomy or its adaptability. Simply saying "yes" to complete the truth-table is misleading. This completion gives two impressions: first, that all of the cases have been accounted for, and, second, that the ability to determine the values for autonomy and adaptability is independent of interaction.

A better characterisation is:

Interaction	Autonomy	Adaptability	Examples
No	Undecidable	Undecidable	?

Of course, I have not considered the multiple types of interaction that I specified earlier.

Interaction type	Name	Inputs	Outputs
I_2	Black hole	Yes	No

As these, too, have no outputs, they would also be undecided for autonomy and adaptability. To account for the other systems types, I_1, I_3, and I_4, in this way would further reduce the table size to five entries.

However, this is not the end of it. I need to return to the original list of agenthood examples and revise it in light of the extra values that I have outlined above. This gives a more comprehensive understanding and complete account of these terms. Interactivity would now have four values, and autonomy, three values. These will take more fully into account the concerns I have specified in the previous section, and this would give 24 possibilities:

	Interaction	Autonomy	Adaptability
1	I_1	S_e	Yes
2			No
3		S_t	Yes
4			No
5		S_i	Yes
6			No
7	I_2	S_e	Yes
8			No
9		S_t	Yes
10			No
11		S_i	Yes
12			No
13	I_3	S_e	Yes
14			No
15		S_t	Yes
16			No
17		S_i	Yes
18			No
19	I_4	S_e	Yes
20			No
21		S_t	Yes
22			No
23		S_i	Yes
24			No

Now combine the concerns expressed about the explanations of the three conceptions, interaction, autonomy, and adaptability, into the table. As interactivity now has four values and autonomy three values, the new table would be reduced to 14 possibilities, as follows:

	Interaction	Autonomy	Adaptability
1	I_1	Undecidable	Undecidable
7	I_2	Undecidable	Undecidable
13	I_3	S_e	Yes
14			No
15		S_t	Yes
16			No
17		S_i	Yes
18			No
19	I_4	S_e	Yes
20			No
21		S_t	Yes
22			No
23		S_i	Yes
24			No

It is possible, of course, to be even more fine-grained about this. It is possible to differentiate the outputs (and inputs) even further to specify types. It is possible to differentiate between the outputs of types of states that would allow or deny revealing the program code, as opposed to the expected types of output that reflect the purpose of the system (in purpose-built systems such as AAs). However, this depth of analysis would go beyond the bounds of this work.

An alternative characterisation of *LoA* is to allow for a *LoA* that hides some inputs/outputs but allows inspection of internal states. [This then looks a lot like the objection to the notion of privileged access, the intentional objection, used in Floridi and Sanders 2003a, Section 3.2 on p. 16.] It seems to me that the problem with this is that the hidden inputs/outputs become unknowable rules for the state transition, but I take it that this is just what Floridi and Sanders are after; if they are unknowable, then an assessment of autonomy and adaptability can be based on particular outputs only. This seems, however, to be rigging the results ahead of time.

Another characterisation is that in which the state transitions occasioned by the inputs/outputs are also excluded from the *LoA*. Setting up a *LoA* to achieve this requires a different *LoA* with which to set up the parameters for the *LoA* required and knowledge of the states to be excluded. This seems to entangle the *LoA* in complexities that are set to render it impossible to use.

All of this discussion of the relationship between interaction, autonomy and adaptability reminds me of one of Arthur C. Clarke's oft-quoted views of technology: "Any sufficiently advanced technology is indistinguishable from magic" (Clarke 1972, p. 139). Any sufficiently described *LoA* is indistinguishable from advanced technology.

To further bolster their claim, Floridi and Sanders provide detailed examples of agents; they cite Webbots and a piece of software called MENACE.

MENACE: In the description of MENACE (a noughts and crosses piece of software), Floridi and Sanders make reference to the program learning. I suspect that, excluding adherents of GOFAI, many might take exception to this use of the term "learning." Normally, people would say that to be able to say that a system had learned something, it (in this case, MENACE) ought to be able to say not only what it had learned, but that it had learned. It would need to be able to answer the question: What has been learned? I do not know if it is necessary to be so upfront in using such a limited used of the term "learning."

In the paragraph beginning with, "This distinction is vital for current software" (Floridi and Sanders 2004, p. 361), they make availability the issue for determining adaptability. There, however, seems to be more to it than this. Along with availability (of outputs), the knowledge-ability of the system doing the evaluation seems to be crucial. With sufficient knowledge of how particular systems work, every (or no) system will be judged as an agent. Is the system doing the judging to suspend what it knows to merely what is at hand (the inputs and outputs of the *LoA*)? But this cannot be so, because of the difficulties that are immediately raised by two related questions: *How is it decided what I am to suspend my beliefs about in order to make the LoA sensible? What am I to use to be able to even start thinking about determining the status of the system being evaluated?* These require answers before the availability issue can be settled. As for the second question, merely being able to recognise a *LoA* must imply a certain level of knowledge of that system as a system and of systems as systems generally. Without that, nothing sensible can be said at all about the (or indeed, any) system. If the *LoA* is to predetermine what I can count as being able to be used in my determination of the system-at-hand's agency, then the whole process seems to pre-determine the outcome. Any sufficiently ignorant determiner would see all (or no) entities as (moral) agents. "Indeed only since the advent of applets and such downloaded executable but invisible files has the issue of moral accountability of AAs become critical" (Floridi and Sanders 2004, p. 361).

It seems that the download ability per se is not the problem, but rather two other things are: the relative ignorance of those affected by their execution (dealt with above), and their reach. The downloading of executable software has been around since the 1950s; it is the now-extensive reach of the downloading that extends its sphere of influence to make it a commonplace concern. Even then, most people would still not think of it as a problem of the morality of the software agent, but rather of the originators of such software. To make the case for the criticality that downloading gives to the AA, it would seem that Floridi and Sanders would need to show that either the reach was within the intentional grasp of the AA or that there was something special in the relationship between reach and accountability. This last part is not clear in their work.

Floridi and Sanders also write: "There are natural LoA's at which such systems are agents." Given my questioning of *LoA*s so far, I claim that there seem to be no *LoA*s that might be called natural, at least not in the sense of the natural world. More explanation is needed to spell out what is meant by this. They are quite right to say that the two *LoA*s are at variance, but there is more at issue than simply the 'open source' versus the 'commercial' view. Again, all of this assumes a particular characterisation (dare I say *LoA*?) of system A, the one doing the comparison.

Webbot: Floridi and Sanders claim that "Since we value our email, a webbot is morally charged" (Floridi and Sanders 2004, p. 370). Surely my valuing my email must have nothing to do with the moral status of a webbot. I, personally, do not value email, while my brother does: does that mean that a webbot is not always morally charged? That cannot be right. A webbot cannot be both morally charged and not morally charged: a contradiction cannot prevail.

In the example of Webbots, the phrase "abstracting the algorithm" (Floridi and Sanders 2004, p. 362) seems almost too convenient. This seems like engineering the selection to get the outcome rather than morality being a consequence of some independent selection of abstraction. Further on, they say, "...we do not have access to the bot's code" (Floridi and Sanders 2004, p. 362). Access is not the only crucial problem. Allowing the evaluator's knowledge is. Access is not necessary if we are allowed to generalise from prior knowledge of bots. Floridi and Sanders' demand must make us necessarily ignorant of how bots work. We are not allowed to use any knowledge that we have concerning bots. How are we to do anything that resembles analysis of the bot without this? It does not seem possible. It also seems that one conclusion that can be arrived at is that the moral agency of another is a function of the evaluator's ignorance.

Floridi and Sanders write that the difficulties humans face resulting from the non-single person (i.e., group) attribution of creating programs can be solved by making AAs morally accountable. I cannot see how this follows. The trail of attribution in the development, testing, implementation, and maintenance of software is indeed long and complex, and some of that trail may become lost in the mists of time and bureaucracy, but this is equally said of human beings. The notion of collective accountability may give the general impression of attributing accountability across this process, but this also does not necessarily follow. Simply saying that extending the class of moral agents to include groups as well as artificial agents, as Floridi and Sanders do, does not mean that such groups are moral agents. Nor does saying that a group is accountable make it accountable. Surely individual AAs must suffer from the same difficulties that humans do and have the same difficulties in relation to groups as humans do if they are to have equal moral status. I do not take up the challenge of any of their claims here.

3.4 Conclusion

Floridi and Sanders' program of ethics for artificial agents depends upon two things: an effective characterisation of agents and a specifiable definition of ethics. Both of these have been claimed by them, centrally through the use of *LoA*s; however, I have found that there are difficulties with each and have suggested where they might be strengthened. In the end, the construction of LoA_2 is too artificial and too simple to count as a natural characterisation of morality.

References

Allen, C., G. Varner, and J. Zinser. 2000. Prolegomena to any future artificial moral agent. *Journal of Experimental and Theoretical Artificial Intelligence* 12(3): 252–261.

Chaitin, G. 1998. *The limits of mathematics*. Singapore: Springer.

Chaitin, G. 1999. *The unknowable*. Singapore: Springer.

Clarke, A.C. 1972. *Report on planet three*. New York: Harper & Row.

Floridi, L. (ed.). 2004. *The Blackwell guide to the philosophy of computing and information*. Malden: Blackwell Publishing, Ltd.

Floridi, L. 2008. The method of levels of abstraction. *Minds and Machines* 18: 303–329.

Floridi, L., and J.W. Sanders. 2001. Artificial evil and the foundations of computer ethics. *Minds and Machines* 3(1): 55–66.

Floridi, L., and J.W. Sanders. 2003. The method of abstraction. In *The yearbook of the artificial*, ed. M. Negrotti. Bern: Peter Lang.

Floridi, L., and J.W. Sanders. 2004. On the morality of artificial agents. *Minds and Machines* 14: 349–379.

Gill, A. 1962. *Introduction to the theory of finite-state machines*. New York: McGraw-Hill Book Company.

Lucas, R. 2009. *Machina Ethica*. Berlin: Verlag Dr. Müller.

Putnam, H. 1975. The meaning of meaning. In *Mind, language and reality*, ed. H. Putnam, 215–271. Cambridge: Cambridge University Press.

Weckert, John. 1986. Putnam, reference and essentialism. *Dialogue* 25: 509–521.

Wittgenstein, L. 1997. *Philosophical investigations*, 2nd ed. Cambridge, MA: Basil Blackwell.

Chapter 4
The *Homo Poieticus* and the Bridge Between *Physis* and *Techne*

Federica Russo

4.1 Physis and Techne in the Digital Era

Very few would deny that the advent of computers radically changed our lives, let alone science and society. Some—notably Luciano Floridi (2008, 2009)—even equates the 'digital revolution' for its importance to the Copernican, the Darwinian, and the Freudian revolutions. The first, putting the Sun at the centre of the universe, radically changed the position of Man and his own perception with respect to Nature. The second, finding common ancestors to various species, vanished the supposed privileged place of Man in the biological kingdom. The third, discovering the unconscious dimension of the mind, made Man realise that he is not fully rational nor transparent, even to himself.

The core change behind the digital revolution is that we are becoming aware of our status of informational organisms *among many others*—an idea that traces back to Alan Turing. In his pioneering paper 'Computer machinery and intelligence' (1950), Turing asked the controversial—perhaps even irreverent—question of whether machines can think and discussed the imitation game as a test for intelligence. Reading Turing some 60 years later in hindsight, we more easily realise that in his arguments there was more than simply seeds for a new area of research—artificial intelligence—but for an altogether different way of looking at intelligent beings (we, the humans) in relation to ourselves, to the environment, and to the (digital) artefacts we are the creators of.

This is the immense change the digital revolution carries forward. We humans lose our privileged place in an anthropocentric world and slowly become aware and

F. Russo (✉)
Center Leo Apostel, Vrije Universiteit Brussel

Centre for Reasoning, University of Kent

Department of Philosophy, University of Kent, Kent, UK
e-mail: f.russo@kent.ac.uk

H. Demir (ed.), *Luciano Floridi's Philosophy of Technology: Critical Reflections*, Philosophy of Engineering and Technology 8, DOI 10.1007/978-94-007-4292-5_4,

accept that we are *informational organisms*, or, as Floridi says, *inforgs*. Being *inforgs* means that we are not, after all, so different from other intelligent engineered artefacts—in fact, Turing was not embarrassed at all in asking whether machines—that is *things*, *artefacts*—can think and be intelligent. As a matter of fact, we share with intelligent engineered artefacts something essential: the informational environment or, as Floridi says, the *infosphere*. The infosphere is the global space of information, which includes the cyberspace as well as classical mass media such as libraries and archives. If the infosphere is the whole space of possible information, then nature belongs to the infosphere too. Thus, recognising that we, intelligent humans and intelligent engineered artefacts, equally share this space brings upfront the need to reinterpret Man's position in reality—that is Man's position in the infosphere.

The strength of Floridi's arguments about the digital revolution is that we don't have to think of post-modern science fiction environments, where humans are de-humanised and AI technology took over. The digital revolution is a revolution that we started living in since the pioneering works in information technologies and that is nowadays blossoming—just think of how many 'digital' actions we perform since we wake up in the morning until we go to bed. The digital revolution, in other words, changed at once our interaction with the external world and our views about who we are. Whilst Floridi argued that such a radical change concerns our role as ethical agents, I will further argue that the radical change concerns as well our role as epistemic agents, in the sense of agents that aim to acquire knowledge about the surrounding world, and as agents that engage in poietic, that is *creative* and *productive*, activities.

More importantly, the digital revolution, according to Floridi, brings up again questions about the relations between *physis* and *techne*, respectively understood as nature and reality on the one hand and as practical science and creation of artefacts on the other hand. The digital revolution, in particular, is increasingly changing the *physis*, in the sense of the 'off-line' world. Our off-line world made of real physical objects is becoming itself part of the 'digital' infosphere because the distinction between 'on-line' and 'off-line' is itself becoming more and more blurred until it will disappear. Information technologies are creating altogether new *e*-nvironments that pose new challenges for the understanding of us in the world.

But those arguments, I will argue, are not confined to the digital revolution. The question of the tension between physis and techne is raised by technology, in general, and in particular by the emerging technologies, such as bio- or nanotechnologies, and consequently by digital technologies too. The digital revolution is in fact a *technological* revolution and as such it encompasses extrovert and introvert changes in our understanding of the world, of ourselves, and of ourselves-in-relation-with-the-world. Whilst Floridi emphasises that the fourth revolution is *digital* and that it therefore affects the position and role of man as *ethical agent*, I want to emphasise that the fourth revolution is a *technological* revolution and that it therefore affects the position and role of man as *epistemic agent* engaging in various poietic activities.

As I shall discuss thoroughly in Sect. 4.3, the revolution technology brings in is a shift in the tools to acquire knowledge about the world. As soon as we understand that *intervening* on nature grants us epistemic access to nature and opens up new possibilities for the creation of artefacts, (pure) science ceases to be the privileged lieu for knowledge. The new configuration is that of a *techno*-science, where the *poietic* aspect is no less important than the *noetic* one.

It is along those lines that, I think, we have to read the considerations that Nordmann (2004) makes about technoscience. Technoscience, he says, is characterised by a shift of focus from representing to intervening, *plus* a change in societal expectations and in the way researchers see themselves. The vocabulary chosen by Nordmann is borrowed from Hacking's well-known *Representing and Intervening* (1983). The choice is certainly not chancy and is in fact well calibrated. Hacking gives us ground to cultivate the idea that the importance of *intervening* on nature lies in the fact that it changed the way we, as epistemic agents, relate to nature. Interestingly, Ihde (1991) even reverses the perspective: he talks about science's embodiment in technology and in the experiment, rather than technology and experiment entering the scientific realm.

Carrier (2004) also investigates the tension and the possible reconciliation between physis and techne, albeit in slightly different terms. He argues for a sort of reconciliation between the two approaches, on the grounds that there is no substantial difference between scientific (theoretical) modelling and modelling in the applied sciences. Carrier's argument ultimately aims to undermine the view of those who claim the alleged inferiority of the applied sciences, on the grounds that modelling is more local in scope. But, the argument goes, more local models do contribute significantly in theoretical research and are not a distinctive feature of applied science.

I would like to further argue that what is special about the emerging technologies is that they are not only making new discoveries, but are altogether creating new environments. Those environments are at once cognitive—in the sense of the space of knowledge—and applied—in the sense of the space of *application* of such knowledge. Nanotechnologies exemplify this situation quite well. On the one hand, nanoscience is discovering that materials have different properties at the nanoscale and at the macroscale. These new properties are opening up possibilities for a new understanding of matters, because the *same* material displays different properties depending on the scale of analysis, as well as for new applications in domains as different as nanomedicine and the food sector. But for this very same reason new ethical challenges arise. The reason, simply put, is the following. There is uncertain and impartial knowledge about the nanoscale *and* at the same time there is strong enthusiasm and élan for new applications, creation of new artefacts etc. The question arises whether there exist *unknown* risks for health and environment. *Unknown*, because the biological activity of nano-materials depends on parameters that are not considered by classical toxicology. This situation leads the various stakeholders (nanoscientists, technologists, policy makers, lay-people, philosophers) to worry about the consequences of licensing the use of nanoartefacts, for instance.

But perhaps the *ethical* worries arising from the emerging technologies ought to be accompanied and even preceded by *epistemological* worries. It is in this sense that, it seems to me, the new environments created by technology put up front again questions about the relation between 'physis', to be passively observed, and 'techne', as a practical and applied science. The gap, as is typically understood, concerns the consequences of our actions on nature. Those consequences may concern nature itself, but also the quality of life of us, who are currently living in this world, and of future generations, who will occupy an environment that we have deliberately altered. This concern for future generations is a natural consequence of the intrinsic projectual character of technology (on this point, see for instance Galimberti 1999, ch. 2). Needless to say, actions are also the results of the ethical principles guiding individual as well as societal behaviour. However, the *poietic* dimension of agents' behaviour now makes traditional ethical approaches unsuitable exactly because they neglect it; agents instead just focus on the behaviour of the individual agent who *happens to be* in the situation she is in. Thus, according to Floridi, *if* the tension between physis and techne can be dissolved, this will be done by a constructionist ethics, rather than e.g. a virtue ethics, where the 'homo poieticus' is herself a creator of the e-nvironment. As Floridi (2009) puts it:

> Fortunately, a successful marriage between *physis* and *techne* is achievable. True, much more progress needs to be made. [...] We should resist any Greek tendency to treat *techne* as the Cinderella of science; any absolutist inclination to accept no moral balancing between some unavoidable evil and more goodness; and any modern, reactionary, metaphysical temptation to drive a wedge between naturalism and constructionism, by privileging the former as the only authentic dimension of human life. The challenge is to reconcile our roles as informational organisms and agents within nature, and as stewards of nature. The good news is that this is a challenge we can meet. The odd thing is that we are slowly coming to realise that we have such a hybrid nature. The turning point in this process of self-understanding is what I have defined above as the *fourth revolution*.

Elsewhere, Floridi suggested that the reconciliation between physis and techne might be provided by the notion of *homo poieticus* (see Floridi and Sanders 2003). The homo poieticus is the ethical agent in the era of technology: she is the *creator* of the situations subject to ethical appreciation. Such a *constructionist* framework goes beyond traditional ethics and is suited to the new environments created by technology. The advantage of a constructionist ethics lies in the fact that, unlike traditional ethics, it does take into account the genesis and the various circumstances that led the agent to be in the situation she is facing. Instead, traditional ethical accounts, whether in the framework of consequentialism or virtue ethics, take the situation as 'given', so to speak. But this, argues Floridi, neglects what is perhaps the most important feature of the ethical agent it the digital era: her *poietic* skills.

In this paper, I also take up the challenge of reconciling physis and techne. The underdeveloped notion of *homo poieticus*, I will argue, is the bridge between physis and techne. Following in Floridi's footsteps, I want to argue that the homo poieticus is not just the ethical agent. The homo poieticus is also the technoscientist, as a creator of crafts and of knowledge, and the philosopher, as a creator of concepts. On the one hand, the technoscientist uses technology both as a means to know the world and as a means to create new 'objects'. Unlike the Aristotelian scientist that passively observes the world, the Baconian technoscientist is a 'constructionist epistemologist' that builds, designs, and models reality to create knowledge. On the other hand, the philosopher, in this perspective, becomes a 'conceptual constructionist': facing new epistemological and ethical environments, the philosopher cannot content herself with applying old concepts or perhaps with adjusting them to the new setting. The philosopher has to integrate herself in this 'poietically enabling' environment and create new modes of thinking.

The paper is organised as follows. Section 4.2 presents the figure of the homo poieticus in Floridi's work on computer ethics. Section 4.3 extends the notion of the homo poieticus first to the technoscientist, and then to the philosopher. Section 4.4 closes the paper drawing general conclusions about the relations between ethics and epistemology.

4.2 The *Homo Poieticus* in the E-nvironment

As mentioned earlier, Floridi introduces the notion of homo poieticus in the context of what he calls the 'fourth revolution', which is the digital revolution. Notably, Floridi is interested in developing a new ethical approach able to cope with the situations that ethical agents, as inforgs, create in the infosphere.

The reason to look for a new approach is that traditional ethical theories all encounter the same problem. Traditionally, ethical discourse focused on what is right and what is wrong to do in a *given* situation. Floridi stressed the point that hardly any traditional ethical approach considers *how* the ethical agent got into the situation she is in. This is why Floridi groups traditional ethical theories under the label 'reactive approaches'. The only aspects that count are the values (in virtue ethics) or the consequences (in consequentialist ethics) of the action taken in a given situation. Nevertheless, the point Floridi wants to make is that behaving morally is not just to be judged *a posteriori* based on values or on consequences. Behaving morally starts much earlier than the moral judgement: it has in fact to do with "constructing the world, improving its nature and shaping its development in the right way" (Floridi and Sanders 2003). Moral behaviour has to do, in Floridi's view, with the *poietic* skills of ethical agents. This poietic dimension is even pushed further (Floridi and Sanders 2003):

> In a global information society, the individual agent (often a *multi-agent system*) is like a demiurge. Her ontic powers can be variously exercised (in terms of control, creation or modelling) over herself (e.g. genetically, physiologically, neurologically and narratively), over human society (e.g. culturally, politically, socially and economically) and over natural or artificial environments (e.g. physically and informationally).

Thus, what is needed to cope with the poietic skills of the ethical agent is a 'proactive approach', that is a 'constructionist' approach to ethics. A proactive, rather than reactive, approach emphasises that the agent plans and initiates action responsibly, thus reducing reliance on 'moral luck'.

Moral luck refers to the problem of morally assessing an agent for facts, factors, or situations that she has no full control of. In fact, on the face of it, it is an acceptable principle, in any ethical theory, that agents should be morally assessable only for what is under their control (Control Principle). However, everyday life shows that this isn't the case—i.e., that we own full control of the situations we are in. Moreover, everyday life also shows that agents indeed undergo moral assessment in such situations. An apparent *impasse* thus arises because, adhering to a narrow version of the Control Principle, we end up in a situation where we cannot assess anyone for anything (for an introduction and discussion on the problem of moral luck, see Nelkin 2008).

A constructionist ethics can overcome the problem of moral luck because, if moral behaviour is but one of the poietic actions of the agent, then there will certainly be at least *some* factors of which the agent had control of and that led her to be in the situation undergoing moral assessment.

The environment created by the digital revolution is a "poietically-enabling environment, which both enhances and requires the development of a constructionist ethics" (Floridi and Sanders 2003). The moral agent in such an environment is, as Floridi calls it, a *homo poieticus*. The homo poieticus focuses not only on the results of her actions in order to use and exploit them, but also on the processes that lead to those results. Thus, she is truly the 'maker', that is the creator and initiator, of both the situation she happens to be in and of the actions she decides to take. She is not simply a *homo faber*—who uses and exploits natural resources—nor simply a *homo oeconomicus*—who produces, distributes, and consumes wealth. In the infosphere, the homo poieticus herself creates and alters digital constructs. This does not necessarily mean being ourselves the creators of some digital artefact such as a computer program, or of a technological device to get connected to the internet, etc. It may simply mean using any object that takes us into the 'online' dimension. Floridi uses the example of following the instructions of a GPS: in spite of appearance, this simple and now so common action has already an online dimension. But there is more than that. As Floridi says, "as a new social space and digital environment, it has also greatly enhanced the possibility of developing egopoietic, sociopoietic and ecopoietic projects" (Floridi and Sanders 2003), that is, as the words suggest, projects about the individual as a persona, about the social environment she shares with other individuals, and about the larger environment she is in.

In Floridi's view, the 'homo poieticus' is a successful way of describing the ethical agent in the 'cyberspace' (as well as in the world 'out there') because it goes beyond the approach of 'situated action ethics' by appreciating the artefacts and the new technology, as well as the creator of these new artefacts. In other words, a constructionist ethics suits the emerging information technology exactly because it puts up front its main characteristics: the *creation* of a special kind of artefacts—the digital artefacts.

Galimberti (1999) insists that the origins of man's poietical skills are to be seen in the intrinsic biological and instinctual incompleteness of man, leading him to develop technological tools and methods to overcome this situation. It is thus in this sense that techne is the very essence of man. The thesis of an instinctual incompleteness of man, leading him to develop other skills to survive in the world, has been anticipated by a number of thinkers from Plato to Bergson, passing through Aquinas, Kant and Nietzsche. The Greeks had illustrated it vividly in the myth of Prometheus. Prometheus steals from Ephesto and Diana technical wisdom and fire and gives them to man in order to supply a lack: contrary to other races, man is naked, barefoot, and defenceless. But Prometheus could not give man the practical and political wisdom, as these were with Zeus.

In the next section I want to argue that there is much more about the homo poieticus. Whilst Floridi focused on the homo poieticus as the ethical agent, I develop this notion further: the homo poieticus is also a technoscientist and a philosopher.

4.3 The *Homo Poieticus*: Technoscientist and Philosopher

4.3.1 The Technoscientist

The Greeks were perhaps the first who tried to study the world *scientifically*, that is independently of religious questions. The Greeks were in fact interested in finding the physical principles governing the *cosmos* (Ficham 1993). Many would agree that Aristotle has indeed been a pioneering scientist, especially in the field of biology. Many others would argue, though, that science—at least in its modern acceptation—could not begin until some 'basic principles' of the Aristotelian method had been discarded. In particular, Aristotle and his scholars at the lyceum carried on scientific investigations through empirical observations and collection of facts.

The idea that the natural world is known by passive observation is in sharp contraposition with the modern conception of science and of scientific method. Arguably, more than in discarding the basic principles of the Aristotelian method, the main change in modern science concerned the introduction of new *tools* to acquire knowledge. One such new tool is *experimentation*. For Aristotle, experimentation is not a means to acquire knowledge but just a means to illustrate knowledge already acquired (for a discussion, see Harris 2005, ch. 1). The scientist, according to Aristotle, aims to establish the 'first principles'—science is *episteme*, namely knowledge of the physis through its contemplation (theoria). On the contrary, science is not *techne*, namely practical or practically oriented science. In other words, science is characterised by *noetic* goals. *Poiesis*, instead, is confined to the arts, to *techne*, and does not allow to reach the upper kingdom of *episteme*.

Let us now make a very long jump forward in time. Since the Scientific Revolution (ca 1550–1700), the natural world is a world that the scientist actively interacts with and manipulates in order to both know and create. The shift is from an 'organic' view of the cosmos, typical of the Greeks and perpetuated in the Middle-Age, to a 'mechanical philosophy' that bright and pioneering scholars such as Francis Bacon, René Descartes, Galileo Galilei and Isaac Newton started to develop. The change has been so profound that 'science' does not just connote 'knowledge' and 'understanding', but embodies, rather than opposes, also *practical* skills (Ficham 1993). It is in fact with Bacon that science becomes a *scientia operativa* (Klein 2008, 2009): to come to know about the world the scientist does not just passively observe it, but she interacts with it. The modern scientist is a *maker*; she performs experiments, namely she actively manipulates factors to find out what causes what

(Ducheyne 2005). Experiments, in Bacon's view, are tools to acquire new information, but also tools to test theories according to Galileo (Ficham 1993).

Making experiments is thus a way to make, build, construct truth—this is in opposition to an ancient truth of physis to be simply discovered. Galimberti also lucidly explaines the tension between physis and techne. He sees a deep difference between the way the Greeks and the Moderns mathematise Nature. He says (1999, p.313):

> In this respect the difference is abyssal: whilst for the Greek mathematics is the order of nature in its making itself manifest (aletheia) to man, for the scientist in the Modern age mathematics is the order that man assigns to nature, forcing it to respond to the anticipated hypotheses.[1]

In sum, there are two major innovations introduced by scholars of the Scientific Revolution: (i) in order to know we need to make, and (ii) what we know is going to be of some practical use. These are, in short, the strongholds of the concept of *techno*science. As a corollary, the *technoscientist*, as I will discuss next, is a *homo poieticus*, that is an epistemic agent that creates both crafts and knowledge.

Let us consider the creation of crafts first. The technoscientist produces the 'objects of technology', e.g. computers, nuclear weapons, medical devices. In general, these are humanly fabricated artefacts. Traditionally, Lewis Mumford proposed a categorisation of technological objects that included utensils, apparatus, utilities, tools and machines (see for instance Mumford 1934). Later on Mitcham (1994) added to Mumford's categorisation also the following: clothes, structures, and automata or automated machines. This list of technological artefacts includes 'tools of doing' and 'tools of making' alike. Needless to say, there are interesting remarks to be made about the distinctions between 'tools of doing' and 'tools of making'. Also, one may debate about alternative categorisations of technological tools. Much can indeed be learned from the phenomenology of artefacts investigating, for instance, their personal or societal effects, or the way they may extend human capabilities and, consequently, alter our experience with the external world (Ihde 1979). But I will not enter those debates here. What interests us the most is that technological objects—crafts—are the products of the *poietic activity* of the technoscientist. In other words, the technoscientist is essentially a homo poieticus. Although Floridi's homo poieticus was essentially a creator of e-nvironments, it is legitimate to extend the notion to the technoscientist because she also *creates*.

But there is another aspect of the poietic activity of the technoscientist that is of relevance here: the technoscientist creates *knowledge*. This, we shall see, is somehow the *trait d'union* between the homo poieticus in her role of technoscientist and of philosopher.

Let us then turn the attention to the creation of knowledge. As before (namely concerning the creation of artefacts), Floridi does not explicitly consider the homo poieticus to be a creator of knowledge. Yet, some insights about the technoscientist

[1] Qui la differenza è abissale: se per il greco la matematica è l'ordine della natura nel suo manifestarsi (aletheia) all'uomo, per lo scienziato dell'epoca moderna è l'ordine che l'uomo assegna alla natura, costringendola a rispondere alle ipotesi su di essa anticipate. (My translation.)

and her poietic activity in constructing knowledge can be found in Floridi's philosophy of technology. More specifically, those insights spur from the kind of epistemology that is part of Floridi's philosophy of information. Floridi is in fact interested in the relations between the natural world and information (Floridi 2010, ch. 2). Such relations will be specified within what he calls *constructionist epistemology*. Let us step back.

Recall, the digital revolution is about our being *inf*orgs in *info*sphere. This means that *information* is key in understanding ourselves, the world, and ourselves-in-relation-to-the-world. It will be worth clarifying what is meant by 'information' in Floridi's philosophy. The first thing worth noting is that information does not merely stand for 'data'. Instead, information, according to Floridi, encapsulates truthfulness, which means that information itself has already a semantic dimension. Of course, that's quite a step, and Floridi (2010, ch. 4–5) offers a number of arguments in support of this strong thesis. Consider now the relation between information and the natural world. The question ultimately concerns the localisation of information: whether there can be information without informee, and whether information can be naturalised in the sense of the semanticisation of data. This is a concern for epistemology, and not a new one. There is a sense in which, in fact, Kant, the German idealists, and the British empiricists were trying to do just that: to understand how we know what we claim we know about ourselves and about the external world (if there is any).

Whether this thesis about the semantic character of information is defendable is certainly an important problem, albeit orthogonal to the issue I am concerned with here. True, in Floridi's account, constructionism is the epistemology *for* information, but arguably what he takes constructionism to be is general enough to be possibly endorsed also by those who do not spouse his account of information. In fact, ultimately, being a constructionist has the following epistemic goal: to hold a particular view of knowledge and of knowledge building. In a constructionist approach, knowledge is the designing and modelling of reality; consequently, we, as epistemic agents, aim to design and model the features and behaviours of reality into meaningful patterns as we experience it.

Let me explain further. To hold a constructionist view in epistemology means, in Floridi's account, to put information on the threshold, as special relation/interface between nature and inhabitants. But this can be generalised further. Constructionism is to be understood in terms of an overall approach to reality. The constructionist epistemology implies an object-oriented treatment of information. Let me phrase this idea in a vocabulary that is perhaps more familiar to the reader. Although there is some objectivity and independence of existence of information (and therefore of the external world), the way we come to know the external reality depends on the agent's modes of modelling and designing it. With the Copernican and Darwinian revolutions, we probably lost our privileged location in the physical and biological realms, but we are still in a position to claim, with Kant, our centrality in the construction of knowledge of those realms.

There is a fairly recent tradition of philosophers and sociologists of science stressing this aspect of construction of knowledge and reality. One of these is Don Ihde (1991). He notices that contemporary science, unlike ancient science, is technologically embodied. In contemporary science, instruments mediate and make possible to acquire knowledge. This is, in essence, the core idea behind *instrumental* realism. Knowledge of the real, in other words, passes through instruments. In the same vein, Ian Hacking (1983) discusses the role of microscopes to see small-size entities. We, as philosophers of science, should indeed worry about the functioning of the microscope because it is the microscope, as an *instrument*, that allows us to find out about the real (micro) world.

Another voice in this 'constructionist choir' is Mario Bunge (1979b). He cogently argues that knowledge, for the technoscientist, is an intermediate goal, a means. The technoscientist mitigates a form of 'epistemological realism'—also shared by the 'pure' scientist—according to which the external world does exist and can indeed be (at least partially) known, with an instrumentalist or pragmatist attitude. Such an instrumental attitude is quite normal, given the objective of obtaining 'practical' results. Thus, the technoscientist and the 'pure' scientist may well be interested in the same scientific object or phenomenon, but whilst the object of study will be a *Ding an sich*—a thing in itself—for the pure scientist, it will be a *Ding für uns*—a thing for us—for the technoscientist.

The idea that for the technoscientist the object of study is a *Ding für uns* can be found already in Heidegger's discussion of technology (Heidegger 1954). Heidegger extends his idea of 'being-in-the-world' (already developed in *Being and Time*) to technology. This becomes 'being-in-the-world-*to-make*', thus emphasising the poietic aspects of human activities. What is more, Heidegger establishes a link between techne and episteme. Heidegger thinks that both techne and episteme 'reveal' or 'disclose' some truth, the difference being in *what* truth they reveal and *how*. The revealing of episteme is revealing a theoretical truth; this is truth of physis 'simply' to be discovered (aletheia). Instead, the revealing proper to techne has to do with poiesis, namely with a bringing-forth, or, in other words, revealing through instrumentality (on this point, see also Galimberti's discussion (Galimberti 1999, ch. 34)). However, Heidegger is against a reduction of techne to mere poeisis of artefacts. In fact, the essence of the technological is in 'enframing', in disclosing meaning through its 'instrumental' sense.

Interestingly, then, *techne* is not anymore in sharp opposition with *physis*. A difference between the two remains and it amounts to the different role the technoscientist, on the one hand, and the pure scientists, on the other hand, give to *episteme*, that is to knowledge. Whilst the latter conceives of knowledge as understanding of the principles regimenting reality *independently* of the use of such knowledge, the former is not only concerned with what can be practically taken out from this knowledge—the artefacts—but also conceives of knowledge as intervention on nature.

4.3.2 The Philosopher

What characterises the homo poieticus is her making, producing, not only (digital) artefacts but also knowledge through technoscience. I want to further argue that 'making' involves also different and, perhaps, higher spheres of the processes of 'making': producing and using thought and ideas.

Again, the seeds are in Floridi's work and hopefully the discussion that will follow will give them manure to grow. Floridi (2010, ch. 1) embraces a particular view of philosophy, namely as *conceptual engineering*: "Philosophy is the art of identifying conceptual problems and of designing, proposing and evaluating explanatory solutions."

In this perspective, philosophical investigation is neither fully logico-mathematical nor fully empirical. This view clearly goes against early stances à la Carnap (1935) and Reichenbach (1951), but also against very recent formal trends in philosophy—see for instance the work of groups in Tilburg, Leuven, or Konstanz, just to mention some scattered over Europe.

Reichenbach (1951, p. 123), for instance, expressed his viewpoint about the need for logical analysis of scientific problems thus:

> It was not until our generation that a new class of philosophers arose, who were trained in the techniques of the sciences, including mathematics, and who concentrated on philosophical analysis. These men saw that a new distribution of work was indispensable, that scientific research does not leave a man time enough to do the work of logical analysis, and that conversely logical analysis demands a concentration which does not leave time for scientific work—a concentration which because of its aiming at a clarification rather than discovery may even impede scientific productivity. The professional philosopher of science is the product of this development.

Although it is sharable that *rigour* is needed both in scientific and in philosophical investigations, another matter is to push this position too much towards a complete reduction of philosophical investigation into logico-mathematical procedures. This seems to be the direction taken by (some) leading scholars in the formal trends in philosophy (e.g., formal epistemology). Witness, for instance, Hannes Leitgeb interviewed for *The Reasoner* (4(4), www.thereasoner.org):

> I just realized I had never considered before whether there was any common thread that runs through the whole of my work. If there is one, then it is on the more methodological side really: I like to apply mathematical methods in order to solve philosophical problems. I call this 'mathematical philosophy'. Very occasionally one has some cool mathematical theorem, and one then looks for the right sort of problem to which it could be applied. But in the great majority of cases one simply comes across a philosophical theory or argument or thesis or maybe even just a clever example, and some mathematical structure presents itself—well, 'presents itself' after a lot of work!

This, needless to say, reminds us of the Leibnizian *Calculemus*. But perhaps there is more than just creating theorems in the activity of the philosopher: the poiesis of *thought*, that is of concepts and of ideas. This is the view of philosophy that Floridi advocates, and that can generally be labelled as *conceptual constructionism*. This position has eminent precursors in the recent history of philosophy. It is arguably in this sense that Bertrand Russell's commented on the use and value of philosophy are to be interpreted. He says (Russell, 1912, ch. 15):

> Philosophy is to be studied, not for the sake of any definite answers to its questions since no definite answers can, as a rule, be known to be true, but rather for the sake of the questions themselves; because these questions enlarge our conception of what is possible, enrich our intellectual imagination and diminish the dogmatic assurance which closes the mind against speculation; but above all because, through the greatness of the universe which philosophy

> contemplates, the mind also is rendered great, and becomes capable of that union with the universe which constitutes its highest good.[2]

Nevertheless, Russell doesn't tell us yet what the philosopher does exactly. Gilles Deleuze and Felix Guattari (1994) are instead much more specific about that. The philosopher, they argue, *creates concepts*. Philosophy is not just contemplation, reflection, or communication. These are activities that any discipline or science can do without claiming to do *philosophy*. Here is the lengthy passage from *What is philosophy?* (Deleuze and Guattari 1994, p. 5–6):

> More rigorously, philosophy is the discipline that involves *creating* concepts. [...] We can at least see what philosophy is not: it is not contemplation, reflection, or communication. This is the case even though it may sometimes believe it is one or the other of these, as a result of the capacity of every discipline to produce its own illusions and to hide behind its own peculiar smokescreen. It is not contemplation, for contemplations are things themselves as seen in the creation of their specific concepts. It is not reflection, because no one needs philosophy to reflect on anything. It is thought that philosophy is being given a great deal by being turned into the art of reflection, but actually it loses everything. Mathematicians, as mathematicians, have never waited for philosophers before reflecting on mathematics, nor artists before reflecting on painting or music. So long as their reflection belongs to their respective creation, it is a bad joke to say that this makes them philosophers. Nor does philosophy find any final refuge in communication, which only works under the sway of opinions in order to create 'consensus' and not concepts.

What the philosopher does is to find new concepts that explain and account for the phenomena around us. Given the ever-changing character of reality, we cannot believe that philosophy finds eternal and ever-lasting concepts. As the world changes, so do the concepts we philosophers create to make sense of it. Paradigmatic examples of concepts created by philosophers in the past are, in the eyes of Deleuze and Guattari, the '*I*' of Descartes, that is the concept of self, or the concept of One and the concept of Idea in Plato's philosophy. Deleuze and Guattari employ the term 'constructivism' exactly to denote this philosophical activity of making up concepts.

Consider now present-day philosophy. Philosophy of information invented concept of *infosphere* and *inforgs*. Philosophy of technology invented concept of *technoscience*. The corresponding sciences could not invent these concepts. The reason is that such concepts are the answers to *philosophical* questions about the surrounding phenomena, not to scientific problems. At best, scientific disciplines can give new names to scientific objects or phenomena, but these aren't philosophically loaded *per se*, or *per* the reflection of the scientist. To give another example, scientists—notably von Bertalanffy (1968)—introduced the concept of 'system' and made a start in what is now called system analysis or systemics; but *philosophers*—e.g. Bunge (1979a and 2000)—created the concept of 'system' to explain a new approach to reality and knowledge.

[2]Quoted from the online version of the book http://www.ditext.com/russell/rus15.html, accessed 4th May 2010.

This 'constructivism'—or 'conceptual constructionism' as Floridi rather calls it—can also be thought of, as mentioned above, as *conceptual engineering*. Consider again the new environments created by technology, in general, and by emergent technologies, in particular. The 'engineering' character of technologies goes beyond the creation of tools and artefacts—it calls for a *conceptual engineering* because the concepts philosophy created in the past are not fit anymore to explain all the novelties we are confronted with. Alfred Nordmann (2004) puts this idea in this straightforward way: "The ontological indifference of the technosciences needs to be complemented by a philosophical concern for the constructions of reality."

The new environments are not just the creation of the digital world, or the discovery of a world at the nano-scale that significantly differs from the world we live in. It is also rethinking the relation between science and society, that is between the scientific community and various stake holders. For instance, Ibo van de Poel (2009) interprets nanotechnology as a 'societal experiment': nanotechnology is not to be done just in the labs in isolation from the world. Instead, the lab is the whole scientific and societal environment, which includes, for instance, public debates between lay-people and regulators. Hence, here is another example of conceptual engineering: the 'old' concept of experiment cannot account for the new environments created by nanotechnologies. A new concept—'societal experiment'—had to be created to cope with these novelties.

This is in line with the philosophical challenge Carl Mitcham posed to various historical reconstructions of technology. Whether the development of technology is told in internalist (that is from the point of view engineers themselves) or externalist (that is from the point of view of humanists interested in the influence of technology on society) terms, what is of utmost importance is what ideas or concepts characterise the 'new' human making.

It will be worth noting, before closing this section, that the layman is a homo poieticus too. Although we can certainly identify the tasks and features of the homo poieticus in her clothes of the specialised ethical agent, or technoscientists or professional philosopher, we shouldn't jump to the false conclusion that poiesis is a feature that just belongs to 'academic' agents, so to speak. The layman is a homo poieticus too, in the way she interacts with world around her and with her peers, in the way she reasons about ordinary decisions and everyday issues. In other words, there is indeed continuity in our activities as homines poietici since we wake in the morning as laymen until we enter our labs of techno-science or of philosophy.

4.4 Ethics Meets Epistemology

So far, I presented the homo poieticus in the clothes of the ethical agent and I argued that she also wears the clothes of the technoscientist (who creates artefacts and knowledge) and of the philosopher (who creates concepts). In this final section I would like to draw some conclusions about what I think is *really* at stake, philosophically speaking, in this reconciliation between physis and techne, through the figure of the homo poieticus.

Let me start with an insightful quote from Carl Mitcham's work. He says that, even in making history of ideas about technology, this should be "the study of how different periods and individuals have conceived of and evaluated the human making activity, and how ideas have interacted with technologies of various sorts" (Mitcham 1994, p. 116).

Now, the homo poieticus allows us to do just that. As a 'maker', the homo poieticus embodies the many aspects of the human making activity: the creation of situations liable to be morally assessed, the creation of crafts and knowledge, and the creation of (philosophical) concepts.

Seen from the eyes of the homo poieticus, technology can be conceived of, with no further tension or contradiction, both as 'knowledge'—that is, as a means to acquire knowledge about technological artefacts as well as natural objects—and as creation of artefacts in the strict sense of the Greek $\tau\epsilon\chi\nu\eta$ or of the Latin *ars*. But in a constructionist perspective, technology can also be conceived of as an activity. Mitcham (1994) lists the following as possible technological activities: crafting, inventing, designing, manufacturing, working, operating, maintaining. Here, the activity may concern the 'action of making' or the 'process of using'.

Once we refer to the *purpose* or *end* to which the technical artefact is used for, this action is ipso facto subject to ethical evaluation. The challenge of ethical theory in response to the rise of technology is not only to enlarge its scope in order to cope with new situations—think of issues raised with regards to the environment (e.g., nuclear weapons) or to the individual (e.g., cloning, transplants), or to the consequences of information society (e.g., individual privacy, corporate security). The challenge is also, as Floridi rightly noticed, to change the ethical theory in order to cope with the roles—technoscientist, ethical agent or philosopher—man has in the era of technology. There is one word that summarises those roles—this is *poiesis*.

The original tension between physis and techne lied in forces apparently pulling in opposite directions: passive observation of the world *versus* active manipulation of it. But technology is to be seen as an opportunity for the agent to better know and act upon the world around, not as the guilty responsible of such tension. Technology asks new questions with respect to 'classical' epistemology. Interestingly enough, many of the questions and worries technology raises (and, particularly, emerging technologies such as bio- or nanotechnology), crucially depend on *what* we know about these emergent spaces of possibilities. Until we don't make clear *how* we can know about the new environments created by technology, any ethical appreciations, especially if anchored to traditional ethical accounts, will be partial and inappropriate.

In other words, if a constructionist ethics is needed (according to Floridi) for the poietic environments created by the digital revolution, a constructionist epistemology is in turn needed for a constructionist ethics (according to arguments hereby given). The reason is that, to say it with Floridi, "the chances of constructing an ethically good x increase the better one knows what an ethically good x is, and vice versa. Constructionism depends on a (satisfactory epistemic access to, or understanding of, the) relevant ontology" (Floridi and Sanders 2003).

Floridi is not an isolated voice in promoting this summit between epistemology and ethics. For instance, Ferrari (2010) urges a contextualisation of the ethical discourse within the ontological, epistemological, socio-economic, and political reflections.

Ferrari's arguments are tested against the specific case of nanotechnology; she is particularly interested in discussing the limits of ethical approaches, such as consequentialists or deontological approaches, that frame all issues in terms of cost-benefit analyses. The consequentialist, for instance, cannot make reliable predictions (due to the high uncertainties at the nanoscale) and therefore can't perform reliable risk-benefit analyses. To this pars destruens, Ferrari (2010) joins a construens pars: "A rigorous unpicking of the ways in which trust informs the work of scientists, affects their social embeddedness, and plays a role in the social construction of technology is still lacking."

Ferrari's overall conclusion is thus that epistemological issues do have a bearing on ethical issues. The main epistemological issue she identifies is, for the case of nanotechnology, the following: "The absence of a commonly accepted definition of nanotechnologies has precise epistemological implications, because it influences the setting and legitimisation of scientific research areas and therefore the scope of the research" (Ferrari 2010). But this situation is not confined to nanotechnologies. Her argument, in fact, generalises to technolog*ies* in that "the setting of goals clearly has ethical implications, because goals and aims are shaped by society and because goals are matters of research policy—in particular through priority-setting".

Floridi, recall, urged us to work towards a successful reconciliation between physis and techne. The stumbling block seems to be, though, the non-neutral character of technology. Galimberti (1999) cogently argues that the non-neutrality also stems from the fact that techne is already the environment we are *in*, not simply the object of our choice. To be sure, the tension between physis and techne arose because the Moderns, by manipulating Nature, overstep its insuperable limits. In the Greek world men cannot dominate the order of Nature but just 'revealing' it. It is for this reason that revealing the truth (a-letheia) of Nature (physis), that is contemplating Nature (theoria), leads to the kind of knowledge that regiments human action and production (praxis and poiesis). This was the origin of the supremacy of theory over praxis in the Greek world. As Galimberti accurately explains again, for the Greeks there cannot be correct technological or political action without knowledge of the immutable laws of Nature.

But the situation has changed. On the one hand, techne, that is poieis, also contributes to acquiring knowledge of the physis. On the other hand, science and technoscience do not discover immutable and eternal truths. Yet, with due amendment, we should follow the advice of the Greeks: sound knowledge of the world positively contributes to make better decisions and to take better actions both in technological and in political contexts.

In sum, a successful marriage between physis and techne, to echo Floridi, is achievable and also utterly desirable. The reason is not only a 'restyling' of the ethical agent with the clothes of the homo poieticus, but also the need of an improved awareness of the technoscientist with respect to her poietic skills. Those two should not travel on parallel tracks that never cross. Instead, they should aim to cross paths to improve our experiences of moral agents and of technoscientists. One may then wonder how to make those tracks cross one another. It seems to me that it is the task of the 'conceptual engineer', i.e. of the *philosopher*, to engage with such a poietic activity.

Acknowledgements I wish to thank Hilmi Demir for organising this volume on Luciano Floridi's philosophy of technology and for encouraging me to contribute to the debate. I would also like to thank Luciano Floridi for discussing with me the core idea of the paper at the very beginning of its gestation. Phyllis Illari was (as always!) kind enough to provide very useful and stimulating comments at the mid-stage draft of the paper. Thanks to the pressing suggestions of Cristiano Turbil, I undertook the reading of the complex work of Galimberti. Finally, financial support from the British Academy is also gratefully acknowledged.

References

Bunge, M. 1979a. *A world of systems*. Dordrecht: Reidel.

Bunge, M. 1979b. Philosophical inputs and outputs of technology. In *The history of philosophy and technology*, ed. G. Bugliarello and D.B. Doner. Urbana: University of Illinois Press. Repr. Scharff R.C. and V. Dusek. *Philosophy of technology. The technological condition. An anthology*. Malden: Blackwell, Chapter 15.

Bunge, M. 2000. Systemism: The alternative to individualism and holism. *Journal of Socio-Economics* 29: 147–157.

Carnap, R. 1935. *Philosophy and logical syntax*. London: Kegan Paul, Trench, Trubner & Co Ltd.

Carrier, M. 2004. Knowledge gain and practical use: Models in pure and applied research. In *Laws and models in science*, ed. D. Gillies, 1–17. London: King's College Publications. http://www.uni-bielefeld.de/philosophie/personen/carrier/Knowledge-Gain.PDF. Accessed 4 June 2010.

Deleuze, G. and F. Guattari. 1994. *What is philosophy?* London: Verso.

Ducheyne, S. 2005. Joan Baptiste van Helmont and the question of experimental modernism. *Physis; Rivista Internazionale di Storia della Scienza* 43: 305–332.

Ferrari, A. 2010. Developments in the debates on nanoethics: Traditional approaches and the need for a new kind of analysis. *NanoEthics*. doi:10.1007/s11569-009-0081-z.

Ficham, M. 1993. *Science, technology, and society. A historical perspective*. Dubuque: Kendall-Hunt Publishing Company.

Floridi, L. 2008. Artificial intelligence's new frontier: Artificial companions and the fourth revolution. *Metaphilosophy* 39(4/5): 651–655.

Floridi, L. 2009. The fourth revolution. *Newsweek*, Japanese ed. http://www.philosophyofinformation.net/publications/pdf/newsweek-article.pdf. Accessed 4 June 2010.

Floridi, L. 2010. *The philosophy of information*. Oxford: Oxford University Press.

Floridi, L. and J.W. Sanders. 2003. Internet ethics: The constructionist values of *homo poieticus*. In *The impact of the internet on our moral lives*, ed. R. Cavalier, 195–214. New York: SUNY.

Galimberti, U. 1999/2009. *Psiche e techne. L'uomo nell'età della tecnica*, 7th ed. Milano: Feltrinelli.

Hacking, I. 1983. *Representing and intervening*. Cambridge University Press.

Harris, R. 2005. *The semantics of science*. London: Continuum.

Heidegger, M. 1954. Die Technik und die Kehre. In *Vorträge und Aufsätze,* 13–44. Pfullingen: Günter Neske Verlag. English translation from Martin Heidegger. 1993. *Basic writings*, rev. ed, 311–341. New York: HarperCollins, Inc. Repr. Scharff R.C. and V. Dusek. *Philosophy of technology. The technological condition. An anthology*. Malden: Blackwell, Chapter 23.

Ihde, D. 1979. *Technics and praxis: A philosophy of technology*. Boston: Reidel.

Ihde, D. 1991. *Instrumental realism. The interface between philosophy of science and philosophy of technology*. Bloomington: Indiana University Press.

Klein, J. 2008. Francis Bacon's scientia operativa, the tradition of the workshops, and the secrets of nature. In *Philosophies of technology: Francis Bacon and his contemporaries*, ed. C. Zittel, R. Nanni, G. Engel, and N. Karafyllis. Leiden/Boston: Brill E-Books. doi:101163/ej.9789004170506.i-582.1. Accessed 02 Feb 2010.

Klein, J. 2009. Francis Bacon. In *The Stanford encyclopaedia of philosophy,* Spring 2009 ed, ed. Edward N. Zalta. Stanford: Stanford University. http://plato.stanford.edu/archives/spr2009/entries/francis-bacon/. Accessed 02 Feb 2010.

Mitcham, C. 1994. *Thinking through technology*. London: The University of Chicago Press.
Mumford, L. 1934. *Technics and civilisation*. New York: Harcourt Brace.
Nelkin, D.K. 2008. Moral Luck. In *The Stanford encyclopedia of philosophy,* Fall 2008 ed, ed. Edward N. Zalta. Stanford: Stanford University. http://plato.stanford.edu/archives/fall2008/entries/moral-luck/. Accessed 4 June 2010.
Nordmann, A. 2004. Collapse of distance. Epistemic strategies of science and technoscience. *Danish Yearbook of Philosophy* 41: 7–34. http://www.unibielefeld.de/ZIF/FG/2006Application/PDF/Nordmann_essay2.pdf. Accessed 4 June 2010.
Reichenbach, H. 1951. *The rise of scientific philosophy*. Berkeley/Los Angeles: University of California Press.
Russell, B. 1912. *The problems of philosophy*. Oxford: Oxford University Press.
Turing, A.M. 1950. Computing machinery and intelligence. *Mind* 59: 433–460.
van de Poel, I. 2009. The introduction of nanotechnology as a societal experiment. In *Technoscience in progress. Managing the uncertainty of nanotechnology*, ed. S. Arnaldi, A. Lorenzet, and F. Russo. Amsterdam: Ios Press.
von Bertalanffy, L. 1968. *General system theory: Foundations, development, applications*. New York: Braziller.

Part II
The Information Revolution and Alternative Categorizations of Technological Advancements

Chapter 5
In the Beginning Was the Word and Then Four Revolutions in the History of Information

Anthony F. Beavers

5.1 A Running Start

In the beginning was the word, or grunt, or groan, or signal of some sort. This, however, hardly qualifies as an information revolution, at least in any standard technological sense. Nature is replete with meaningful signs, and we must imagine that our early ancestors noticed natural patterns that helped to determine when to sow and when to reap, which animal tracks to follow, what to eat, and so forth. Spoken words at first must have been meaningful in some similar sense. But in time the word became flesh (corpus) and dwelt among us, as "inscription" (literally, to put into writing) inaugurated the dawn of human history. This did not happen instantly. One place to enter the story is with clay tokens to represent trade transactions that in time became accounting tablets and, then, the world's first literature (Enmerkar and the Lord of Aratta, The Epic of Gilgamesh, etc.) and codes of law (The Codes of Ur-Nammu, Lipit-Ishtar, Hammurabi, and so forth.) This event happened around the north shore of the Persian Gulf sometime in the 4th millennium BCE and was enshrouded in mystery as the role of the scribe trained in the art of inscribing and deciphering signs belonged to the priest (Deibert 1997). With the sanction of religion, writing gave birth to "civility" (literally, life in the city) and defined the line between "history" and "pre-history," the latter being a term designating everything that happened before. There is little doubt that the invention of writing was significant and that it deserves recognition as the first revolution in the history of information. Life as we live it today would have been impossible otherwise.

Innovations in writing technologies happened with significant effects, but at various points in the history of information, changes in technology were so dramatic that they reshaped the course of human history in radical ways. The revolution in printing

A.F. Beavers (✉)
Philosophy and Cognitive Science, The University of Evansville,
Evansville, IN, USA
e-mail: afbeavers@afbeavers.net

H. Demir (ed.), *Luciano Floridi's Philosophy of Technology: Critical Reflections*,
Philosophy of Engineering and Technology 8, DOI 10.1007/978-94-007-4292-5_5,

is well-studied; the invention of the printing press and movable type (c. 1450) has been credited as the catalyst for the Reformation (sixteenth–seventeenth centuries) and for allowing the Renaissance (fourteenth–seventeenth centuries) to take hold, both as necessary contributors to the Enlightenment (seventeenth–eighteenth centuries), which gave birth to the modern state and innovations in philosophy and science (Martin 1993; Deibert 1997; Eisenstein 2005). A ripple effect followed the printing press requiring reassessment of the theological enterprise that redefined our understanding of the human being's place in the world and the cosmos, as we went from being an *imago dei* (a divine "imprint" made in the image of God) living in nature, God's creation just outside the Garden of Eden, to human individuals set afloat in a solar system, though quite able and endowed with curiosity and reason.

More transformative still was the revolution in information technologies that began in the middle of the nineteenth century. The invention of the Daguerreotype (1839) signaled the birth of practical photography; and other mechanical and electrical technologies including the telegraph (1836), the telephone (1877), the phonograph (1878), radio (1906) and television (1926) made a multiplicity of informational media move quickly, crossing spatial and temporal boundaries at an alarming rate to bring a world of people closer in the span of a few short years. The rise of the modern corporation and, of course, international, world-wide warfare are tied inextricably to this information revolution, since they could not have emerged without them, along with tools to allow friends and family members to migrate across geographical locations while remaining "in touch."

Of recent interest and often credited as the start of the information age is what we might call the "digital revolution" that began with Alan Turing (1937) and firmly took hold with the popularization of the PC in the 1980s. It accelerated the flow of multimedia information so far beyond what was possible in the previous era that even information visionaries like Thomas Edison and Alexander Graham Bell could not have imagined its extent, though, as we will see, they anticipated it nonetheless. More so, the introduction of computers into communications technologies added another dimension to this history by introducing automated information processing. No longer was informational technology restricted to the mere storage, transmission and retrieval of information; machines could be built to manipulate it as well.

We live in this context today. Inter-networked digital technologies afford communications between human and artificial information processors (both "inforgs" in Floridi's language) that interact together in a collective space (the "infosphere") to produce a collective body of information that is archived for easy retrieval. Of course, these technologies have produced their own variety of toys and with them mechanisms for several forms of social interaction that range from the trivial (though not un-importantly) entertaining to the educationally, and even interpersonally, complex. No doubt, something major is happening around us informationally by the addition of automated digital information processing to the technological affordances of previous generations. Sitting at the start of what will no doubt be an unimaginable transformational revolution involving everything human and historical, it is impossible to know now what all of it can mean. But we see its effects emerging as the geopolitical scene explodes into a global arena populated with multi-national corporations richer than many countries and where the mechanisms of civil (and uncivil!) control rely significantly on the politics of information flow, all the while comprehending it through the lenses of computer-mediated information technologies and interacting with each other via email, text message, chat client, Twitter, and other social networking sites such as Facebook.

A transformation of this magnitude must certainly qualify as a revolution, a fourth one in the history as I have here outlined. For the sake of clarity in what follows, I name them the (1) Epigraphic, (2) Printing, (3) Multimedia, and (4) Digital Revolutions, making no claims to have discovered them, since each has been studied in extreme detail. In what follows, I will comment on each revolution in turn before offering a discussion spawned by Floridi's notion of "the Fourth Revolution" (see 2008, 2009, 2010, for instance), which corresponds to the last I have enumerated here. Though we name the fourth in common, Floridi's three previous revolutions are designated differently. I say this without criticism, because he intends to draw out the implications of the "Fourth Revolution" in different relief. That is, he largely situates his revolutions "in the process of dislocation and reassessment of humanity's fundamental nature and role in the universe" (Floridi 2009, p. 156). Thus, he is primarily concerned with shifting identities (of both self and world) across revolutions and the philosophical implications of such. My comments are of a more historical nature. Nonetheless, because this reflection is offered as broad commentary on the context in which Floridi situates the "Fourth Revolution," it is important to say something about his taxonomy. Perhaps it is best here to let him speak for himself:

> Science has two fundamental ways of changing our understanding. One may be called extrovert, or about the world, and the other introvert, or about ourselves. Three scientific revolutions have had great impact in both ways. They changed not only our understanding of the external world, but also our conception of who we are. After Nicolaus Copernicus (1473–1543), the heliocentric cosmology displaced the Earth and hence humanity from the centre of the universe. Charles Darwin (1809–1882) showed that all species of life have evolved over time from common ancestors through natural selection, thus displacing humanity from the centre of the biological kingdom. And following Sigmund Freud (1856–1939), we acknowledge nowadays that the mind is also unconscious and subject to the defence mechanism of repression. So we are not immobile, at the centre of the universe (Copernican revolution), we are not unnaturally separate and diverse from the rest of the animal kingdom (Darwinian revolution), and we are very far from being Cartesian minds entirely transparent to ourselves (Freudian revolution). (Floridi 2009, p. 156)

To be clear, I do not doubt the historical reality of these revolutions and the meaning that Floridi attaches to them, even though we must recognize that any such talk is pretty coarse grained (and Floridi does, as I do of my own views here). However, we could just as well have added the "Marxist Revolution" into this mix, citing Marx's conception of human beings as workers situated in a network of bureaucratic relations in the midst of industrial and economic transformation and the incredible efficacy it enacted on the geopolitical stage. This would make Floridi's "Fourth Revolution" a 5th, and possibly a 6th or 7th, depending on the how one carves out history. There is also the philosophical question of whether the named revolutions have come and gone or whether they continue to fight it out in the effort to reinterpret who we are (see Floridi 2008). (Consider, as an example, the battle that continues between creationism and evolution in the United States.) These are mere quibbles, since it is clear that Floridi enumerates his revolutions to provide a context for characterizing what is happening today as a result of life within the infosphere. A taxonomy of every historical revolution that has influenced our understanding of human identity and its context is not his immediate concern. (To be sure, this would be an impossible project, in any case.)

Floridi, of course, is not blind to the fact that the information revolution could be said to begin with writing, noting that this historical usage is "not what is typically meant by the information revolution" (2010a, p. 4). Nevertheless, casting the Digital Revolution against the backdrop of these others (the Copernican, Darwinian, Freudian, etc.) lends focus on what to target in analyzing the information age; so perhaps something complementary can be said, if Floridi's "Fourth Revolution" were to be plotted on the trajectory of the history of information flow itself. My hope here then is to resituate this central concept in Floridi's work for just a moment to help fill out the context for the philosophy of information. To this end, I will present a short (over-generalized and abridged) characterization of each revolution as I have laid them out, and then offer a bit of discussion. The next section presents caricatures that I hope true enough in their generalities to set the stage for comment.

5.2 Four Revolutions in the History of Information

5.2.1 The Epigraphic Revolution

When speech takes to writing, it transcends the moment to make its mark in space. Whether this occurrence is a recipe for remembering or forgetting, as Plato questions in his *Phaedrus*, the event signals the spatialization of temporal information and the emergence of an early form of hard storage useful because it off loads information from a brain into a shared environment. Its elegant simplicity is almost unfortunate, since it easily leads us to overlook its magnitude; in fact, only fairly recently has research on the impact of cognitive technology (e.g., Norman 1994; Clark 1997, 2001) made this significance clear. Marks of some sort serve as "stand ins" for (or representations of) words, things or ideas that are etched onto a surface that preserves them for however long. A technique governs this art that in essence inscribes temporal streams of thought into a spatial arrangement in the act of writing itself to be temporally resequenced later in the act of reading. The precise spatial arrangement is unimportant, whether proceeding from the top of the "page" to the bottom, from left to right, right to left, or alternating back and forth in tracks like those left by a plow, so long as the technique of reading follows the proper order for deciphering signs.

Other technicalities (literally) are quite important. The encoding strategy (whether using pictographs, ideographs, logographs, a syllabary or letters of an alphabet) is critical, because it determines the granularity of information that can be encoded, regulates variability in information compression, and impacts the spread of literacy. The materials upon and by which such symbols are imprinted are equally critical, since these affect the preservation and transmission of information and, in so doing, also affect the spread of literacy. Much of the transformation in information technologies during this revolution, in fact, are best understood as innovations concerning one or the other of these.

In the Near East, Cuneiform, the earliest known writing system, encoded information by pressing cone-shaped marks with a wedge-shaped stylus into wet clay and then baking them into tablets. Initially, it was a pictographic system that slowly evolved into a syllabary and was used to encode several languages, until it died out around the beginning of the A.D. period (Green 1989). Though not all ancient writing systems evolved from pictographs to alphabets, Egyptian Hieroglyphics did as well (Fischer 1989). The Phoenician alphabet, developed much later in mid-eleventh century BCE, however, coded for sound from the beginning and was carried by merchant trade into the Mediterranean to form the basis of the Hebrew, Aramaic, Greek and Latin alphabet, ultimately becoming the modern alphabets we use in the West today (Logan 1986).

Using symbols to code for sound is informationally efficient since it lessens the load on our cognitive abilities and, perhaps more importantly, *immediately allows anything that can be said to be written and read.* Additionally, by combining a small number of signs to represent a large number of words, reading and writing become easier to master, thereby encouraging the spread of literacy, and, according to some, engendering civilization itself (Logan 1986).

Equally important to this early spread of information were the materials involved in writing. This is partly because textual information is transmitted (at this point in the story) by physically transporting texts, a process best served when texts are most mobile and durable. Clay tablets are heavy and easily broken, so, in time, writing with ink on papyrus will create lighter more transportable texts. (The fact that papyrus was only available in the Nile valley may explain in part why Western civilization firmly took hold around the Mediterranean, despite its origins in Mesopotamia.) But both the brittleness and scarcity of papyrus led to the development of parchment, which could be processed anywhere animal hide (sheep, cows, goats, rabbits and even squirrels) was available (Deibert 1997). Parchment is more durable than papyrus and, thus, more safely transported, enhancing both the transmission and preservation of information. Other innovations happen concerning both encoding strategies and materials that are important to the story, but here omitted in the interest of saving space.

Technological innovations either offer affordances that did not exist prior or improvements on existing ones. This fact is precisely what makes them innovative. Understanding affordances in this context is a matter of specifying what a given technology allows or permits that was not possible before. Accordingly, technological innovations are best thought of as necessary, and only sometimes sufficient, conditions for social, scientific and technological change. A durable, transportable means of transmitting speech through text affords several things of importance, though in the broadest terms, it expands the scope of information flow over temporal and spatial (and thus historical and political) boundaries. Whether *polis* follows *logos*, or *logos polis*, civility is irrevocably tied to the spread of information. Where one goes, so does the other; and as the lines of textual dissemination go farther and faster, polis grows to empire. The Epigraphic Revolution is thus tied to the age of civilization. Soon, however, Christianity will learn the power of the word, and as people learn to worship it, the Church will rise to curator of ancient wisdom. The city will decline to reawaken toward the end of the Middle Ages, in the thirteenth century, about the time that paper is introduced into the West from China and the great Medieval universities were founded. We still live in the wake of this reawakening.

5.2.2 The Printing Revolution

The Renaissance began in the fourteenth century before the printing press (c. 1450), advocating a new humanism and supplying a need for texts. Even long before the Renaissance, books were already on the scene (Diringer 1982). Though the revolution in printing follows a spark that it therefore could not have ignited, it nonetheless can be credited in large part for contributing to the Enlightenment, including innovations in philosophy, politics, mathematics and science that brought with them a new worldview and a new sense of self-awareness (Deibert 1997). It definitely facilitated the Reformation which depended on the quick duplication and the widespread dissemination of texts (Deibert 1997; Edwards 1993; Eisenstein 2005). So, when the fifteenth century opened, inventors and an industry were ready in wait to respond with what might best be described as the *mass production of writing*. They moved quickly too. Citing Saxby (1990) and Febvre and Martin (1976), Deibert (1997) aptly describes the situation: "About 20 million books were printed before 1500 in Europe among a population at the time of about 100 million. This number of books, produced in the first fifty years of printing, eclipsed the entire estimated product of the previous thousand years" (p. 65). He goes on to note that "Febvre and Martin estimate that 150 million to 200 million were then produced in the next hundred years."

Of course with the demand for books, an industry immediately responded. Deibert continues:

> By 1475, printing workshops had been established throughout the Rhineland, and in Paris, Lyons, and Seville. By 1480, printing centers had sprouted through all of Western Europe … in all to 110 towns.... By 1500, the number of towns … had risen to 236. By the sixteenth century, western Europe had entered a new communications environment at the center of which were cheap, mass-produced printed documents emanating from the many printing presses stretched across the land. (pp. 65–66)

Deibert's depiction of this printing revolution is eclipsed by the language of exabytes, quantities so large we have no practical sense of how big they actually are. But in their day, these numbers are significant. To put this in context, the population of Europe at the time of the printing press was one quarter the size of Facebook's today.

Thus, one can perhaps surmise that by 1600 there were approximately two books in circulation in Europe for every literate and non-literate person, nothing like what we will see in terms of the information explosion of today, but significant nonetheless.

It was especially significant in terms of resituating authority and creating a spirit of individualism. By becoming the primary vehicle through which the Protestant Revolution would take hold, the Printing Revolution challenged the hegemony of the Catholic church. Equally importantly, as Lawhead (2002) points out, is that it engendered a sense of *epistemic* Protestantism as well. Just as with regard to theology Protestantism provided the faithful with a direct line to God, human beings were resituated with regard to the study of what was the case. *Individual* minds were now conceived as having direct access to a truth that could be discovered by following the proper methods. The scientific revolution unfolded in this light, and with it, the sense of rationally-enlightened individualism that would support the rise of our modern democracies. Coupled with the rapid increase in texts published in the vernacular, a new sense of national identity also emerged (Deibert 1997).

In broad terms it seems fair to say that by the eighteenth century it was more fashionable to be a well-informed individual than a child of God, or, at least, that God had been redefined as a divine architect whose essence could be read directly off of "the book of nature," in which case being a child of God meant being a well-informed inquirer in pursuit of truth, metaphorically "enlightened" no longer by mystery or divine inspiration, but by reason. The appearance of two texts bear witness to this transformation even in their titles: John Toland's *Christianity Not Mysterious: or, a Treatise Shewing That There Is Nothing in the Gospel Contrary to Reason, Nor Above It, and That No Christian Doctrine Can Be Properly Call'd a Mystery*, first published in 1696, and Matthew Tindal's *Christianity as Old as Creation; or, the Gospel as a Republication of the Religion of Nature*, published in 1730. In the beginning was the word, and in the emerging religion of the Enlightenment it was *printed* in nature itself and *republished* in the form of *scripture*.

The printing press, and indeed the printing metaphor itself, will thoroughly take hold before the eighteenth century closes, spreading literacy, a new authority in a new institution of authorship, and a collection of enlightened minds, empowered and able to govern themselves as informed citizens of democratic states. Indeed, as a result of the Printing Revolution, the word was now set free. Though several will try in subsequent generations, there will be no taking it back, and as free inquiry, individual invention and experimentation carry us through the next century and physics transforms into mechanical and electrical engineering, the flow of information itself will be industrialized. We still live in this era of industrialized information flow.

5.2.3 *The Multimedia Revolution*

The Multimedia Revolution started with a distant sound beeping out in dashes and dots, taking letters that originally code for sound and matching them to other audible patterns that could be easily sent over a wire. Just two tokens could represent every letter, readily affording the transmission of writing over distances. This event is significant because it decoupled the flow of information from the exigencies of transportation technology. Where prior the transmission of text was by physically moving it around, now it could move on its own independent of the courier, caravan and wagon cart. Before this revolution is finished, technology will increase the speed of transmission in ways never before imaginable, transcending the wires in time to take to the airwaves, sending moving text, pictures and sound directly to our living rooms thanks to the marvels of radio and television (Winston 1998).

The history of technological innovation during the Multimedia Revolution is convoluted and complex. Even trying to describe it primarily in terms of the reach of information would exceed the space allowed, since the industrial revolution industrialized information flow itself, providing a sudden escalation in the development and spread of information-based technologies. Some of these (along with their approximate date of invention) include: Telegraphy in 1836; The Daguerreotype in 1839; The Telegraphic Printer in 1856; The Stock Ticker in 1863; The Telephone in 1877; The Phonograph in 1878; The Light Bulb and the Photophone in 1880; Wireless Telegraphy, Wax Cylinder Phonography and the Motion Picture Camera, all in 1891; The Rotary Telephone in 1898; Radio and Teletype in 1906; Television in 1926; Electric Phonography in 1927; and Magnetic Tape in 1928. Innovation continued into the second half of the twentieth century with Cable Television in 1948; Cassette Tape Recorders in 1958; Touch Tone Phones in 1963; Color Television in 1966; and the VCR in 1969.

Though innovation in information technologies occur to this very day (altered greatly by the digitization of information and the sudden popularity of the personal computer in the early 1980's), even early on in the Multimedia Revolution major effects were already being felt (Beavers and Sigler 2010). Just 34 years after the invention of the telephone, a full length history of it appeared. Herbert Casson's *History of the Telephone* in 1910 paints a vivid picture of the social changes engendered by its arrival on the scene. He writes:

> What we might call the telephonization of city life, for lack of a simpler word, has remarkably altered our manner of living from what it was in the days of Abraham Lincoln. It has enabled us to be more social and cooperative. It has literally abolished the isolation of separate families, and has made us members of one great family. It has become so truly an organ of the social body that by telephone we now enter into contracts, give evidence, try lawsuits, make speeches, propose marriage, confer degrees, appeal to voters, and do almost everything else that is a matter of speech. (p. 199)

When we look back from the perspective of today, it might initially seem that the trajectory of technologies that culminate in our networked world was accidental. But the inventors behind this revolution were conscious of what they were doing, what was happening around them as a result and where we were headed with regard to information technology. Of the ten affordances that Edison promoted with the invention of the phonograph, using it to record music is named fourth. Distance education, or at least asynchronous learning outside the presence of a teacher, is indicated in his list. More importantly is what we might call the Edison/Bell vision of an information network. Enumerated last, Edison notes that one affordance of the phonograph is "connection with the telephone, so as to make that instrument an auxiliary in the transmission of permanent and invaluable records, instead of being the recipient of momentary and fleeting communication" (Edison 1878). We can easily see here a system of hard storage accessible over telephone lines, a point emphasized more poignantly by the fact that Turing noted in 1946 that his ACE computer could also be connected to the telephone system (Hodges).

Though initially tied to wires, almost immediately, Bell and others were working on wireless telephone transmission. A variety of techniques were tried; Bell's favorite invention, the photophone, invented in 1881, for instance, could send signals 200 yards on a beam of light (Bell Family Papers 1862-1939), thereby anticipating modern fiber optic information transmission. Furthermore, long before the turn of the century, the promise of a global communications network with and without wires was in place, so much so that in a lecture at the Imperial Institute in 1897, W. E. Ayrton made an apt prediction:

> There is no doubt that the day will come, maybe when you and I are forgotten, when copper wires, gutta-percha coverings, and iron sheathings will be relegated to the Museum of Antiquities. Then, when a person wants to telegraph to a friend, he knows not where, he will call in an electro-magnetic voice, which will be heard loud by him who has the electro-magnetic ear, but will be silent to everyone else. He will call, 'Where are you?' and the reply will come, 'I am at the bottom of the coal-mine' or 'Crossing the Andes,' or 'In the middle of the Pacific'. (Fahie 1900, p. vii)

Regular cell phone service still might not reach to the bottom of a mine shaft or to the top of the Andes, but Ayrton's prediction was still correct in its generalities. The world was about to change, and these early inventors knew it even before the twentieth century began. Bell's introduction of helpful services not only answered a need for the telephone in society, but soon people would wonder how they ever lived without it. By mid-century the backbone and the vision of an information superhighway was firmly in place awaiting the digitization of information. Something of global significance was about to happen. We now live in the early days of this transformation.

5.2.4 The Digital Revolution

The statistics on the reach of information because of the development of digital technologies and, in particular, the Internet, are staggering. Citing a variety of sources, Floridi paints a decent portrait. So, let us again let his picture speak for itself:

> To have some simple, quantitative measure of the transformations experienced by our generation, consider the following findings. In a recent study, researchers at Berkeley's School of Information Management and Systems estimated that humanity had accumulated approximately 12 exabytes of data in the course of its entire history until the commodification of computers, but that it had produced more than 5 exabytes of data just in 2002: 'print, film, magnetic, and optical storage media produced about 5 exabytes of new information in 2002. Ninety-two percent of the new information was stored on magnetic media, mostly in

> hard disks. [...] Five exabytes of information is equivalent in size to the information contained in 37,000 new libraries the size of the Library of Congress book collections' (Lyman and Varian [2003]). In 2002, this was almost 800 MB of recorded data produced per person. It is like saying that every newborn baby came into the world with a burden of 30 feet of books, the equivalent of 800 MB of data on paper. This exponential escalation has been relentless: 'between 2006 and 2010 [...] the digital universe will increase more than six fold from 161 exabytes to 988 exabytes.' (2009, p. 154)

Indeed, in the current age, the reach of information continues to accelerate so quickly that talk about any one or two people or any few technologies would be too incidental to be informative. Furthermore, the speed of transformation is so rapid that it is almost impossible to write about information technology at the level of specifics. Someone starting a book on Facebook today, for instance, would have to worry about whether the network dynamics that it supports would still pertain by the time the book was finished. This situation is especially pressing when it comes to legislation concerning the regulation of information flow. Even a cursory inspection of the current political landscape shows that the laws simply cannot keep up. As a consequence, this is all I intend to say about the specifics of the Digital Revolution save for some analytic points offered in the discussion below.

5.3 Discussion

5.3.1 Unifying and Differentiating These Information Revolutions

The tracks cut into history by the current exposition are way too broad, even if viewed only from the perspective of the history of information. The control of information in each of these ages by civil, religious and economic authority means that a *politics of information* and, equally, an *economics of information*, must be taken into account in understanding these historical transformations, along with the role that the computational sciences (math, logic, computer science and computer engineering) exercised in processing information and advancing human understanding. Outside of informational phenomena, a variety of other scientific and technological changes must also be considered. The transportation industry itself continues to support the circulation of information, as it did early on. Today, planes, trains and automobiles afford easy changes in physical presence as people come together across the globe to visit, speak and exchange ideas. And, of course, the history of educational institutions themselves and their curricula is more significant than many are inclined to acknowledge. Even so, a broad outline of these information revolutions in terms of the history of information technology tells a salient part of the story.

Where there is change, something must remain the same, or we are dealing with entirely different phenomena. A revolution implies a change and thus occurs in the wake of the one that came before, preserving something of what was there as one epic unfolds into the next. Thus, revolutions should be thought of as overlapping waves, rather than a sequence of different eras. This seems explicitly clear in the case of information revolutions. The fact that I am sitting here writing text through the medium of a digital computer while using a variety of computational tools to access text for research connects me to the Epigraphic and Digital Revolutions. Reading and writing have yet to vanish, and human beings still think through the vehicle of words. That I'm writing this at 11:30 p.m. in my study lit by bulbs indicates that I remain bound to the Multimedia Revolution, the television on with the sound down and airing news of the Gulf oil spill, as I listen to a (digitized) Strauss Opera on iTunes. Furthermore, that this "paper" will be disseminated through the vehicle of the publishing industry still shows vestiges of the Printing Revolution. But these superficial traces barely touch the unifying elements that tie these revolutions together.

These common elements are founded on old ideas, though transposed into the language of the Digital Revolution, they can sound quite new. This is unfortunate, because it lends the appearance that we are reading the future back into the past when, in fact, we are not. Whether coded digitally and sent over the airwaves or coded alphabetically and pressed into a tablet, information is encoded, stored, transmitted and received. These basic elements thus comprise the unifying components of continuity across epochs, until the Digital Revolution adds what might appear at first as a new technological affordance, namely, information processing. (I use the word "appear" here for reasons that will be clear in the next section, even though *mechanical* information processing is indeed reserved for the advent of automated computational devices.) Though this is not insignificant, the primary substrate for the changes from one revolution to another concern then the kind of information that can be stored and transmitted, the speed of information transmission, its preservation, and its reach. Indeed, these elements allow information to transcend the moment to make its mark in space and time, thereby allowing it to cross temporal and spatial boundaries.

By contrast, what changes are the specific techniques and technologies that allow these elements to have their play. Thus, with the invention of writing we see a technology for off loading information into the environment. As other changes in information encoding and improvements in the materials for information storage are made, the speed of information transmission and its reach increase. Minor technological improvements (with major historical consequences) occur until the invention of the printing press, which affords a sudden escalation in the speed of information flow and its reach, because mass produced text allows information to travel along various routes in parallel inexpensively. Multiple copies of a text stored in various places, of course, also affect the preservation of information. As we move through the Printing Revolution, this escalation in the reach of information is inextricably tied to the collapse of the Medieval world and the rise of the Enlightenment, which brought with it new understandings of self and world.

The industrialization of information flow that began at the end of the nineteenth century represents yet another sudden leap in the speed of information transmission and its reach, but this time with machines that could also move pictures and sound, not just text. It also decoupled the mobility of information from the transportation industry.

Telephone, teletype, radio and television, significantly ushered in a new world order only on the basis of which could something like a World War be possible. It also allowed a new kind of communicative presence between persons (and nations!), both synchronous and asynchronous, that brought interlocutors together without making them present in the flesh.

As the technologies of the Multimedia Revolution start moving digitized information and new digital machines emerge, we find ourselves once again at the beginning of an unfathomable leap in the availability of information, the speed of its transmission and its reach. To overstate the case just slightly, massive amounts of information are globally ubiquitous, though respect requires that we acknowledge a new division between the information rich and the information poor. The Digital Revolution affords such easy mobility of information that one-on-one audio-visual communication via tools like Skype, private news sources in the form of blogs with international readerships, and the fact that anyone anywhere can make a movie for all the world to see are quickly becoming omnipresent, so quickly, in fact, that it is impossible for governmental legislation and scholarly analysis to keep up. Even so, does the transition from the Multimedia to the Digital Revolution represent a mere difference in degree, more information moving faster and farther, or is something different in kind also going on?

5.3.2 Technological, Scientific and Cognitive Co-incidence

The study of technology as it impacts culture is primarily a concrete affair. This is because sometimes technologies lead and sometimes they follow other cultural developments. In looking at the impact of technology, then, in one sense it is inappropriate to generalize, each case being unique into itself. However, this is not the end to understanding this story if our revolutions can be plotted on a common trajectory, that is, on a continuum of development regarding the same set of technological affordances or some other underlying commonality. So far, they have concerned the kind of information that can be stored and transmitted, the speed of information transmission, its preservation, and its reach, with information processing arriving with the computer. Thus, it looks at first glance that we have a discontinuity at this point with the Digital Revolution, since it adds rather than merely develops an affordance. But this is not entirely correct. Seeing that this is so requires broadening the perspective of information technologies to include their function. Such technologies are in fact only useful in their use, not as ornaments, and in this regard one critical import is that they connect us together, which means that at bottom information techniques and technologies are always instantiated in a network of informational relations. Thus, one way to plot the trajectory of the history of information revolutions is to look back at the evolution, not merely of particular technologies, but also of the networks they enable. Characterizing information technologies as embedded in networks along with the human beings who use and communicate through them provides one unifying factor to set the historical context of our current Digital Revolution.

Prior to the Epigraphic Revolution, speech and gesture connected us to each other through the mediation of sound and vision. These early networks, very much physical, mechanical and biomechanical phenomena, connected brain to brain by way of sound and light. Without writing, however, communication happened within the physical proximity of respondents, even if oral networks allowed word to be passed along. Two things are of critical importance here. Even from the start, information processors did exist on these early networks, not in the form of technologies, but in the human beings that used them. Thus, information processing does *not* first arrive with the Digital Revolution, though the invention of technologies to automate this task does. Second, these early networks were bound by the limits of space and time to set informational boundaries around the tribe and its environment. In this context, writing emerges as a form of hard storage, buffering information physically in materials ("off line," so to speak). By its very nature, text affords asynchronous information flow, while providing ready memories that individuals and generations can look back to later for themselves. From the network perspective, distal physical relations now become informationally relevant. As literacy spreads and materials improve, so too does the speed of information transmission and, with it, the boundaries of the tribe expand first to the city then the empire. Information networks, in other words, collide and assimilate giving rise in time to the *cosmopolitanism* of the Hellenistic period.

Additionally, insofar as information technologies connect human information processors, it should not be surprising that as networks expand, invention and knowledge production increase. Indeed, from the cognitive perspective, even on the level of the individual human processor writing enables discovery and invention (Clark 1997), thereby increasing the general intelligence of human beings. Add to this fact faster lines of dissemination across a wider range of informed human information processors, and we should expect an escalation in our cognitive capacities both as individuals and as collectives, and indeed this is precisely what happens.

Replication also affects the speed of transmission, since the same information can travel simultaneously by different routes. The mass production of writing in the Printing Revolution thus could be characterized as an explosion in the general size and scope of information networks, as better informed human processors endowed with faith in reason are set free to think on their own, though within the confines of common languages and ideas provided by the collective. Whether for better or worse, the Multimedia Revolution transcends the text, and in so doing, invites a new kind of easily-acquired public literacy in the form of motion pictures, television, radio and so forth. Indeed, already by the third and fourth decades of the twentieth century, the transformation is well underway, enabled by new encoding strategies and the development of public "airwaves" that allow single streams of audiovisual transmissions to travel instantaneously to anyone wanting to "tune in."

Limited by what was primarily a one-way communicative relationship, *mass* media largely situated information providers and information consumers asymmetrically, even though viewers and listeners could write or call in to a television or radio show or perhaps communicate on a widespread scale with others by capturing the media's attention. Even though several information networks were largely asymmetrical, they nonetheless collided with other networks, once again expanding the reach of information, but with a cost. The situation was aptly summarized by Emmanuel Levinas in 1982, who characterized society at the time as one …

> whose boundaries have become, in a sense, planetary: a society, in which, due to the ease of modern communications and transport, and the worldwide scale of its industrial economy, each person feels simultaneously that he is related to humanity as a whole, and equally that he is alone and lost. With each radio broadcast and each day's papers one may well feel caught up in the most distant events, and connected to mankind everywhere; but one also understands that one's personal destiny, freedom or happiness is subject to causes which operate with inhumane force. One understands that the very progress of technology—and here I am taking up a commonplace—which relates everyone in the world to everyone else, is inseparable from a necessity which leaves all men anonymous. Impersonal forms of relationship come to replace the more direct forms, the 'short connections' as Ricoeur calls them, in an excessively programmed world. (p. 212)

Some of us who were inhabiting the academy at that time lamented or praised the end of "logocentrism" that inaugurated a postmodern worldview in which the representation replaces the presentation and in which the forces of dissemination empowered Hermeneutics, Deconstructionism, Post-structuralism, Critical Theory and a host of other ways to approach the communications environment of the day. None of us were ready, it is fair to say, for the onslaught of multimedia, two-way, synchronous and asynchronous communications between individuals and groups that would come with the Internet and that would allow individuals to interact informationally with the collective. Indeed, the same year that Levinas offered the description above, *Time Magazine* named the computer as "person of the year," noting that …

> in 1982 a cascade of computers beeped and blipped their way into the American office, the American school, the American home. The "information revolution" that futurists have long predicted has arrived, bringing with it the promise of dramatic changes in the way people live and work, perhaps even in the way they think. America will never be the same. (Friedrich 1983)

Nothing could have been further from the truth, as we now all know, and not only for America, but also for the world. The Digital Revolution had now begun, and in the context of even historical time, it immediately exploded (within 20 years) into an information network of global proportions uniting human and automated information processors, thereby significantly rearranging the communicative playing field.

In terms of the network perspective I am taking here, interactivity changes everything, and in the emerging world of the Internet, an arena in which all information is, in principle, retrievable from anywhere and in which any two people or a community can communicate instantaneously, it is making a staggering difference at a rate beyond our ability as humans to comprehend. It is difficult to say what it all means, to determine whether it was destined from the start, and to say where it will end, but there is no doubt that it matters more than any of the three revolutions previously mentioned in the evolution of our species, even though it depends on each previous revolution in important ways. For the first time, with the Digital Revolution *our species can relate interpersonally through the mediation of machines that process information along the way* and thereby affect who relates to whom, which facets of our social life and interests will develop, what kind of economic and political action we may take, and our sense of self (or selves, as the case may be). Thus, something different in kind does arrive with the Digital Information. Consequently, if the previous revolutions altered so greatly the shape of human history, there can be no doubt that this one will do so with greater force, thereby, as with the past, raising foundational philosophical questions and inviting new methodologies for addressing them. The stage is thus set for historically contextualizing the philosophy of information.

5.3.3 Philosophical Entanglements, or Historically Contextualizing the Philosophy of Information

Philosophy is a self-reflective discipline in the sense that as it moves forward it always remains cognizant of its presuppositions. It is also self-aware of the communications environment in which it unfolds and how to use and manipulate the informational tools at its disposal. This fact is already apparent in the study of rhetoric and composition that emerged with classical Greek philosophy and the use of oratory and disputation in Hellenistic and Medieval philosophy. It seems fair to say that philosophy begins both enamored and cautious about the *logos*. Plato's famous comments in the *Phaedrus* bear direct witness to this observation, but the concern is present in his dialogues more generally and in Aristotle's awareness of the power to persuade through language in his *Rhetoric*. Thus, even though philosophy emerges in the wake of the Epigraphic Revolution it begins with explicit concern over the power of the word in general. By the time we reach the Medieval period, philosophy seems content in thinking that philosophical problems can be settled by quoting the masters and debating the meaning of their words. Philosophy comes to a standstill, in other words, in the scholasticism of Christianized classical philosophy hiding in the form of theology set to the service of the Church. During this period, confessing (professing, proclaiming—choose your favorite word) the truth was more important than finding it. After all, we already had it on good and ancient authority.

Characterizing "scholasticism" generally as an "internal, negative force" in philosophical development, Floridi (2011) notes that "it gradually fossilizes thought, reinforcing its fundamental character of immobility and, by making a philosophical school increasingly rigid, less responsive to the world and more brittle, it weakens its capacity for reaction to scientific, cultural, and historical inputs, [and] divorces it from reality..." (p. 11). Scholasticism, this time intended in the specific sense of its historical Medieval manifestation, seems precisely to have reached this point. Intolerant, unable to innovate and respond to an emerging spirit of inquiry, it found that the ultimate resolution to a philosophical problem was to burn the "heretic" at the stake. Even before the invention of the printing press, it had reached its breaking point, and a new philosophy was about to emerge, almost as if ideas were just waiting to find a way to travel from mind to mind. The *Republic of Letters*, a community of "enlightened" individuals in the seventeenth and eighteenth centuries that crossed national boundaries (and, indeed, the Atlantic ocean), took up arms in the form of writing and publishing, making use of the Printing Revolution and giving birth not solely to modern science and modern philosophy, but also to learned societies and academic journals. The creation of the *Royal Society* in 1662 fostered the spirit of individual inquiry according to the doctrine of *epistemic* Protestantism mentioned above that afforded individuals direct access to the truth in spite of ancient authority. This spirit is aptly present in philosophers of the Early Modern period, such as Descartes (1596–1650), Spinoza (1632–1677), Locke (1632–1704), Leibniz (1646–1716), Berkeley (1685–1753), Hume (1711–1776) and Kant (1724–1804). Private thoughts, and with them the notion of privacy more generally (private property, individual and inalienable rights, etc.), gave birth to the notion of society as a community of individuals engaging in collective action within a new kind of public state, the modern democracy. At its core was the idea that the collective was defined by the individuals who lived within it.

Hegel (1770–1831) challenged this picture, looking for a more integrated relationship between the individual and the "system," culminating, at least according to Kierkegaard's reactive reading in *Fear and Trembling* (1843), in a doctrine that defined individuals in terms of the collective rather than the reverse. In between, Samuel Morse set to work in 1825 in search of a quick means for communicating information over distance, and the Multimedia Revolution was about to begin. Its effects would be felt on both sides of the Atlantic as the vision of a long distance international communications network would become a reality. Philosophically, the presence of a networked conception of humanity was visible in two forms, one positive, advocating a new communitarianism, as in Marx (1818–1883), and negatively, advocating emancipation from the herd, as in Nietzsche (1844–1900). In epistemology, similar effects were apparent in a reaction against Cartesianism and, particularly, against the notion that knowledge could be validated on the basis of independent thought. In America, pragmatism overtook the quest for privately-validated truth in the works of Peirce (1839–1914), who situated truth as a public agreement among a community of inquirers, while, on the European Continent, the Vienna Circle (founded in 1922) advanced a doctrine of logical positivism that would constrain meaning itself to empirical verifiability and, thus, to public visibility. Wittgenstein's posthumously-published *Philosophical Investigations* (1946/1953) famously argued that there can be no private language the very year before Heidegger's *Letter on Humanism* (1947/1993) asserted that the very distinction between private and public is itself problematic, even as he tried to rescue philosophical thought from a techno-scientific conception inherited from Husserl, who sought to make philosophy an exact science.

Though twentieth century philosophy is notoriously characterized as divided between the "Continental" and "Analytic" traditions, in the context of the networked global communications environment sketched here, they seem more concerned with a similar set of issues rather than different ones, even though they disagree on method. As we move past the initial shock of the "telephonization of city life" described by Casson above, the so-called "Analytic/Continental Divide" is starting to look less substantial and more like pointless arguing, not in the form of careful and sustained discussion between individuals, but between two (until recently) separate networks of inquirers that do not speak the same language even when talking about the same things. In this light, we must wonder whether accepting rather than struggling against a new networked conception of humanity might jar philosophy lose from the constraints of its recent scholasticism. To invoke Floridi (2011) once more:

> ... philosophy is indeed like a phoenix; it can flourish only by constantly re-engineering itself. A philosophy that is not timely but timeless is not a *philosophia perennis*, which unreasonably claims unbounded validity over past and future intellectual positions, but a stagnant philosophy, unable to contribute, keep track of, and interact with, the cultural evolution that philosophical reflection itself has helped to bring about, and hence to flourish. (p. 12)

We find here not a philosophy of evolution, but the notion that philosophy is evolutionary, that it belongs to a community of inquirers, who as responsible processors of information, disseminate their findings to build an information commons beyond the comprehension of any single individual, yet, in hyperbolic terms, accessible as needed to all, not a *Republic of Letters* but a networked community of informants. We are, to situate this in the language of the Digital Revolution, information processors who read from and write to a common tape and who will, in time, find each other as needed and when relevant, thanks to the mediation of socially networked computer technologies.

In advocating the philosophy of information as a new *philosophia prima*, Floridi sets out on a new frontier, "not by putting together pre-existing topics, and thus reordering the philosophical scenario, but by enclosing new areas of philosophical inquiry—which have been struggling to be recognized and have not yet found room in the traditional philosophical syllabus..." (p. 24). From the perspective of this paper, Floridi is not merely calling for a new philosophy suited to an old communications environment, but among the first to respond within the constraints of a new one. What will philosophy look like as we become aware of our place as *inforgs* within the *infosphere*? What indeed will our presence in the *infosphere* do to the history of philosophy? It is far too soon to say. But in a world where the speed of informational change is so rapid that legislation and analysis cannot keep up, we will either adapt to new methods of inquiry and new informational tools or let the forces of technological change roll over us (or perhaps, worse yet, both.) We are undergoing something dramatic, and we do not yet know what. Perhaps this very imperative will necessitate the transformation of philosophy in the Fourth Revolution that digital technologies both afford and require with the philosophy of information at its foundation.

Acknowledgments I wish to thank Dick Connolly, Christopher Harrison and Brent Sigler for their help with research on this paper, and, of course, Luciano Floridi, for providing something provocative to which I could react.

References

Beavers, Anthony, and Brent Sigler. 2010. Mechanists of the revolution: The case of Edison and Bell. In *Proceedings of the VIII European conference on computing and philosophy*, ed. Klaus Mainzer, 426–430. Munich: Verlag Dr. Hut.

Casson, H. 1910. *The history of the telephone*. Chicago: A. C. McClurg and Co.

Clark, A. 1997. *Being there: Putting brain, body and world back together again*. Cambridge, MA: MIT Press.

Clark, A. 2001. *Mindware: An introduction to the philosophy of cognitive science*. Oxford: Oxford University Press.

Deibert, R. 1997. *Parchment, printing and hypermedia: Communication in world order transformation*. New York: Columbia University Press.

Diringer, D. 1982. *The book before printing: Ancient, medieval and oriental*. New York: Dover.

Edison, Thomas. 1878. *North American Review*. U. S. Library of Congress. http://memory.loc.gov/ammem/edhtml/edcyldr.html. Accessed Aug 1 2010.

Edwards, M. 1993. *Printing, propaganda and Martin Luther*. Berkeley: University of California Press.

Eisenstein, E. 2005. *The printing revolution in early modern Europe*, 2nd ed. New York: Cambridge University Press.

Fahie, J. 1900. *A history of wireless telegraphy, 1828–1899, including some bare-wire proposals for subaqueous telegraphs*. London: William Blackwood and Sons.

Febvre, Lucien, and Henri-Jean Martin. 1976. *The coming of the book: The impact of printing 1450–1800* (trans: David Gerard). New York: Verso. (Orig. pub. 1958.)

Fischer, H. 1989. The origins of Egyptian hieroglyphs. In *The origins of writing*, ed. W. Senner, 59–76. Lincoln: University of Nebraska Press.

Floridi, L. 2008. Artificial intelligence's new frontier: Artificial companions and the fourth revolution. *Metaphilosophy* 39(4/5): 652–654.

Floridi, L. 2009. The information society and its philosophy: Introduction to the special issue on 'The philosophy of information, its nature and future developments'. *The Information Society* 25(3): 153–158.

Floridi, L. 2010. *Information: A very short introduction*. New York: Oxford University Press.

Floridi, L. 2011. *The philosophy of information*. New York: Oxford University Press.

Friedrick, Otto. 1983. The computer. *Time Magazine*, January 4th. http://www.time.com/time/subscriber/personoftheyear/archive/stories/1982.html. Accessed 14 Feb 2011.

Green, M. 1989. Early Cuneiform. In *The origins of writing*, ed. W. Senner, 43–57. Lincoln: University of Nebraska Press.

Heidegger, Martin. 1993. The letter on humanism. In *Martin Heidegger: Basic writings*, ed. David Krell, 213–266. New York: HarperCollins. (Orig. pub. 1947.)

Hodges, Andrew. *The Alan Turing Internet scrapbook*. http://www.turing.org.uk/turing/scrapbook/ace.html. Accessed 4 Aug 2010.

Kierkegaard, Søren. 1983. *Fear and trembling* (trans: Howard Hong and Edna Hong). Princeton: Princeton University Press. (Orig. pub. 1843.)

Lawhead, W. 2002. *The modern voyage: 1400–1900*, 2nd ed. Belmont: Wadsworth.

Levinas, Emmanuel. 1989. The pact. In *The Levinas reader*, ed. Seán Hand, 211–226. Cambridge, MA: Blackwell. (Orig. pub. 1982.)

Logan, R. 1986. *The alphabet effect: The impact of the phonetic alphabet on the development of Western civilization*. New York: William Morrow and Company.

Lyman, Peter, and Hal Varian. 2003. *How much information? 2003*. http://www2.sims.berkeley.edu/research/projects/how-much-info-2003/. Accessed 14 Feb 2011.

Martin, Henri-Jean. 1993. *The history and power of writing* (trans: Lydia Cochrane). Chicago: The University of Chicago Press. (Orig. pub. 1988.)

Norman, D. 1994. *Things that make us smart: Defending human attributes in the age of the machine*. Cambridge, MA: Perseus Books.

Saxby, S. 1990. *The age of information: The past development and future significance of computing and communications*. London: Macmillan.

The Alexander Graham Bell Family Papers at the Library of Congress: 1862–1939. U. S. Library of Congress. http://memory.loc.gov/ammem/bellhtml/bellinvent.html. Accessed 1 Aug 2010.

Tindal, Matthew. 1730. *Christianity as old as creation; or, the Gospel as a republication of the religion of nature*. Google Books. Accessed 14 Feb 2011.

Toland, J. 1696. *Christianity not mysterious: or, a treatise shewing that there is nothing in the gospel contrary to reason, nor above it, and that no Christian doctrine can be properly call'd a mystery*. Google Books. Accessed 14 Feb 2011.

Turing, A. 1937. On computable numbers, with an application to the Entscheidungsproblem. *Proceedings of the London Mathematical Society* 2(1): 230–265.

Winston, B. 1998. *Media technology and society—a history: From the telegraph to the internet*. New York: Routledge.

Wittgenstein, Ludwig. 1953. *Philosophical investigations* (trans: G.E.M. Anscombe). Oxford: Blackwell.

Chapter 6
I Mean It! (And I Cannot Help It): Cognition and (Semantic) Information

Valeria Giardino

6.1 Introduction: We Are Inforgs in an Infosphere

To introduce Luciano Floridi's theses, I will start from what I believe is his own starting point: defining the role and the challenges of philosophy in the contemporary world. In his writings, Floridi presents to his readers a scenario that is very familiar to any human being who is a member of the contemporary society and pursues everyday all the typical activities of that society. It is before our eyes: in the last decades, the world has changed dramatically and so fast that also relatively young people have witnessed some of these changes in person. The metamorphosis is still in progress: it is easy to predict that in the following years the world will continue changing and evolving. The question now is: towards what will these changes bring our world and us? Moreover, are we ready for such a new world and are we aware of what is happening at all?

It is at this point of the story that philosophy enters the scene, offering the conceptual tools that are necessary in order to answer to these questions. Some critics could think that an 'old' discipline such as philosophy has nothing to say about the dramatic transformations that are happening today. As a consequence, it cannot have a role neither in finding solutions to the new challenges that this new world presents us nor in making any prediction about what is going to happen next. The same critics could think that philosophy has nothing to offer since other kinds of expertise are needed today: the contemporary world is calling for people who are able to speed up these changes, for example pushing the new technologies to their limits and then beyond them, or creating tools that would allow for better interactions between humans and machines. Floridi shows that these critics are wrong, and, being a philosopher myself, I think he is right.

V. Giardino (✉)
Institut Jean Nicod (CNRS-ENS-EHESS), Paris, France
e-mail: Valeria.Giardino@ens.fr

H. Demir (ed.), *Luciano Floridi's Philosophy of Technology: Critical Reflections*, Philosophy of Engineering and Technology 8, DOI 10.1007/978-94-007-4292-5_6,

If the challenge of philosophy today is still conceptual analysis as it has always been, then philosophers should try, in Floridi's words, "to capture the new *Weltanschaung* that might be dawning on us" (Floridi 2007, p. 3). Although this does not mean operating directly on the world, it is still an important task to make sense of what is happening to us – as cognitive agents – and to the environment in which we cognitively act. The world today has been re-ontologized, that means that the categories that we have inherited from our ancestors to speak about the world have changed. New artifacts and new technologies have appeared and – to some extent – we are now 'human' in a different way; through the lens of conceptual analysis we acquire this awareness. Philosophy keeps pursuing its ever-lasting objective: to make sense of the world and of us as agents within the world.

Let us consider an example. Our concern is the everyday world, so I will describe what I see here and now. I am working at this book chapter, my face partly reflected by the computer screen. Please note that I am writing something that will be – hopefully – published, in a physical book as well as in an electronic book accessible on the Internet by submitting the correct username and password. I am sitting now in Hall 3 at the Orly airport, waiting for my flight, my notebook and one of my books are next to me, my computer is on my knees and my credit card in my wallet, ready to be used in case I want to buy a half-an-hour connection to the Internet. Finally, if I am not forgetting anything of relevance in this context that I have stuck in my pockets before going out – my USB-flash drive maybe? – I have with me my two mobile phones: an old one with an Italian sim card, and a smart-phone with a French sim card by which I can connect to the Internet, even more easily than with my computer, since my monthly fixed-rate allows downloading a certain amount of information from the Internet.

A book, a notebook, a computer, an electronic card with virtual money on it, two phones and finally a human being: me. That are many questions that philosophy could ask about this very common situation. A first and very fundamental question, which sounds very philosophical indeed, could be the following: in this scene, where am I in the end? The strategy that Floridi adopts to answer to questions such as this is to look at this scenario by focusing on one specific feature: its *informational* space. In his interpretation, the scene gets informationally colored: I become an *informational* agent, and the objects around me are framed following my cognitive activity in the way they respond to my cognitive – and informational – demands. As Floridi has suggested, we are beyond a Newtonian image of the world, made of 'dead' cars, buildings, furniture, clothes, which do not interact with us and cannot communicate, learn or memorize. There is something we share with all these 'new' objects: our common environment. In fact, we are all part of what Floridi defines as the *infosphere*: me, the new world and the objects that make it possible and real, can all be re-conceptualized in informational terms. The environment we are living and acting in today is drastically different from the one that past generations have been experiencing, since the infosphere is becoming more and more *synchronized*, *delocalized* and *correlated*. It is more synchronized because the very notion of time has changed: the new informational processes are faster and can happen in parallel. It is more delocalized because the very notion of space has changed: information is everywhere – for instance, there are more and more archives of all kinds of information accessible from the Internet. It is more correlated because the very notion of what an interaction is has changed: as informational beings, we are almost always interconnected. Time, space and interactions as we knew and learnt them are thus dramatically evolving into a 'different' time, a 'different' space and finally into a 'different' concept of what acting and reacting amount to.

There is even more to say: it is not only the environment that has changed, but we have changed too. As Floridi suggests, "we are probably the last generation that will experience a clear difference between onlife and online" (Floridi 2007, p. 9), since the onlife of the infosphere as a whole will always be online: at some point, there will be no difference between processors and processed, online and offline, and all interactions will become equally digital. As a consequence, a new form of agent is emerging, a hybrid (multi)agent, partly artificial and partly human. In the scene described, I am such a hybrid agent, equipped as I am with my laptop, my phones, and all my, let's say, technological extensions. I am an *inforg*, not merely an organism but an informational one.

At the end of the twentieth century, a new myth emerged in films and novels: popular science kept telling us that, at some point in the future, a new being would have appeared, whose body would have had partly human features and partly mechanical ones. Today, the myth of cyborgs – half human-half machine beings completely identical to us on the surface – has faded. We realize now that some transformations had taken place indeed, but not in our body, as it was predicted at the dawn of the Artificial Intelligence program. The transformations occurred through the re-ontologization of our environment and of ourselves: we have not evolved into cyborgs as we thought but into *inforgs*. Our body has not changed in any uncanny way but we have found ways of augmenting our mental and informational capacities. More drastically, our environment and we have gone through a process of re-ontologization that has changed our way of seeing the world and ourselves forever.

6.2 Two Possible Scenarios for Philosophical Inquiry

In this context, the task of philosophy is to discuss, to understand and to anticipate any further transformation. In my view, the challenges are mainly of two kinds: two scenarios of problems are open before philosophers.

The first scenario is *ethical*. As Floridi suggests, we have to realize that the infosphere is "a common space, which needs to be preserved to the advantage of all" (Floridi 2007, p. 9). The risk, which Floridi refers to as being more than a mere possibility but a concrete event that will come true soon, is that a chasm will create between those who can be in the infosphere and those who cannot, between insiders and discriminated outsiders, between an informationally rich environment and others who are doomed to be informationally poor. As Floridi claims, this "will redesign the map of worldwide society, generating or widening generational, geographic, socio-economic and cultural divides". Moreover, this gap will not be reducible to the distance between industrialized and developing countries, since it will cut across society (Floridi 2002). In this respect, our categories must be once again revised: the contemporary society is preparing the ground for tomorrow's digital favelas. To this picture, I would add that also within the informationalized portion of the society, an abyss will separate those who have access to *all* the information from those who have a partial or worst a controlled access to it. Another important ethical issue concerns the very notion of Self, which assumes new features in the infosphere.

The second scenario is instead *epistemological*. The question is: how are these transformations affecting our way of perceiving the world and ourselves as agents? In which respects do the intrinsically limited powers of our mind get augmented when we become inforgs? Are we revising our criteria for something to be considered as knowledge? And finally, do we access meaning differently in the infosphere? In the following sections, I will discuss some of these questions.

The ethical and the epistemological are the two main scenarios opened to the philosophical analysis, not to mention another scenario that lies in between the first two and rises both ethical and epistemological issues: *education*. As Floridi points out in many passages, we are constructing a new environment that will be inhabited by future generations. It is not something far away from us: it concerns our children. In Floridi's words, at the moment we are *e-migrants*, since the *Umwelt* as we knew it is being absorbed by the infosphere. But this situation will not last long. In fact, future generations will be different from us because they will be digital natives and not digital immigrants: our children will be born *in* the infosphere and therefore they will recognize themselves from their birth as inforgs. The very crucial question is how this change will affect their way of learning and their criteria for what is reliable knowledge. For example, consider the discussion about the so-called 'wisdom of the crowd' (Surowiecki 2004): if we look at new Web tools such as Wikipedia, we know that in most cases they are considered reliable. This is possible because the number of people contributing to them is so high that it annihilates the potential errors or imprecisions due to one or more individuals. Are educational systems and institutions ready for this kind of transformations? What does an inforg-child need to learn and to know to prepare for her future life in an informational society? Once we will be able to answer to these questions, a further challenge would be to discuss and to define what the infosphere requires as the most appropriate and effective tools for teaching.

Of course, the scenarios I have just described are all intertwined and do have an influence on one another. I would say: everything and everyone is interconnected in the infosphere. The world we are experiencing today is more and more permeated by information; this information comes in different forms and formats and is diffused through the new technologies. We currently make use of these technologies that are becoming more and more familiar to us and get continuously improved. It is an epochal change, a fourth revolution, as Floridi claims.

If the challenge of philosophy is to analyze how this revolution has changed our understanding of the world and of ourselves, my challenge in this article will be to claim that some of Floridi's suggestions should be partly revised and further discussed. In the remainder of the article, I will present the four revolutions Floridi talks about, and I will claim that there are other revolutions that can be considered in the history of human culture. Some of them are interesting in the perspective of discussing the reshaping of our new environment and of our new selves in the infosphere. I will discuss an ambiguity in Floridi's use of the term information and propose to consider his fourth revolution as the Second Information revolution. To solve this ambiguity, I will distinguish between information and *semantic* information, which implies meaning and understanding. If I am right, this distinction will make new difficulties emerge, in dealing with the new picture of the world and possibly in providing predictions. Finally, I will present some questions that emerge once we consider humans' cognitive capacities in accessing meaning on the background of the new context, the infosphere.

6.3 One, Two, Three… Many Revolutions: Human Culture

6.3.1 Looking Back in Time: Are There Other Revolutions?

According to a classical analysis, in the sixteenth century the Copernican revolution has changed our conception of ourselves and of our place in the world. When the Earth lost its place at the center of the Universe, we were displaced from it as well. More generally, we were not anymore the reason for that Universe to exist and to move around us. As a consequence, our world evolved into a new sort of world, and thus was re-ontologized: from being the center of the Universe, it transformed into a humble planet that constitutes only one of the elements in a complex system of physical relations. The Earth is not special; on the contrary, it is analogous to its brother and sister-planets that are in the same way attracted by the Sun. Not only had our world transformed: at the same time, we had evolved into a new sort of human beings. We had become mere inhabitants of one of the planets that are part of the Solar system.

Yet, despite this first re-ontologization, we retained our place at the center of the animal kingdom. It was a second and more recent revolution, the Darwinian one, which removed us also from that illusionary spot. In his theory of the evolution, Darwin showed that human beings are not so different as they thought from animals: as a species, we are primates, namely we are at the end of a long chain of mutations and transformations and we share common ancestors with other animals. Once again, the world around us as well as ourselves had transformed into something new, because we were forced to re-ontologize it.

After this second revolution, there was still one thing that was under humans' control: our own mind. Human beings were still at the center of their rational game, and, as Descartes taught, they had a conscious access to their ideas by introspection. But this perspective too was doomed to be revised, because of the third revolution, the Freudian one, which brought human beings to the discovery that the mind has an unconscious side. The consequence was that a portion of our Self became inaccessible to us.

This is not the end, since the recent transformation of the environment has shown that we need to re-ontologize further our picture of the world and of ourselves. What revolution are we experiencing now? What is revolutionary about the scene depicting me working at the airport with my computer on my knees?

According to Floridi, we are not – or at least not only – experiencing a computer revolution. It is not sufficient to acknowledge the widespread diffusion of computational devices to describe what is happening today. Think once again at the scene I described: there is a computer, it is true, but there are also mobiles and the possibility of connecting to the Internet as well. So, if not a computer revolution, are we experiencing a digital revolution? Once again, Floridi's answer is negative: what about the success of enterprises such as Amazon, which are giving to books – and to e-books – a new renaissance? Therefore, following Floridi, there is just one possible answer: the revolutionary element in the new scenario is information and the role it plays in it. Things evolve into energy and into information, and what matters are the changes in the life cycle of this information. Going back to me sitting in the airport hall, what matters there is the information flowing around the scene and, through me, across the different devices near me and into the Internet. It is the twenty-first century and we are part of the Information revolution. We have coined a name for the society we are living in, permeated by computer science and ICTs: the *information* society. But what about us? We are informational beings and the world has turned into an informational world. The Information revolution is the forth revolution, and is happening now. Moreover, the Information revolution has its hero as well: Alan Turing. His work, or better what was *in nuce* in his work, has changed forever our understanding of the world and of ourselves as cognitive agents.

Though I am in general sympathetic with Floridi's rational reconstruction, I would argue that in the course of human cultural evolution, it is possible to individuate other crucial steps in the transformation of our ontology, *before* the Copernican revolution. As I will show, once an evolutionary perspective is assumed, our engagement in symbolic activities appears as being crucial for our cognition. For this reason, I will claim that the Information revolution Floridi refers to is in fact the Second Information revolution; moreover, according to some views, it can be considered as a degeneration of another revolution: the Cognitive revolution.

Let me consider first other topical moments in the evolution of human culture and, more specifically, in the evolution of cognitive artifacts, before Floridi's first revolution. Though very back in time, it is unquestionable that cognitive artifacts have played a major role in the shaping of our world and of us as cognitive agents. To clarify, I am not arguing that Floridi is not aware of the relevance for our cognitive history of the several innovations that were introduced each time that a new cognitive technology was created. What I am suggesting is rather that we might assume an evolutionary perspective and consider two very important events. First, the time when human beings began to communicate by means of a *language*; secondly, after that, the time when they invented *writing*, and thus began not only to produce words but to share them in a kind of public format that could be stored in archives. Both those steps were crucial in the evolution of human cognition, since they revolutionized human beings' access to meaning: new channels became available to communicate and to make sense of the world around us and ourselves.

Take numerical cognition as an example. From what has been shown in the empirical research, humans are equipped with some spontaneous representations: whatever their education and their culture, they are able to make simple comparisons between numerosities. Humans, together with some of their evolutionary precursors and other animals, are able to represent precisely numerosities up to 4, and, for larger numerosities, they approximate (Dehaene 1997). This given, as some authors suggest, once number words are acquired, our representational powers get crucially improved (Frank et al. 2008). In fact, numerals play a fundamentally *compressing* role for what regards our more spontaneous representations. Experimental data show that a subject who does not master the number word system will not be able to track numerosities across time and space. This is the case of the speakers of the Amazonian monolingual hunter-gatherer tribe Pirahã, who have a limited inventory of words for numbers. Although they show the same spontaneous representations of Western controls, their performances in matching tasks get inaccurate when the tasks involve different spatial organization or memory. This suggests that number words can be conceived as a *cognitive technology*: once available, they add a second and most of the time preferred route for encoding and processing information. The same could be claimed for other cognitive activities as well, such as color recognition (Gilbert et al. 2006; Uchikawa and Shinoda 1996; Winawer et al. 2007) and navigation (Hermer-Vazquez et al. 1999). In fact, also in distinguishing colors and in orienting in space, humans have at their disposal from very soon some spontaneous representations; it is only afterwards that these representations get integrated by an appropriate and public code. These codes serve as cognitive technology like in the case of numerals, because they constitute a useful and effective cognitive tool. This does not mean that once this code is acquired, the more spontaneous representations are lost; on the contrary, when the appropriate code is suppressed or not useful, speakers perform in the same way as speakers of languages who do not possess the relevant technologies. In this perspective, other cognitive technologies that offered new possibilities for our cognition can be acknowledged. Let us consider for example writing, which improved the possibility of sharing words – and therefore our ideas, our opinions, our knowledge – with other members in our community, across time and space. A further improvement has been in particular alphabetic writing, that solved the problems of ideographic writing by offering a reliable and public code in order to visualize human speech.

My approach is in line with the idea that cognition is 'distributed': as Hutchins (1995a, b) explains, cognitive events are not encompassed by the skin or skull of an individual. If we look at human cognitive activity 'in the wild', we discover at least three interesting kinds of distribution of cognitive processes: the cognitive processes can be distributed (i) across social groups, (ii) in the coordination between internal and external structure, be it material or environmental, and finally (iii) through time, in such a way that the products of earlier events can transform the nature of later events. We must consider these kinds of distributions if we want to understand human cognition. In fact, as I suggest here, the invention and the use of cognitive artifacts as scaffolding structures for our reasoning are involved in the organization of our functional skills into cognitive functional systems (Dror and Harnad 2008). Human beings, despite the limitations of the cognitive systems we know they are born with (Kinzler and Spelke 2007; Spelke 2004), were able to develop new practices and new cognitive strategies to augment the powers of their minds, showing an extraordinary capacity in creating tools that would help them in the processes of both describing the world around them and acting upon it. Some of these tools had an intrinsically cognitive function, which allowed them to enhance recognition, communicate, economize their cognitive resources, and provide faster and accurate transitions from premises to conclusions. Therefore, language is special because it is cognitively primary, but not *so* special in the end. Our relationship with language as well as with complex mathematics is analogous to our relationship with chess. As Tomasello (1999, p. 208) claims, the cognitive skills involved in language, complex mathematics or chess "are products of both historical and ontogenetic developments working with a variety of preexisting human cognitive skills, some of which are shared with other primates and some of which are uniquely human".

6.3.2 Symbolic Activity and Cognitive (Info)Artifacts

What then to say about these important steps in the evolution of cognitive artifacts if we consider Floridi's view?

First, I want to point out that my objective is not to suggest that every introduction of a new cognitive artifact such as numerals or alphabetic writing should be considered as a revolution in our way of relating to the world and to ourselves as cognitive agents. What I want to claim is rather that the task of creating new technologies to find new ways of improving our more ancient and spontaneous cognitive capacities and communication skills has started a long time ago; we have *always* urged to produce semantic information, by all possible means: speaking, writing, printing books, painting pictures, shooting films… One objection to this idea could be that there are also other reasons why these forms of externalization of our thoughts have been introduced. For example, there may had been esthetic reasons, for instance in order to arouse emotions or evoke pleasure, or social reasons, for instance in order to affect action or promote collaboration. I am not denying this; though, I want to focus on the cognitive and communicative reasons for which they had been created: my aim is to point out that to some extent we have been living in an informational environment all along. In fact, our culture deals by nature with information and pursues the realization of newer and newer means to reach the world and the others around us.

The consideration of these crucial moments in the cognitive history of human beings helps thus reshaping Floridi's infosphere: the infosphere encompasses informational cognitive agents – the inforgs – and informational cognitive tools – the *info-artifacts* – that are a *new form* of cognitive artifacts. In fact, humans have been creating and inventing cognitive artifacts all the time; nevertheless, the most recent artifacts they invented, beginning with the Information revolution, have proved to be drastically different. They are not only cognitive tools at our disposal, but also seem to have an onlife of their own: not only they memorize, as writing does to some extent, but they also learn, and most of all, respond to us. Think about Plato's objection to the use of writing: written words speak as though they make sense, but if one asks them an explanation for what they are saying, they will go on telling the same thing for ever, over and over again (Plato 1997). The new tools, instead, interact with us, in a dialogue-like exchange, and this is not metaphorical talk since it faithfully describes our everyday interaction with them.

This brings us to another issue, which concerns the possibility of considering an older Information revolution. Let us accept that from the beginning we have been creating cognitive tools; one could think that, as a consequence, from the beginning we have been living in some kind of *proto*-infosphere; with this environment as a background, we were *proto*-inforgs: we were doing our best in sharing information and finding new ways of improving our communication, and yet we were lacking the necessary technology – the one that Turing provided – to make this information flow all around as it happens today. Though, this reconstruction is ill posed, since it is given from the point of view of what has happened 'next'.

A more faithful reconstruction would rather show how the history of our cognition has been deeply influenced by the fact that from the very beginning we engaged ourselves in symbolic activities, and that these activities have become, in a long historical and cultural process of creation and selection, more and more complex. As Deacon (1997) observes, our ancestors found a way to create and reproduce a simple system of symbols. Once available, these symbolic tools quickly became indispensable. This was indeed a revolution in the ontology of information through the billions of years of the evolutionary process, from the time when living processes became encoded in DNA sequences: "because this novel form of information transmission was partially decoupled from genetic transmission, it sent our lineage of apes down a novel evolutionary path – a path that has continued to diverge from all other species ever since" (p. 45).

If this is correct, then I propose that the revolution Floridi talks about is the 'Second' Information revolution. From sequences of DNA to cultural transmission (First Information revolution), from cultural transmission to online transmission (Second Information revolution). Therefore, Floridi's fourth revolution is a revolution not because information is now everywhere – since we have always urged for it, but because now it is conveyed and spread by genuinely new artifacts: *info*artifacts.

It is undeniable that the new tool available, the Internet, is revolutionary. Nevertheless, we could ask whether it has really qualitatively changed our way of accessing information. Actually, though information is all around, it does not seem to imply that we will take advantage of it. In fact, we are driven by our choices – most of the time biased – and Internet surfing is no exception. Our own interests guide us, and we employ the available technology to select the information according to them. Therefore, despite the Internet revolutionary character, old mechanisms of information diffusion risk to be replicated: people – even once they have become inforgs – tend not to engage themselves in 'occasional encounters' with information, but have predetermined preferences.

Moreover, Floridi often talks about evolution, but I am not sure that the 'evolution' label can be applied to the changes in us or in our artifacts. If he intends this term in a Darwinian sense, then it would be difficult to think that our cognitive artifacts did evolve – and are still evolving – thanks to a series of random mutations, as Darwinian evolution prescribes. It is not so much an issue of random mutations here, but rather a matter of how our spontaneous and precocious capacities have played a role in the characterization of the new technologies. For what concerns us as e-migrants, we have not really evolved when we transformed into inforgs, since our cognitive capacities have not intrinsically changed: if we want to say that the environment and our artifacts have evolved into a new form of scaffolding structure for our cognitive capacities, then we should also claim that this happened because our original cognitive capacities have constrained their evolution, quite like our biology still constrains all our possible mutations.

There are two further questions. First, what about the new and the future generations? Will they develop some kind of new and different cognitive capacities in the interaction with the new info-artifacts? Will these capacities be transmitted to their offspring? Secondly, would these new technologies find ways of evolving in isolation from humans, by their own random mutations? What could determine such an event?

I hope to have shown that to some extent we have been inforgs all the time. The revolution we are experiencing today is the consequence of our urge for information pushed to the limit: our environment has become informational, and this will change our primitive relationship with information. First only the DNA was conveying information, then our symbolic activity made us capable of transmitting information to the future generations through culture; today, information is everywhere, going through the environment and through the people in it. Still, in this picture, there is an ambiguity in the term 'information' that needs to be solved. In order to do that, in the next section I will distinguish between information and *semantic* information, which implies meaning and understanding.

6.4 Cognition and Semantic Information

6.4.1 Humans as Semantic Engines

In the DNA double helix, as well as in Turing machines, information is conceived as a code, a string, and has not anything to do with meaning or understanding. By contrast, *semantic* information requires meaning and understanding. Floridi claims that, by re-ontologizing ourselves as inforgs, we recognized how significantly *but not dramatically* different we are from smart, engineered artifacts, since we have, as they do, an informational nature. But what kind of information is Floridi talking about when he refers to 'informational nature' in the two cases? Are we referring to the same notion of information? If we are not, as I believe it is the case, then there is still a sense in which we indeed are dramatically different from responsive artifacts, as Floridi himself seems to admit in a recent paper (Floridi 2009).

In the previous section, I discussed the reasons why I believe we do not refer to the same kind of information in the two cases: we as humans have a relation with information – semantic information – that is different from the relation machines have. Let us go back to the cognitive artifacts we use, and to symbols alternative to words available to us in order to communicate and externalize thought, such as gestures or diagrams. We invented them – maybe by taking inspiration from nature – in order to convey information, both to ourselves and to the others (Tversky 2005 together with Kessell and Tversky 2006). Given the fact that we are now living in the infosphere, is there a reason why we should renounce to them? Would they become useless or would they disappear? I do not think so. We all keep on gesturing or drawing diagrams, despite all the informational powers of computational devices. Do machines make gestures or draw sketches? Despite the advancement in technology, at the moment they cannot.

A great number of cognitive activities imply the use of such cognitive tools, which are widespread. Let us consider the case of mathematics. Avigad, in trying to define what mathematical understanding is, claims that ascriptions of understanding are best understood in terms of the possession of certain *abilities* (Avigad 2008). The individuation of such abilities is in fact a central issue for automated formal verification, since in order to improve the human-machine interaction it is necessary to consider how different sorts of agents will have different strengths and weaknesses. For example, in the process of looking for a mathematical proof, humans commonly show to spontaneously refer to diagrams, most likely because they are good at recognizing symmetries and relationships in information when it is so represented. The same cannot be said for machines, which can keep track of gigabytes of information and carry out exhaustive computation where humans would be forced to rely on little tricks. Avigad's suggestion is that our theories of understanding should then be relativized to the particularities of the relevant class of agents. They are all informational agents, but they treat information differently.

If our aim is simply to claim that machines as well as humans have access to information and are cognitive agents of some sorts, then we have to conclude that from this point of view they are not different. Yet, they are still dramatically different in the way they *deal* with information. In a recent paper Floridi shows to be aware of this difference: he claims that humans are the only *semantic engines* that have been so far available in the universe: we produce meaning and we have always produced it (Floridi 2009). By contrast, artificial agents are "*syntactic engines*, cannot process meaningful data, i.e. information as content, only data at a lower- or higher- level. … Humans are the only semantic engines available, *the ghosts in the machines*" (my italics). Though, this seems to contrast with his previous claim that after the fourth revolution we re-ontologized ourselves in such a way that we have become of the same nature of machines. Humans and machines may well have the same informational nature, but it appears now that only humans process semantic information, while machines are syntactically powerful. They possess two different abilities. This is far from new and may seem to echo Searle (1980) and his *Chinese Room* objection to Strong AI. Nevertheless, my aim in this article is more modest than Searle's: I simply want to discuss the possibility of distinguishing among different cognitive subjects having different cognitive capacities, and I do not want to take any stance on the nature of consciousness.[1]

There has been, especially in the 1990s, an enormous effort in AI as well as in computer science to create semantic machines or a Semantic Web; nonetheless, the results have not met the target yet.[2] Up to now, machines are still only syntactic engines, and we know why – we are the 'ghosts' in them, in the end! – but what to say about our own powers? What are the conditions behind our being semantic engines? This question was, as I will show, at the origin of what has been defined the Cognitive revolution.

6.4.2 Meaning-Making and Meaning-Flexibility

I will consider now Bruner's (1990) point of view on what he defined the Cognitive revolution, taking place in the 1950s. We know that it is the same revolution Floridi refers to; nevertheless, Bruner gives a different interpretation to it. According to Bruner's reconstruction, the aim of that revolution at the beginning was to discover and describe *formally* the meanings that human beings were able to create out of their encounters with the world. The objective in the long run was to set forth hypotheses about which *meaning-making* processes were implicated in humans' cognitive activity. As I have already discussed, human beings engage themselves in symbolic activities to construct and make sense of the world and of themselves as well. Bruner's hope was that such a revolution, as it was conceived at its origins, would have lead to the collaboration of psychology with its sister interpretative disciplines such as the humanities and the social sciences. It is only a collaboration of this kind that can allow the investigation of such a complex phenomenon as meaning-making. But the happy ever after did not work out. In fact, in Bruner's opinion, the emphasis began shifting from the *construction* of meaning to the *processing* of information, which are profoundly different matters. The notion of computation was introduced and computability became '*the*' good theoretical model; this brought far from the original question – the revolutionary one – which was about the conditions of our meaning-making activity, whose answer would have

[1] Moreover, I take the distinction between syntax and semantics from my work on the philosophy of mathematical practice, and on the limits of the foundationalist approach to mathematics.

[2] See Floridi (2009) for an interesting discussion on the contrast between the Semantic Web and the Web 2.0 enterprises.

explained our semantic power. In Bruner's words, "information is indifferent from respect to meaning. In computational terms, information comprises an already precoded message in the system. Meaning is preassigned to messages. It is not an outcome of computation nor is it relevant to computation save in the arbitrary sense of assignment" (p. 4). For this reason, the Cognitive revolution "has been technicalized in such a manner that even undermines that original impulse" (p. 1): it has become the (uninteresting) Information revolution.

Meaning is thus different from information because it does not come *before* the message, but it originates through the message itself and the sharing of the message. In fact, public meanings are the result of a *negotiation*. According to Bruner, the picture is a new transactional contextualism, since the realities that people construct are *social* realities, negotiated with others and distributed between them. It is 1990 and Bruner claims: "the social world in which we lived was, so to speak, neither 'in the head' nor 'out there' in some positivistic aboriginal form. And both the mind and the Self were part of that social world. If the cognitive revolution erupted in 1956, the contextual revolution (at least in psychology) is occurring today" (p. 1). Knowledge is situated-distributed, and this happens not only because knowledge has a cultural nature, but also and most of all because our knowledge *acquisition* has a cultural nature. Moreover, knowledge has also a social nature, because it gets socially constructed (Berger and Luckmann 1966). Human beings are semantic engines, and they engage themselves in meaning-making and meaning-negotiating. For this reason, meaning is flexible: as Bruner says, we show a 'dazzling' intellectual capacity for envisioning alternatives.

Let me focus on the contrast between this spontaneous flexibility in meaning-making and classic information theory, according to which a message is informative if it reduces alternative choices. The assumption that we need a code in order to reduce alternative decisions curiously parallels what happened for mathematics at the beginning of the twentieth century, when the logical paradigm became 'the' good theoretical model. As Grosholz (2007) discusses, mathematicians as well as scientists bring together disparate registers of discourse into rational relation. Nevertheless, logical positivism made the dimension of the growth of scientific knowledge almost impossible to see, since it unrelentingly attempted to impose a homogeneous discourse on science, not considering the constitution of the meaning of the terms used in scientific propositions. The epistemological ideal that was put forward by Russell, Carnap and others, was to unify mathematics and science into one formalized theory that had to respond to the demand for a homogeneous idiom as a vehicle for deductive reasoning. If the form of reasoning alone must transmit truth, then all the terms must be stable and 'alike'. This demand was too strong: the foundationalist program failed in its pretension of replacing the mathematical language with formal logic. As well as information in Turing's sense is not meaning, formal logic is not mathematical discourse. Both meaning and mathematical discourse are not characterized by being homogeneous, but by implying different interpretations and different perspectives on the same object as well as on ourselves as cognitive agents.

Consider the use of diagrams in particular, and the way they externalize thought in order to facilitate reasoning. Why are they so effective? If we try to answer to this question by considering the way it would be possible to extract propositional information from diagrams, we would be on the wrong track. There is no unique and apparent propositional content that can be extracted at different times from a diagram. Rather, diagrams are subjects to both *physical* constraints – since they are two-dimensional physical objects – and *conceptual* constraints – because they require a user to interpret them. Topological relations, for example, are very basic spatial relations such as proximity or enclosure that would not change in a diagram if the diagram were printed on a rubber sheet and the sheet were stretched or twisted (Willats 1997). Nevertheless, the recognition of such spatial relationships must be accompanied by interpretation, so that the diagram can be used to the aim of obtaining new conclusions within a specific theory. These two constraints are integrated in the way the diagram is reproduced and manipulated: a diagram is thus interpreted *dynamically* and informal inferences take the form of physical transformations. In fact, the rules of diagrammatic representations are normally externalized as procedures, and, as a consequence, what must be learnt in order to master a diagrammatic system is not abstract rules, but instructions on how to act on the diagrams and to read and interpret them correctly. To sum up, the correct interpretation of a diagram gets intimately connected to the *systematic actions* that are performed on it (Giardino 2010).

I want to argue then that representations must be considered in the way they are used in the meaning-making process. It would also be possible to think, as Walton (1990) suggested, that some of the things in our environment prompt our imagination in order to broaden our imaginative horizons: "imagining is a way of toying with, exploring, trying out new and sometimes farfetched ideas. Hence the value of luring our imaginations into unfamiliar territory" (p. 22). Children games express paradigmatically the ability we show in coping with meaning flexibility and meaning-making, also in situations that are completely new. This is a strong characteristic of our cognitive capacities: we look for meaning and create meaning where we don't find it. And in many cases this cognitive primitive capacity is not apparent in the subsequent stabilization of explicit and formal rules to constrain information.

In his work in developmental psychology, Tomasello (1999) has pointed out to a feature that could be one of the conditions for our meaning-making capacity. According to his studies, we are not only cognitive agents but most of all we are *intentional* agents. The human symbolic activity, of which language is a direct manifestation but not the only one, derives from the joint attentional and communicative activities that the understanding of others *as intentional agents* engenders. Once we recognize the other as an intentional agent, then we are ready as individuals to take an outsider's perspective on our own behavior and cognition, and as a consequence we engage ourselves in a representational redescription, iteratively re-presenting in different representational formats what our internal representations represent (Karmiloff-Smith 1992). Cognition becomes then more systematic: we become capable of using knowledge in a more flexible way in a wider array of relevant contexts.

In the same spirit, Bruner proposes that it is precisely this 'push' to construct narratives that determines the order of priority in which grammatical forms are mastered by the young child. The child is not only reporting: she is trying to make sense of herself and of the world around her, to structure her experience. The achievement of the capacity of giving representational redescriptions and providing narratives is not simply a mental achievement but an achievement of social practice that lends stability to the child's social life.

6.5 Negotiating Meaning in the Infosphere as Inforgs

As I claimed in the previous section, when we consider *semantic* information and meaning, we see that we are still the only semantic engines in the universe. This is a belief that any revolution has not (for the moment!) proved to be false. Our meaning-making capacity is a core issue of our cognition and is related with other capacities as well, such as the creation of cognitive artifacts, the flexibility in assigning meaning, the tendency of seeing the world and the others in intentional terms, and finally the ability to redescribe the world and the others by means of different kinds of representation. Machines, at least up to this point, do not share all these cognitive capacities, which are connected to semantic information. This given, Floridi is right in pointing our that the machines that we are using today are dramatically different from past cognitive artifacts, because they are cognitive *info*artifacts, and for the first time they interact with us, responding to our demands and to our needs. As a consequence, the real challenge for a philosophy of information today is to see in which way our already available capacities of cognitive artifact-making, meaning-making, meaning-flexibility, shared intentionality, and representational redescription are to be affected once we get absorbed by the infosphere.

Let us consider first meaning-making. If public meaning is always constructed, since it is the result of a negotiation, will meaning be negotiated differently in the infosphere? Delocalisation, synchronicity and correlation: how will these features not only change the very notion of time, space and interrelation, but also affect the process of meaning-negotiation? If information – though not necessarily semantic information – is everywhere in the infosphere, then the process of transaction involved in the construction of meaning will surely be influenced by its flow. But how?

Similar considerations can be made for flexibility of meaning. If semantic information is flexible, while information in computational terms is unambiguous, will our capacity of constructing newer and newer perspectives on things and on ourselves in the meaning transaction process lose strength? A crucial issue will be to try to understand to which extent the transformations of our environment and of our technology will have consequences on our interpretation of the world and of ourselves. Shall we become less flexible in meaning assignment because of the influence and the omnipresence of information? Once online and onlife will be one thing, what will happen to meaning and information? Will our capacity of negotiating meaning be reduced?

Another crucial issue to be considered is the role of intentionality in the infosphere. Following Tomasello, sharing intentionality is a basic requirement for human cognition. If we accept his analysis, what about then our relationship with technological responding artifacts? I have shown that we have already re-ontologized them as agents, which are informational as well as we are, but what about the possibility of attributing intentions to them? If we believe we can attribute intentions to them, then it would be really difficult to claim that they are different from us. If we believe we cannot, then we still have to check whether their omnipresence has transformed our very attributions of intentions and our capacity of assuming the other's perspective. If information is everywhere, we should be able to develop ways of understanding where the relevant information is, namely the one that has been given to us with a communicative intention. Will we be able to do that?

The same considerations can be made for representational redescription and narrative in the infosphere: what shape will they take in an environment that is always interconnected and always on line? How will the child structure her experience in the infosphere in order to lend stability to her social onlife?

A final worry could be whether we should accept a sort of technological determinism, according to which the technology will fix any social problem, or on the contrary assume a kind of social determinism of knowledge, in which case the inforgs will be identified as the ones who define the new technological scenarios.

These and more questions arise when we take into account the difference between information and meaning and we project it into the infosphere. In my view, they are the most crucial issues that philosophy of information should be concerned with, in order to provide predictions and to prepare the inforgs to what will happen next, to them and to their world, in their infosphere.

6.6 Conclusions

Let me go back to the scene at the beginning (thought in reality as you might expect it has changed: hopefully I am not anymore waiting for my flight in Hall 3 at the Orly airport). There is the computer, the mobile phones, the Internet and me. I am living in the infosphere, and I am an inforg, that means a hybrid being partly human partly artificial, always – or almost always – on line. I have an informational nature, but, differently from my computer and my phones, this informational nature derives from the fact that I am a symbolic being. I am the ghost in my smart phone and in all the other info-artifacts I use. As Deacon (1997) pointed out, though I share the same earth with millions of living creatures, I also live in a world that no other species has access to, a world full of abstractions, impossibilities, paradoxes. As other members of my species, I was born with some cognitive systems ready to work, but they are limited – and I am limited as well, since I will not last forever. Though, thanks to cultural transmission, I can overcome my limits: I have at my disposal all sort of cognitive artifacts that are ready for me to be used and shared, the more powerful of which is language. Moreover, after Floridi's forth revolution, which I define as the Second Information revolution, information permeates all the surroundings, in such a way that I am always connected: my time, my space and my interrelations with the others and with the very new and responding infoartifacts have changed. I am a new form of symbolic being: it does not mean that my cognitive abilities have changed but that the ontology of the world around me and of myself has changed. What will my representational description of my situation of 'e-migrant' be?

The challenge of philosophy is to answer to questions such as this. Floridi has uncovered a domain of research, the infosphere, which was calling for a new theorization and a conceptual analysis. The methodology he suggests to adopt is to reason in terms of the re-ontologization of our world and the re-discussion of our beliefs. My suggestion is that any further step in this direction requests the collaboration of what Bruner calls 'the interpretative disciplines', such as psychology and the human and social sciences. At least two scenarios are open to the investigation, an ethical and an epistemological one, and issues in education have emerged as well.

In this article, I tried to show that a particularly interesting aspect to discuss is the role, in this picture, of semantic information, which is the expression of a symbolic activity that has up to now been shown to be specifically human. Will one day a fifth revolution come that will take away from us also this ultimate illusion? That day, will our own technology design intentional and semantically powerful machines? At the moment, we do not know. The task of philosophy of information is to provide the appropriate framework that would allow us to make useful predictions in order to prepare the future generations and ourselves.

Acknowledgements I want to thank the group working on Public Representations at the Institut Jean Nicod for all our useful discussions on similar topics, and in particular Elena Pasquinelli and Giuseppe A. Veltri who read a preliminary version of this article. The research was supported by the European Community's Seventh Framework Program ([FP7/2007- 2013]) under a Marie Curie Intra-European Fellowship for Career Development, contract number no. 220686—DBR (Diagram-based Reasoning).

References

Avigad, J. 2008. Understanding proofs. In *The philosophy of mathematical practice*, ed. P. Mancosu, 317–353. Oxford: Oxford University Press.

Berger, P.L., and T. Luckmann. 1966. *The social construction of reality: A treatise in the sociology of knowledge*. Garden City: Anchor Books.

Bruner, J. 1990. *Acts of meaning*. Cambridge, MA/London: Harvard University Press.

Deacon, T.W. 1997. *The symbolic species*. New York/London: W. V. Norton Company.

Dehaene, S. 1997. *The number sense*. New York/Cambridge (UK): Oxford University Press/ Penguin press.

Dror, I.E., and S. Harnad (eds.). 2008. *Cognition distributed: How cognitive technology extends our minds*. Amsterdam: John Benjamins.

Floridi, L. 2002. Information ethics: An environmental approach to the digital divide. *Philosophy in the Contemporary World* 9(1): 39–45.
Floridi, L. 2007. A look into the future impact of ICT on our lives. *The Information Society* 23(1): 59–64. An abridged and modified version was published in TidBITS.
Floridi, L. 2009. The semantic web vs. web 2.0: A philosophical assessment. *Episteme* 6: 25–37.
Frank, M.C., D.L. Everett, E. Fedorenko, and E. Gibson. 2008. Number as a cognitive technology: Evidence from Piraha language and cognition. *Cognition* 108(3): 819–824.
Giardino, V. 2010. Intuition and visualization in mathematical problem solving. *Topoi* 29: 29–39.
Gilbert, A.L., T. Regier, P. Kay, and R.B. Ivry. 2006. Whorf hypothesis is supported in the right visual field but not the left. *Proceedings of the National Academy of Sciences* 103: 489–494.
Grosholz, E. 2007. *Representation and productive ambiguity in mathematics and the sciences*. Oxford: Oxford University Press.
Hermer-Vazquez, L., E.S. Spelke, and A.S. Katsnelson. 1999. Sources of flexibility in human cognition: Dual-task studies of space and language. *Cognitive Psychology* 39: 3–36.
Hutchins, E. 1995a. *Cognition in the wild*. Cambridge, MA: MIT Press.
Hutchins, E. 1995b. How a cockpit remembers its speeds. *Cognitive Science* 19: 265–288.
Karmiloff-Smith, A. 1992. *Beyond modularity: A developmental perspective on cognitive science*. Cambridge, MA: MIT Press.
Kessell, A.M., and B. Tversky. 2006. Using gestures and diagrams to think and talk about insight problems. In *Proceedings of the 28th Meeting of the Cognitive Science Society*, ed. R. Sun and N. Miyake. Mahwah: Lawrence Erlbaum Associates, Inc.
Kinzler, K.D., and E.S. Spelke. 2007. Core systems in human cognition. *Progress in Brain Research* 164: 257–264.
Plato. 1997. *Phaedrus* (trans: Alexander Nehamas and Paul Woodruff). In *Complete works*, ed. John M. Cooper and D.S. Hutchinson. Indianapolis/Cambridge: Hackett Publishing Company.
Searle, J.R. 1980. Minds, brains, and programs. *The Behavioral and Brain Sciences* 3(3): 417–457.
Spelke, E.S. 2004. Core knowledge. In *Attention and performance: Functional neuroimaging of visual cognition*, vol. 20, ed. N. Kanwisher and J. Duncan, 29–56. Oxford: Oxford University Press.
Surowiecki, J. 2004. *The wisdom of crowds: Why the many are smarter than the few and how collective wisdom shapes business, economies, societies and nations*. New York: Doubleday.
Tomasello, M. 1999. *The cultural origins of human cognition*. Cambridge, MA/London: Harvard University Press.
Tversky, B. 2005. Visuospatial reasoning. In *Handbook of reasoning*, ed. K. Holyoak and R. Morrison. Cambridge: Cambridge University Press.
Uchikawa, K., and H. Shinoda. 1996. Influence of basic color categories on color memory discrimination. *Color Research and Application* 21: 430–439.
Walton, K.L. 1990. *Mimesis as make-believe. On the foundations of the representational arts*. Cambridge, MA/London: Harvard University Press.
Willats, J. 1997. *Art and representation: New principles in the analysis of pictures*. Princeton, NJ: Princeton University Press.
Winawer, J., N. Witthoft, M.C. Frank, L. Wu, A.R. Wade, and L. Boroditsky. 2007. Russian blues reveal effects of language on color discrimination. *Proceedings of the National Academy of Sciences* 104: 7780–7785.

Part III
Applications: Education, Internet and Information Science

Chapter 7
What Happens to Infoteachers and Infostudents After the Information Turn?

Elena Pasquinelli

Elena Pasquinelli is post-doc researcher at the Department of Cognitive Studies of the Ecole normale supérieure (Paris) and scientific coordinator of the Group Compas, a Think tank dedicated to the contributions of cognitive science to education in the era of digital technologies. Her research concerns the cognitive and conceptual aspects of the interaction with virtual and fictional worlds, the role of technologies such as video games and mobile phones on learning, and the theoretical aspects of a science of learning and of human-technology interaction based on cognitive science and on evidence.

7.1 Introduction

> *The information revolution has changed the world profoundly, irreversibly and problematically, at a pace and with a scope never seen before. It has provided a wealth of extremely powerful tools and methodologies, created entirely new realities and made possible unprecedented phenomena and experiences. It has caused a wide range of unique problems and conceptual issues, and opened up endless possibilities hitherto unimaginable.* (Floridi 2003)

Philosopher Luciano Floridi describes our era as the result of an *information revolution*. The information turn has made of us *inforgs* (connected information organisms) evolving in the *infosphere*: a place where distinctions between learning from digital on-line – as opposed to physical, off-line one – interactions and contents

E. Pasquinelli (✉)
Department of Cognitive Studies, Ecole normale supérieure (Paris), Paris, France

Groupe Compas – Education, Technologies, Cognition, 29, rue d'Ulm,
75005, Paris, France
e-mail: elena.pasquinelli@gmail.com

H. Demir (ed.), *Luciano Floridi's Philosophy of Technology: Critical Reflections*, Philosophy of Engineering and Technology 8, DOI 10.1007/978-94-007-4292-5_7,

are less and less relevant. Let us imagine walking in the street with our mobile phone in our pocket (not a huge leap of imagination, in fact). Someone calls from far away, we answer and engage in a conversation about a strange art object we are looking at, right in front of us; a picture of the mysterious object is soon taken, and sent to the phone-friend. The phone-friend, tickled by curiosity, searches the Internet for street exhibitions in our town. Meanwhile, we approach the object, and find a code; we then point the camera of our smart-phone onto the code, and an artist appears next to the mysterious object – on the screen of our phone, of course - ready to explain the meaning of the artwork, and to guide us – GPS activated – through an entire maze of no-more so mysterious objects of art that are physically installed in town and through another maze of artworks that the same artist has created with digital tools: representations that are activated by special codes disseminated in the town and that we see on the screen of our telephone, when we point the camera on the real spot. By simply using a smart-phone one can experience that "*The digital is spilling over into the analogue and merging with it*" (Floridi 2007, p. 64), and that the real world is part of the infosphere (the picture you sent to your phone-friend). This is why the infosphere is "*now vast and infinite*" (Floridi 2007, p. 62), ICT (Information and Communication Technologies) being "*among the most influential factors that affect the ontological friction in the infosphere*" (Floridi 2004, p. 63). Friction is the resistant force to the flow of information within a certain region of the infosphere; when friction is low, information freely circulates in a way that makes inforgs – as inhabitants of the infosphere – not necessarily savvy, but at least *informed*: they have no right to claim ignorance, and they know that others know. Mobile phones have done much to reduce *friction*. They are so portable, *always in* (the pocket) and *always* (switched) *on*, that they are much more similar to glasses for short-sighted people, than to sophisticated ICT. But they *are* sophisticated ICT. This fact transforms those who wear them in nicely sophisticated *ITentities* with troubles in sight. Troubles mainly concern ethical issues, such as the risk that the digital divide – the unequal distribution of information technologies, hence: of friction in the infosphere – will generate new populations of "excluded" across and within societies.

> *As a consequence of such re-ontologization of our ordinary environment, we shall be living in an infosphere that will become increasingly synchronized (time), delocalised (space) and correlated (interactions). … Although this might be read, optimistically, as the friendly face of globalization, we should not harbour illusions about how widespread and inclusive the evolution of information societies will be. The digital divide will become a chasm, generating new forms of discrimination between those who can be denizens of the infosphere and those who cannot, between insiders and outsiders, between information rich and information poor. … But the gap will not be reducible to the distance between industrialized and developing countries, since it will cut across societies.* (Floridi 2010, p. 9)

At the same time, developing countries are showing a great deal of ingenuity in exploiting the potentialities of ICT so as to create economical and educational otherwise absent possibilities. Forms of mobile banking in Kenya and in other African countries (Greenwood 2009), as well as educational mobile practices in South Africa and India – that I will illustrate later in this chapter – even suggest a possible counter-colonisation of ingenuous ICT practices from developing to developed countries.

Despite the potential effects (and side-effects) of the massive introduction of ICT in our daily life, the information revolution has barely modified the way we teach and learn (at least in school). Rather than happening, the information revolution is invoked, funded, measured, asserted as a goal for the wealth of the nations. However, the pace of its implementation is not the same as in business, daily social life, banking, and administration. Why? And how can ICT change the educational scenario? The first part of the present chapter discusses why education seems to be at least in part recalcitrant to the information revolution. The second part of it discusses the potential effects of the information turn on education. An exhaustive list of the best practices with ICT at school is beyond my scope, but maybe also beyond feasibility: new educational uses are invented daily for tools that were initially not conceived for education. I will thus discuss what is likely to happen to education when the Fordist model of the classroom – walls separating the school as *the* special place for learning, tables and chairs so that everybody has her own specific place, the teacher as instructor, the pupils as passive listeners – is breeched by mobile phones, video games, and, of course, the computer.

7.2 Why the 4th Revolution Hasn't Revolutionized Education, Yet

In an interview of 2004, MIT educational guru Seymour Papert proposed the following thought experiment: imagine a country where a sophisticated civilization has arisen, and where philosophy, arts, and sciences flourish in spite of the fact that nobody has ever had the idea of writing (Papert 2004). The moment arrives when paper and pencils are invented; rapid and huge transformations happen in the domains of commerce, as well as of science, and someone asks: why not education, as well? This starts a debate: should we begin with one pencil per class, or three pencils per class, or wouldn't it be better to create special classes with mountains of pencils? It is clear, Papert comments, that teachers could do interesting things even in these circumstances; but this scenario has nothing in common with the role of paper and pencils in our civilization. Yet, our society assigns to computers a crucial role in the capacity of operating with knowledge, e.g., scientists do work with computers. Only kids aren't entitled to do the same. Is it because computers are fragile objects, or is it because the "*infostudents – and infoteachers scenario*" (students and teachers inhabiting the infosphere) entails a radical change in the way we conceive education and learning?

Things have naturally changed since 2004: schools are more and more equipped with computers, as well as with other digital devices especially designed for teaching. UK is one of the leading countries for the diffusion of ICT in schools, from primary to high schools. ICT widespread in schools mostly consist in computers (1 computer per 6 pupils in primary schools), electronic interactive whiteboards, but also game consoles; more rarely mobile phones. The number of computers naturally increases when moving from primary towards higher education, and the same is true for Internet access and bandwidth (Rudd et al. 2009). Meanwhile, teachers who have no digital literacy *at all* are becoming rare (less than 7% in Europe in 2006). And yet, the information revolution has not happened: a rough picture of the use and different distribution of computers in schools – which is not limited to the gap separating developed and developing countries - of the challenges of an evolving digital literacy, of the ambivalence towards new forms of interaction made possible by mobile phones, video games, wikis and other forms of social networking, shows a relative resistance of the world of education in terms of information friction.

7.2.1 The Recalcitrance of Education to ICT Penetration: A Rough Picture

In spite of the wide diffusion of ICT tools, even the very optimistic report produced in 2009 by Becta (the late British agency for the introduction of technologies in education) admits that there is still room for improvement, for instance in what concerns the *use* of otherwise widespread new technologies for interactive and engaging forms of learning and teaching that go beyond the projection of presentations on electronic whiteboards (Rudd et al. 2009, p. 26). In other words, even in technologically advanced contexts such as the educational British system, the *use* of digital tools is not as developed as one could hope. Electronic whiteboards and computers can still be *used* as traditional tools. This is maybe explained also by the fact that infoteachers are still far from being an established reality:

> *In the following five countries, more than 5% of all teachers are not using computers because they say they see "no or unclear benefits": Germany (10.5%), Latvia (8.6%), France (7.5%), Belgium (5.8%) and the Czech Republic (5.5%). There exists a strong correlation between this scepticism and lack of motivation to use ICT in class and the age of teachers: the older the teachers, i.e. the longer they are teaching, the more likely they are to lack motivation for ICT use in class because they do not see benefits in its use for pupils.* (Korte and Hüsing 2007, p. 22)

Moreover, the distribution of digital resources is not uniform. Let us take the situation of Europe in 2006:

> *The clear European leaders are Denmark (27 computers per 100 pupils, 26 of which are connected to the internet), Norway (24 computers per 199 pupils/23 internet connected), the Netherlands (21/20) and the UK (20/19) and Luxembourg (20/18). The figures in these countries are significantly higher than the European average of 11 computers per 100 pupils (of which 10 are internet computers). Almost all new member states belong to the group of laggards which include countries such as Latvia, Lithuania, and Poland; however Portugal and Greece also find themselves in this group of countries, with 100 pupils having to share only 6 computers.* (Korte and Hüsing 2007, p. 20)

The part of the picture devoted to the use and distribution of ICT would not be complete without considering the situation of developing countries. It is true that these are the best candidates for becoming *digital slums* (Floridi 2010); but developing countries can also surprise us, and inspire education in unpredicted ways. ICT for education is a major concern for international organizations supporting development in poor countries – such as the World Bank – and it has become one of the topics of developing countries policies. However, as stressed by Kozma (2008), policies and changes in the classroom practice (where classrooms exist, or are attended) can significantly diverge. Accurate policies can crash against digitally illiterate teachers, or pre-existing educational programs based on rote learning. It should be added that ICT policies are expensive choices, and must be justified against results, especially when developing countries are at stake. Outcomes expected from the introduction of ICT in education should hence be stated in *measurable* ways, and *actually measured* in order to monitor their effects (Wagner et al. 2005). Maybe one good reason why information and communication technologies haven't revolutionized education, yet, is because persuasive *evaluations* of the capacity of ICT to enhance, or to transform education, are still lacking.

It should also be reminded that, unlike other forms of literacy, digital literacy is an evolving competence. The 2010 version of the Horizon report (the annual issue of a research project established in 2002 with the aim of identifying emerging technologies likely to have a meaningful impact on the following 5 years' education, training and research) describes the situation as follows: everybody agrees on the importance of digital literacy, but training in digital skills is still rare in education programs; this lack is made more salient by the continuous transformation of the technology, which changes the very notion of literacy. As opposed to learning to write and compute, digital literacy is in fact always evolving, so training quickly becomes obsolete (at least training focused on tools).

> *This reality is exacerbated by the fact that as technology continues to evolve, digital literacy must necessarily be less about tools and more about ways of thinking and seeing, and of crafting narrative.* (Johnson et al. 2010, p. 7)

E.g., the Horizon report 2010 indicates the following four key trends for the period from 2010 to 2015: first, the pervasiveness of information, which challenges educators to revisit their capacity in sense-making and credentialing information that is *everywhere*. Second, the desire and possibility to work and learn *wherever and whenever*, to access information *just in time* and *on demand*; this second trend is potentially disruptive for the distinction between formal (school) and informal learning, and is especially made possible by ubiquitous computing. Pragmatically: by the development of mobile phones and mobile learning, and by the decentralisation of the IT support, we are becoming more and more used to the idea of browser-based software independent from a specific hardware device, and this is the third trend. The fourth trend is: *collaboration*. This looks more like wishful thinking, when it comes to education, but the idea is that (some) schools have created an environment and a climate in which students and teachers work together toward a common goal. So, if these are the trends for the next 5 years, what do the technologies to keep watch over (the emerging technologies that present remarkable potential from an educational perspective) look like? Mobile computing devices (e.g. smart phones) and open content are expected to reach mainstream use in the next year; electronic books and augmented reality accessible to everyone should hit education in 2–3 years; and finally, gesture-based computers and visual data analysis are foreseen to have an impact on education in a 5 years long term perspective. Definitely, digital literacy cannot be bound to computer related skills, but becomes a matter of gaining an *attitude* towards the opportunities (and side-effects) represented by new media technologies and practices. A situation that could represent another reason for the recalcitrance of education to the information turn: because of the necessity of continuous updating, and of the acquisition of a general attitude on the side of (info)teachers.

Among new media and practices, are also included controversial items such as mobile phones and videogames. The active engagement and diffusion of videogames among learners – representing an opportunity for education for more engaging experiences – is absent from the Horizon Report 2010, but strongly present in the reflection about educational technologies; e.g., the proceedings of the 2006 Summit on Educational Games sponsored by the Federation of American Scientists (FAS 2006) open with an enthusiastic endorsement of the introduction of videogames in education:

> *Modern video and computer games offer a rich landscape of adventure and challenge that appeal to a growing number of Americans. Games capture and hold the attention of players for hours as they struggle to operate a successful football franchise, help Romans defeat the Gauls, or go through the strict regimen of Army basic training in virtual landscapes. People acquire new knowledge and complex skills from game play, suggesting gaming could help address one of the nation's most pressing needs – strengthening our system of education and preparing workers for 21st century jobs.* (FAS 2006, p. 3)

Both mobile phones and video gaming are hence foreseen (by different communities) as potentially disruptive technologies for learning and education. Their diffusion forces training in digital literacy to evolve. At the same time, the two are strongly fought by many teachers, and parents – not the indiscriminate, compulsory use of videogames, or the bad habits that teens seldom/often show with mobile phones, but their very existence and use by kids and pupils. So, while Mobile Learning (or Mobile Computing) becomes a domain of research – structured by a community of practice, a series of conferences, an association, a number of national and international projects involving developed countries as well as developing ones[1] – mobile phones are banned from schools in a number of countries (BBC 2005; Bremner 2009). Health issues and misbehaviour (from cheating to bullying) are the reasons adduced for the ban, not adverse effects on learning. The fate of educational videogames is less dramatic, but still controversial. The success of videogames for simulating

[1] E.g.: The International Association of Mobile Learning: http://mlearning.noe-kaleidoscope.org/; Handheld learning conference: http://www.handheldlearning.co.uk/; MoLeNET: http://www.molenet.org.uk/; MobileActive: http://mobileactive.org/

military and other 'serious' situations has reached the world of training, vocational training and education and produced a domain of studies called Game-Based Learning. Like Mobile Learning, Game-Based Learning gives rise to a number of conferences and projects, and the ground for a certain number of educational products (the diffusion of both is much more evident in the UK than in other countries, due to the activity of several organizations).[2] It is however not easy to evaluate the effective gain in learning that Game-Based Learning or Mobile Learning produce: controlled test are an absolute rarity in classrooms; studies on the positives effects of (video)gaming mainly concern visuo-motor coordination (Byron 2008; Mitchell and Savill-Smith 2004). Again, the *lack of evidence*, of proper measures – and more generally: of methods for evaluating the effects of technologies as complex as videogames on skills as complex as those required by schooling – could be one reason for the slow penetration of this technology, when added to the fact that videogames have raised strong, negative reactions. In the US, the National Institute on Media and the Family[3] (a private association) has been conducing a strong fight against video games, arguing from studies such as Gentile (2009) and Anderson et al. (2006) on videogame addiction and on the arousal of violent behaviours (the data only refer to immediate reactions after the game, long-lasting effects not having been measured). Supporters of videogames at school, oppose to the naysayers' view that studies on videogame addiction do not prove any causal effect of video games on negative behaviours, but just show that – in a minority of children – negative schooling attitudes and an excessive use of videogames occur together (Prensky 2006; Gee 2007a, b). However, the lack of a large, solid, shared body of evidence certainly undermines both the positive and the negative attitude. A gap exists between trends and penetration that is reinforced by the difficulty of updating skills on an evolving domain and the absence of assessments capable of proving opportunities, and measuring side-effects.

The purpose of this chapter is not to take a stand in the debate, but to show that when it comes to school education and to young learners the introduction of digital technologies is far from being neutral and technologies raise strong suspicions and resistance. What about more common tools: blogs, Wikipedia, and all the manifestations of the Horizon Report's trend number one, in a nutshell, the pervasive circulation of information? In educators' talk, we found the same ambiguity that affects games and mobile phones. Wikipedia is perhaps the most controversial issue with its "cut and paste" easy solution to homework. On par with searching a solution on the web during in-class exams, or receiving tips via the mobile phone, cut and paste in homework research and composition is perceived as a form of cheating (Bulstrode 2008; Johnson 2007). Cheating is in fact one of the bad habits attributed to technology. It is true that it becomes quite easy to answer pre-shaped, factual questions with Google,

[2] E.g.: Games Based Learning initiative, The Consolarium, LTS Scotland: http://www.ltscotland.org.uk/ictineducation/gamesbasedlearning/index.asp; Games Based Learning conference: http://www.gamebasedlearning2010.com/; Educause http://www.educause.edu/EDUCAUSE+Review/EDUCAUSEReviewMagazineVolume39/GameBasedLearningHowtoDelighta/157927

[3] http://www.mediafamily.org/

or Wikipedia. But the question is: Why blame technology, and not (at least also) the questions? It is a fact that not all questions are easily answered by copying Wikipedia entries, and that just asking students to *write* an entry would block them from copying it (and teach them about the process of writing and modifying entries). But this example shows that introducing and using new technologies in education might involve an additional change of attitude that goes beyond adjusting to evolving technologies: a change in the goals of education and in the understanding of how learning occurs. Before dealing with this issue in the second part of this chapter I will analyse the ways the information revolution could take for overcoming the recalcitrance of education.

7.2.2 The Way Up and the Way Down

One consideration arising from the quick tour we have taken in the (promised) land of infoteachers and infostudents is that formal education seems to be recalcitrant to the information revolution, or at least to approach this revolution with the pace of a slow penetration, and a huge amount of doubts. Another consideration is that policy makers seem to be totally sold to the idea that the 4th revolution should/will change school.[4] The information turn thus is really a *promised* land for educational policies, e.g. for the European Commission.

> *The Commission's policy of "information society for all" (European Commission, 2000, 2004) emphasizes the need to bring every business, school, home, and citizen into the digital age. One goal of the policy is to promote digital literacy that would provide students with new skills and knowledge that they will need for personal and professional development and for active participation in an information-driven society.* (Kozma 2008, p. 1086)

Changing educational systems from a *top-down perspective* is however a hard job. Educational systems are big machines, which have their habits and resist change. Moreover, even when policies change and rational considerations exert pressure, practices do not necessarily follow. In his influential book *Mindstorms*, Seymour Papert tells a story about how difficult it can be to change a system from inside, to unravel established habits, and ask people to move to new ones, even if the new ones are more effective or less expensive from some point of view (but are they proved to be? The problem of evidence, and proper measures cannot be underestimated). The introduction of the QWERTY keyboard on which I typed this chapter is the result of a solution to a perfectly contingent problem. In the time of old typewriters (computers ancestors in text writing), typing two letters one after the other could produce errors if the two letters were adjacent. So, it was decided to set letters

[4] E.g.: Becta: http://www.becta.org.uk/; European Schoolnet: http://www.eun.org/web/guest;jsessionid=C88D79E4E3EE7B1A7E8583DF559DF3D6; National Education Technology Plan: http://www2.ed.gov/about/offices/list/os/technology/plan/2004/site/edlite-default.html; InfoDev: http://www.infodev.org/en/index.html

that are commonly close in our words very distant from one another: professional courses in typewriting had adopted the QWERTY model, and developed training that fitted with that particular keyboard. Very soon, better machines were produced, and the original QWERTY model became useless. Nonetheless, since then nothing has changed, and we are still struggling with our illogical keyboards, and trying to explain to our kids why their so technologically-advanced computers and gesture sensitive, born-for-natural-interaction devices do not allow to quickly identify letters. The story of education could be the same as QWERTY keyboards, resisting change coming from top-down claims for rationality, efficiency, cognitive functioning. This is a good reason, Papert says, for believing that change will come from *outside* the system of education. In 1980 he prophesied that the day every child and adult would possess a computer, learning would undergo a seachange, and schools would have to follow (Papert 1980).

It is on these premises that the One Laptop Per Child (OLPC) was born, claiming that each and every child in the world should possess a computer, especially kids from developing countries.[5] The day this will happen the very idea of teaching and learning and of teachers and learners as we conceive of them will dissolve (and a big amount of social injustice will dissolve too, overcome by knowledge and global participation). For this reason, low cost (the goal – not quire reached – was to keep the cost under 100 $), low power, robust laptop computers (called "XO") have been designed, and given to about 1.6 millions of kids in the world (with governments paying for the computers). The XO is delivered with programs not directly aimed at learning, but rather at creating and interacting: each laptop is in fact connected with the XO laptops of the area so as to allow distance collaboration and sharing of the contents that kids are able to create with their personal computer. At the heart of the OLPC project hence lie the idea that a quantitative factor can translate into a qualitative revolution (a revolution hitting both education and poverty), and the view that learning is a constructive process: *children* are the agents of change, once they become active in their learning, and in teaching as well – for instance teaching their parents to read and write, as it happens in Peru. When this occurs, the information revolution has a major effect in blurring the boundaries between teaching and learning, as *infostudents* become *infoteachers*. But does this happen? In 2009, OLPC has met a big objective: a contract with the government of Uruguay to bring a green computer to every child in the country. In March 2010, Rwanda's government decided to endow every Rwanda kid from 9 to 12 with a XO laptop. Despite all this, many consider OLPC a failure (Nussbaum 2007; Dukker 2007). The number of XO sold to governments has not reached the expectations that could make it economically viable – big 'clients' such as India and China have not followed the OLPC sirens. Additionally, OLPC computers require maintenance and have to travel in difficult conditions, requiring a large and distributed organization and lots of diplomacy.

---

[5] The director of the project is Nicholas Negroponte, but Seymour Papert and Alan Kay (all three from MIT) are amongst the educational theorists and computer scientists recognized as being at the same time inspirers and supporters of the initiative that has seen light in 2005. OLPC: http://laptop.org/en/

Above all, even the OLPC project has somehow taken the *top-down path to the information revolution*, rather than the bottom-up path. Customers (the kids) have been involved only at the later stage of the project; the laptop, its programs and concept, have been delivered as ready-to-use – unsolicited – "gifts". Everybody knows how it feels to get a birthday gift one didn't ask for. A big surprise, but well, we so badly needed or wanted that other beautiful whatever-it-is. This could be in part the story of OLPC. Just in part because OLPC remains an inspiring and influent project, and because meanwhile the OLPC initiative has contributed to lower the price of laptops in a meaningful way. Nevertheless, some critics have expressed the idea that OLPC designers should have spent more time in India, Africa, South America villages, observing uses and needs of the local populations, namely children; and influent experts in ICT in education have contrasted the OLPC top down-model with a *truly* bottom-up approach: the steady, spontaneous multiplication of mobile phones in developed countries as well as in developing ones (Trucano 2009).

7.2.3 The Bottom-Up, Practice-Bound Stance

This is an example of the bottom-up rising of an educational project involving sophisticated but widespread technology. In 2003 a South African mobile provider launches a free instant messaging service supporting both texts and multimodal messages: MXit.[6] Penetration of mobile phones in South Africa is high, but messages are expensive. MXit thus becomes a hit, especially among teens and pre-teens. The success immediately calls for restrictions at school: MXit is seen as a drug eroding school results and social behaviour.

> *Schools are calling for tough new rules to curb the use of a cellphone instant messaging service that is becoming an obsession affecting pupils' work. And one school has already a support group for pupils who are addicted to the service, designed by a South African and called "Mxit"... Immaculata High principal Kubeshini Govender, where pupils addicted to Mxit are getting support, said: "Mxit is a drug. The learners become dependent on it." ... Some chatted all night and were tired in class the next day. Others spent most of the day on Mxit.* (Keating and Williams 2006)

It is at this point that Laurie Butgereit of Meraka University launches an educational project which exploits the strengths of MXit (its diffusion, the existence of strong practices amongst teens) in order to address the need teens have for a better education in mathematics. The project (called "Math for MXit" or "Dr Math") consists of one to one remote assistance in mathematics via SMS texts (Butgerteit 2007). The tutor (a student in mathematics or engineering) sits in front of a computer and remotely answers the questions asked by students having trouble with their

[6] Mxit: http://www.mxitlifestyle.com/

homework; he never solves their problems directly, but guides them step by step to the solution, and to an understanding (as we can read from the transcriptions of the interactions). This model cannot but remind us of the way forums work. Results of the evaluation of Dr Math project are still to come, but users' comments are positive and a new project has been launched (Imfundo Yami Imfundo Yethu) which involves a Finnish organisation, the South Africa Department of Education, Nokia, and 260 Grade 9 and 10 learners from six schools for producing controlled evaluations (Vecchiatto 2009).

What do we learn from the Dr. Math (on MXit) case?

> *Math on MXit takes advantage of the fact that teenagers are* already *using MXit to communicate with their friends.* (Butgereit 2007)

First, the educational activity provided by Math on MXit is not the reproduction of something that exists in traditional education: one-to-one tutoring being a very desirable but expensive situation (Bloom 1984). We also know that African families are hungry for tutors for their kids, but these tutors are often amateurish and expensive. We hence thus a clear *need* on the educational and social side, which traditional systems can hardly fulfil. Secondly, rather than inventing new practices, and trying to make them popular, Dr Math *colonises* existing, common practices with educational purposes. In a perspective that is coherent with the information revolution described by Floridi, the principle of colonisation consists in *grafting* educational purposes into the ecology of the infosphere. When students go to school they are taken away from the "real world": the infosphere, with its practices and its ecology made of phone cells, messages, and a wide variety of ways of producing and sharing information. Once back home they are inforgs again (they start again mxing, gaming, surfing the net). Colonisation represents an ecological approach to bringing the information revolution in the domain of education.

What would the opposite scenario look like? Something like this: a brand new technology that students (and also teachers) do not know how to use is added to the classroom. Moreover, this unpractised technology does not bring a new function into the educational panorama, but it is limited to the *electrification* of pedagogical activities, tools and roles that can be very well realized in more traditional ways (Casati 2009). In other words, technology is used as a modernizing paintbrush or as a form of electrification of books, teachers and blackboards.

To conclude, the 4th revolution is yet to reach education, for several reasons among which we can cite: the lack of appropriate and shared evaluations of effects and side-effects, the difficulty of upgrading to continuous changes in hardware and practices, the challenge to educational habits. Moreover, injecting technology is not enough and changing educational habits is such a hard job – prone to the mistake of simply adding some digital make-up onto traditional activities. The way for the 4th revolution to come and invest education could be then be better represented by a form of *ecological colonisation of existing, widespread technologies and practices*. A double colonisation, since this model comes from brilliant ideas spreading *from* developing countries: will innovation in education be the place for a counter-colonisation?

7.3 What Is It Likely to Happen to Education When the Information Revolution Happens?

Even if it is true that education is recalcitrant to the information turn, it is still reasonable to question whether the perspective of ICT spreading to education is a desirable scenario. There is in fact no consensus on this issue. Some fear the deepening of the digital divide which afflicts developing countries, and which adds its negative effects to those of a number of other gaps. Some (defined by Seymour Papert as "critics") think that the extension of new technologies is not desirable at all, even in the perspective of developed countries (Papert 1980). The critics believe that computers can make a difference, but that this difference is not worth pursuing. So, for example, some critics propend to accept that video games have an effect on young minds, and that video games can "teach", but that what they teach is violence, short attention span, immediate gratification (Anderson et al. 2006). Or that reducing the friction in the circulation of information can endanger education, because it creates a sort of "information overload"; e.g., president Barack Obama has recently rallied technophobics (according to The Economist) asserting that information can become a distraction for learners:

> *With iPods and iPads and Xboxes and PlayStations – none of which I know how to work – information becomes a distraction, a diversion, a form of entertainment, rather than a tool of empowerment.* (The Economist 2010)

President Obama has not condemned ICT as a whole (it would have been rough for someone who has made an exemplary use of the Internet, and is still making an unprecedented use of Twitter and YouTube). According to The Economist's analysis, infopresident Obama's speech implicitly contains a distinction between good, empowering information and bad, distracting information. Still, the discourse was addressed to students of Hampton University, and the quoted sentence could thus be interpreted in the following alternative way: information is not good or bad in itself; yet, when educational environments do not limit the pervasiveness and free circulation of information (when the infosphere becomes frictionless) it becomes difficult to attend to the information proposed by the teacher in the classroom. It is not untrue that sending SMS messages, consulting YouTube or even Wikipedia is incompatible with the Victorian model of the classroom, where a teacher speaks to listening pupils. But this is not the only, possible scenario.

In addiction to critics, Papert introduces two other categories of attitudes towards ICT in education: "optimists" and "sceptics". Optimists believe that computers can make a *qualitative difference* in learning; it is not just a matter of improving instructional teaching and school education, but of empowering individuals to choose the way they want to learn by creating learning tools that can be used outside schools: ICT *augments* education, in the sense that it changes education into something, which can benefit from the entire infosphere. What grounds the optimistic attitude is the view that learning is a cognitive process, which goes beyond dedicated instruction (school): people learn from their experience, all their lives; children learn from their environment and culture. Changing the furniture of the environment, changing the tools and habits that are part of the culture, also changes the way we learn, and think. Floridi would say that this produces a re-ontologisation of the learning environment, a transformation of its intrinsic nature (Floridi 2007). For instance, infolearners will ask for different schools that correspond to their way of learning, and to their idea of knowledge: knowledge which is accessible anytime, anywhere; knowledge which is constructed by multiple, interconnected intelligences; and knowledge which is gained through active patterns of search, hence meaningful from the searching individual's perspective. In the framework of this massive change information overload no longer is a problem, because the very structure of educational contexts, methods, and aims is transformed by the expansion of the infosphere. On the opposite side, sceptics do not expect the presence of computers to produce a massive change in how people learn, and think; according to them, all that ICT can do is to *enhance* instruction (as opposed to augmentation), by providing a means for better teaching in schools. Interactive whiteboards can be considered as "enhanced" blackboards, which allow teachers to display multi-modal contents (images, videos, charts) and to save exercises and notes; this is a lot more than can be done with a traditional blackboard, but it does not represent (or at least, not necessarily) a revolution in how students learn, and teachers teach.

The three categories described by Papert do not belong to the same "natural kind". Critics and optimists both believe that ICT will produce a radical change on learning and thinking, but they evaluate differently the *desirability* of its effects. Sceptics do not hope or fear, they just don't *believe* (or *estimate*) the information turn to represent a massive change for education. We obtain two axes along which different positions can be aligned. E.g., (Aviram and Talmi 2005) draw a matrix along two axes they call *approaches* and *attitudes*. Approaches range from the assumption that technology can be subsumed under the traditional school and curriculum – and that its introduction has a qualitative but not a "revolutionary" effect – to challenges to the very notion of school as a physical space and to its aims. The transition from one extreme position to the other is represented by seven beliefs: a. that computers should simply be present at school as they are everywhere else; b. that technology should serve curricular purposes, by becoming a discipline (computer science) or by taking advantage of ICT for teaching the subject matters included in the current curriculum (e.g. sciences or maths); c. that new technologies are part of a change in the way contents are taught/learnt at school (for instance, through more constructive and interactive methods); d. that the whole organization of educational spaces and time, roles and curricula is changed by the advent of ICT; e. that school disappears in favour of remote and even virtual schools; f. that ICT in education is part of a deep cultural revolution ; g. that change should be shaped by values (Aviram and Talmi 2005). The cultural approach that characterises f. fits particularly well with the philosophy of information proposed by Luciano Floridi, because it recognises that ICT has a re-defining (re-ontologising) impact on our way of living and thinking about things, and because it acknowledges the fact that the educational revolution is part of a deeper revolution that has transformed Western culture.

> *The cultural approach is quite rare in discussions on ICT and education. Those who rely on it are mainly academics, intellectuals or futurists. The approach remains unknown to many teachers, and even to many academics. Adherents of the cultural approach maintain that educationists should be aware of the revolutionary, defining nature of ICT, and strive to adapt the education system to the new culture. Such adaptation could take diverse routes. One may judge the rising postmodern culture favorably and recommend radical changes in the school structure in order to adapt it to the new 'human situation' (what we call below the 'radical' attitude). Conversely, one might judge it unfavorably and opt for preserving and strengthening the existing structure of education (the conservative attitude).* (Aviram and Talmi 2005, p. 171)

As for the second axis, Aviram and Talmi distinguish five attitudes, which can be driven by goals: those of i. agnostics, ii. conservatives, iii. moderates, iv. radicals and v. extreme radicals, ranging from those who do not care about what the impact of ICT would or should be, to those who "*believe that ICT is a Trojan horse inside the base of the prevailing educational system, and that the latter will not (and, quite often, should not) survive it.*" (Aviram and Talmi 2005, p. 172).

In what follows I will illustrate some examples of what Trojan horses could look like, and their potential effects on the Victorian school. The process of bringing the effects of the information revolution into formal education is slow because formal education has created special places for learning, and these special places tend to keep learners separated form the world. Trojan horses can enter the heart of the educational system: school; but they can also settle in the periphery of the citadel and slowly change the perception of what education is (i.e., re-ontologise education).

7.3.1 Trojan Horse No 1: Computers

"Hole in the wall" is an initiative aimed at slums and poor villages in India, imagined and realized by Sugata Mitra (now professor of Educational Technology at the School of Education, Communication, and Language Sciences, Newcastle University). In 1999, in Kalkahi (a poor borough of New Delhi) a real hole was made in the real wall separating the NIIT (the learning solutions corporation Mitra was working with) from the adjoining slum: a computer was slipped into the hole, for free use. Children came, spontaneously, and started using the computer to look up information on the Internet (videogames and CDs have also been employed in further settings). Many skills were required, which the kids did not possess yet: to use a mouse, to understand how a web page is structured, and most of all to read English – a major issue for education in India, where English is the mandatory requirement for the access to higher education. The observation of kids operating with the computer, coming back day after day, getting better at digital literacy, and collaborating, came to reinforce the pedagogical stance that Mitra has since then identified as "Minimally Invasive Education" and "unsupervised learning": learning that develops from the natural exploratory activity of children, especially when children are brought *together* and interact with an object, which is able to deliver *information* in different shapes (Mitra and Rana 2001). This same model has been exported from India to Cambodia, to Africa, and even to UK, as a project named "Self Organised Learning Environments" (Mitra 2009).

> *"Minimally invasive" refers to the least possible, negligible, or the minimum help required by the child to initiate and continue the process of learning basic computing skills. This minimal amount of help from other children at the MIE learning station is necessary and sufficient for the children to become computer literate. This "help", which is the fundamental aspect of MIE, could be from peers, siblings, friends, or any other child familiar with computers. Children are found to collaborate and support each other. The learning environment is characterized by its absence from adult intervention, openness and flexibility. Children are free to operate the computer at their convenience, they can consult and seek help from any other child/children, and are not dictated by any structured settings.* (Mitra et al. 2005, p. 3)

The method is meant to apply whenever there are no real teachers at hand, at least of good teachers (because they will not accept work in remote parts of developing countries; as well as of developed ones). Many solutions have been put in place, in the world, to compensate for this absence – special books, radios transmitting courses to the classroom (and the physical teacher), educational TVs, open universities and what is called e-learning, or learning at distance (through CDs, the Internet or even mobile phones) - all with a common denominator: being addressed to the individual learner, or to the individual learner as immersed in a typical classroom structure (as when learners look and listen to radio and wait for questions posed by the teacher or take their exercises on a mobile phone) (Trucano 2005). In Minimally Invasive Education this modality is challenged twice: first, learners become teachers for other learners (they *peer-teach* each other in groups); and, second, learners *search* information, instead of receiving it as a form of instruction, or of test (which is not to say that instruction and tests are not useful and effective). MIE has turned out to be effective, at least for the acquisition of computer literacy: children collaborating around a computer reach levels of digital literacy, which are comparable with those acquired by the means of traditional classroom instruction (but it should be acknowledged that they normally spend more time interacting with the computer than children using the computer at school) (Mitra et al. 2005). This means that a minimal investment (much less that one computer per child) could make the difference in terms of digital divide (which has been cited at the beginning of this chapter as a major ethical preoccupation in the information age). Moreover, the Hole in the Wall experience points to an issue, which is deeper than the positive effects of MIE on digital literacy and divide: a *different way of learning* is made possible – or at least made easier – by the fact of living with other inforgs in a rather frictionless infosphere (friction being reduced in this case by the presence of just one computer). This form of learning is *self-directed*, *collaborative*, and *independent from formal structures* and settings.

Would it have been possible to achieve the same result before the information turn? In other words: is the fact that information can freely circulate a necessary condition for this form of learning to exist? A thought experiment can shed light on this question: let us imagine a group of children wandering among the scrolls of the Ancient Library of Alexandria; certainly, they did access large quantities of information; however, those scrolls could not respond to their actions: they would not close in response to a bad search. Computers do. Sugata Mitra describes the discovery of one of the first "Hole in the wall children": he touches touching the screen in a certain way and sees the page disappear, another appear; the kid then goes back and forth in search of new reactions from the machine. Tools of the information turn do react to learners' actions (and to states of the world, if they have a GPS, or whatever kind of sensor) with a change in their informational content. They can thus become part of a dialogue in a way books, radios, cinema (and even just reading a Wikipedia page) cannot afford.

Hole in the wall and MIE is however a rather extreme form of *no-schooling*, confined to places with no school to choose as an alternative. Few parents in Paris, London, Rome, New York, Singapore, Tokyo, would choose to send their children playing with a computer in the street, *rather than* going to school. But some children, even in these big cities, do not want to go to school: they quit, disengage, suffer school phobia, or illness. Is there an alternative to bringing them back to the classroom? Forms of no-schooling or virtual schools have been tested: learners do not meet physically, do not collaborate in presence, but only at distance, and via the computer; they receive some form of follow-up which is somewhat more "invasive" than Minimally Invasive Education. Notschool project, originally founded by Stephen Heppell, aims at reengaging students in the learning process, without imposing a school environment: learners have access to chat rooms, mentors (one for six learners or researchers), and a virtual community.[7] In 2007 the project included 1,000 learners, with 96% obtaining some form of accreditation. The principle behind Notschool is the principle that education can reach learners everywhere: they do not need a *physical* space called school (which does not mean that schools should not exist). What makes it possible is the existence of a continuous flow of information, and of high bandwidth.

7.3.2 *Trojan Horse No 2: Mobile Computing*

We have seen how computers can transform learners into infoteachers and breach the walls of schools (re-ontologisation of authority and space). Schools are also time-organisers, defining which is the moment for learning, and which for entertainment (not during the lesson: information overload), for play, and for socialising. Does the information turn also affect (re-ontologise) our perception of time and, in particular, the idea that there is a time for learning and a time for doing other things? I think this particular re-ontologisation could depend upon the spread of two practices, which I have cited as influential on education in the first part of this chapter: mobile learning and serious gaming. Let us consider them in turn.

[7] Not School: http://www.notschool.net

Mobile learning is a recently minted label in the domain of technology-based education; it refers to the use of mobile technologies, such as mobile phones, PDAs, and even portable game consoles, or, more deeply, to the notion of *mobility*. Let us go back to the scenario described in the introduction: the mobile phone is used for taking pictures of the real world, digitalizing them, and plotting information of different sorts against real objects. There no fixed time for doing this: the user chooses when to take her phone out of her pocket. This is why it is said that mobile learning is *anytime*, and not only *anywhere* (Ally 2009). This *mobility over time* can represent a great advantage for learners who have strong temporal constraints: working people trying to learn a new language or skill while going to their job, travelling people; but also students going to school and wanting to learn more in between home and school. E.g.: a teacher in India, attending a 1-week course in order to improve her capacity to teach English to her pupils, can also have the same course, and the exercises she needs to practice, on her smartphone, and bring them with her in the small village she has been assigned to. Once there, she will be able to consult the lessons, and revise them through the exercises, at any moment she needs it; she will also be able to send SMS for checking the answers, or asking questions to her tutor.[8] A kid in an isolated village of India, with no Hole in the wall kiosks around, can still be involved in a project like Millee[9]: smartphones are given to kids who can use them at any time for playing special games for learning (serious games; the contents of the game reproduce games that kids "physically" play in their village), as well as build projects in schools (kids take pictures, create their own contents, and show them at school). The incredible growth of mobile phones (even smart phones) in developing countries is then a positive premise for the spread of an educational use of mobile phones, for both children and for adults (Traxler and Kukulska-Hulme 2005). Adult population is a major target for educational actions aimed at reducing the gap between developed and developing countries, in terms of access to knowledge (general and digital literacy). Mobile phones could represent one of the most powerful (in terms of penetration and opportunities) Trojan horse for this population to access the infosphere and, through the infosphere, to raise literacy in developing countries (in all those contexts where literacy is the limit). This is also because mobile phones are flexible tools, in the sense that they can serve at least three kinds of uses. First, as computers, mobile phones allow accessing and producing digital information: taking and sending pictures (and videos), recording and listening to sounds and music, playing games, reading and writing, accessing the Internet as passive and as active users (Prensky 2005). Let us imagine a student spending some time in London in order to perfect her fluency in English. Of course, she has joined a "traditional" course, but she also has a mobile phone. With that, she walks in the street, where she impolitely listens at conversations: she catches words, and phrases she does not know, and she notes them on her phone, using it as a log.

---

[8] English in Action (Open University): http://www.englishinaction.com/

[9] Millee: http://www.millee.org/

Or better, she searches for their definition on Google, she writes SMS to course mates and tutors and ask their advice on the particular problem she has met (Kukulska-Hulme 2009).

Second, mobile phones sense the environment, and respond to it: bar codes scanners, GPS, compasses, accelerometers gyroscopes embedded in smartphones are sensors that allow multiple applications: from augmenting reality with digital contents[10]; to write on the phone just writing words in the air (Agrawal et al. 2009); or to write on a projected keyboard so that even physical action in the world becomes digital information and command for the machine (Maes and Minstry 2009). Even walking can become digital information, with an iPod (if one wears Nike shoes).[11]

Third, mobile phones are tools for communicating (via voice, texts, and images) with other individuals, and also with machines (Sharples 2005). One can receive tips about lessons, be put in a network with other learners that are interested in the same topics, by automatic systems for managing networking and administration in a University campus (Brown 2008). In some cases individuals and machines can be combined in such a way as to become one indistinguishable information station. Imagine being in a remote village of Cameroon, and badly needing to know who is the richest man in the world, or what is the price of tomatoes today, or which is the right pesticide for your dying plantation. Imagine not having access to Internet, at least not directly, but owing a mobile phone (and some credit), and a number to call, where an operator takes the question, searches the Internet, and provides the answer. A smooth flux of information, a frictionless infosphere is established up to the remotest corners of the planet, if one can make a call, even in the absence of the computer functionalities. This is the lesson of Question box project: a service for calling an operator who searches the infosphere, from dedicated phones distributed in Indian villages, or from one's own mobile phone in Cameroon.[12]

The three uses just mentioned can affect the way we conceive access to information, knowledge and education, in a way that goes beyond the *anytime, anywhere* refrain.

First of all: learners can access information when it is really needed, when it is meaningful. Accessing information *just in time* has potentially large consequences, which go beyond education. Knowledge does not necessarily need to be stored in mind, when one knows where to find information and how, and if one is *confident* that one will be able to access it at any moment. Mobility is hence a premise for considering ICT as a form of *cognitive extension* (Clark and Chalmers 1998), for instance a memory extension, which is not so different from "internal" or "brain" memory. As memory, mobile phones are always with us, always on. This does not mean that there is no difference between internal processes and extended ones, but that ICT tools can be used as cognitive tools with an effect on cognitive actions and performances. It is obvious that the fact of possessing mobile phones wouldn't have changed the necessity for people living in Ray Bradbury's Fahrenheit 451 world

[10] E.g.: Layar http://www.layar.com/

[11] Apple-Nike: http://www.apple.com/ipod/nike/

[12] Question box: http://questionbox.org/

to learn entire books by heart, in order to save them from burning. Mobile phones can burn, too (actually, we could imagine a future where memory can be selectively erased; but it is the future).

Secondly, learning can happen *in context*. While other media tend to set a separation between the learner and the physical world, and between digital information and physical objects, mobile phones allow a perfect integration of the two: they take the learner out of the box (Van der Klein 2008). Thus, in mobile conditions context can affect learning in two complementary ways: objects raise questions, which learners can answer through the help of digital information (augmented reality mode); and objects provide answers to questions raised by digital information (augmented digital representation mode). An educational project developed by Waag, a Dutch company, illustrates the second mode: grouped in small teams, young learners follow a quest in the Medieval streets of Amsterdam; they walk in search of monuments in order to answer the problems raised by a video game for mobile phone; at the same time, they stay in contact with residential teams searching the Internet with computers.[13] Mobile phones thus allow a form of experiential education, as Dewey described it in the last century: where knowledge is acquired through experience (active exploration), and connects to the learners' experiences (interests, motivation, life) (Dewey 1997).

7.3.3 Trojan Horse No 3: Serious Games

Experience is a key word for the vision of education purported by John Dewey, a vision never fully realized. Lack of appropriate means could explain it. So, let us see what happens when new technologies are employed to make experience possible. In the 1990s, a group of researchers of Vanderbilt University, coordinated by John Bransford, launched a long-term project devoted to designing and testing a method for the learning of sciences, which would comply with Dewey's considerations about experience, and with the notion of inert knowledge as introduced in 1929 by another philosopher: Alfred North Whitehead (CTGV 1990). Whitehead had claimed, in front of his colleagues, that Victorian school provides students with a form of knowledge, which is not used for anything but for responding to tests (Whitehead 1929). This knowledge is inert, because schools teach broad, but not deep, and because they disconnect knowledge from the reasons of its existence, from the contexts of its application. But in Whitehead's view, as well as in Dewey's, pieces of knowledge are nothing but tools, which help people coping with the world. This is also the perspective adopted by Bransford and colleagues (and they certainly are not the only ones) in proposing an anchored instruction method for learning maths, pivoting around the videotaped adventures of a fictional character: Jasper Woodbury (CTGV 1990). Jasper finds himself driving a boat or a plane, and facing

[13] Frequency 1550: http://freq1550.waag.org/

problems of fuel, distance, time. At the end of the movie, students are asked to plan the solution to the quantitative problems Jasper is faced with, and to compute. Instruction is thus anchored to "real" contexts, and mathematical tools serve to solve "real" problems.

Twenty years later, this same approach is proposed in the framework of serious (video)gaming: not only are concrete problems posed to learners in the context of the representation of a certain situation; but learners are asked to find the solution and to implement it directly in the game (something that was impossible with non-interactive technologies like videotapes). Serious games, as well as simulations without gaming (the difference being that simulations have no winners, no reward and no competition), have spread in a number of domains: military training – including the simulation of social interactions with civil populations – surgical training – including training on virtual frogs in school using dissections – the training of pilots – also to earn a (real) licence to fly civil planes, with no other experience than driving military planes. Going back to Jasper Woodbury, games have been *designed* for teaching and learning biology, physics, history, and mathematics (Prensky 2005). Commercial Off-The-Shelf games (COTS) are currently used in schools for stimulating children to write and imagine scenarios, for inviting them to collaborate around the organisation of events, for increasing efficiency and speed in elementary computation, and for all those learning activities, which inventive teachers can imagine from diverting commercial products from their original aim and colonising them with educational purposes (again) (Felicia 2009).

Naturally, the idea that play is important for children, and even for learning, is not new: it is not a product of the information age, or of the videogame industry. Historically, the first theories of play were purely descriptive and aimed at finding a role for play in the human development (from the end of the nineteenth century). The normative idea that play should be exploited for learning is more recent; among others it has been asserted by Maria Montessori, and is still implemented by schools inspired by her vision. Still more recently this same idea has been revived by the advent of videogames, and has given birth to what is called Game-Based Learning, or better: Digital Game-Based Learning (Prensky 2005; Gee 2007a, b). It has even been asserted that modern videogames (whatever their original scope) are machines for learning: players must learn how to play in order to enjoy the game; if the game does not facilitate learning, then the designer is out (Gee 2007a, b). This strong constraint would be the reason why videogames embed very efficient pedagogical principles: learners feel like active agents because they make things happen; learners form expertise by practicing skills until they are nearly automatic, then having those skills become insufficient to face new situations in a way that makes it necessary to think and learn anew; learners are put into fish tanks that are similar to real situations for their structure, but without the dangers and excessive complexity of the real world: only certain variables are selected and stressed ("*With today's capacity to build simulations, there is no excuse for the lack of fish tanks in schools*": Gee 2007a, p. 39); players do not start from the manual, but from playing the game and then going to read the manual for knowing more ("*Game manuals, just like science text books, make little sense if one tries to read them before having played the game*": Gee 2007a, p. 38); hence learners can start from experience rather than from general definitions and principles; and, naturally, learning and pleasure are joined together.

> *Pleasure and learning: For most people these two don't seem to go together. But that is a mistruth we have picked up at school, where we have been taught that pleasure is fun and learning is work, and thus that work is not fun.* (Gee 2007a, p. 10)

So, at the same time, games (digital or not, but it is a fact that the discussion has been revived by so-called Digital Game Based Learning) question the distinction between time for learning and time for pleasure, and make it possible or at least easier to challenge the idea of education as the *transmission to the new generation of bodies of information and skills that have been worked out in the past* (Dewey 1997), because digital fish tanks are ideal tools for experiencing simplified, models of reality that are designed for pedagogy.

New technologies, or better: the way they are *practiced* in some exemplary cases, challenge some of the tenets of a model of schooling and education – a model which probably is not realised in its complete form in any school of the twenty-first century, but that is present in our vision of education, positively or negatively.

7.4 Conclusions

In the preceding sections, I have identified the Victorian school with a number of characteristics: a dedicated space (separated from other social enterprises and physical places), a dedicated time (the time for learning), well-defined roles (one teaches, the others learn), and contents (inert knowledge). I have shown that all these characteristics are challenged by practices that have become possible *after* the information revolution, even if this does not mean that they will be transformed.

The pervasive flux of information is a potential Trojan horse into the traditional structure of education. Firstly, ICT practices spreading in developing countries, and in especially "deprived" conditions (in terms of educational systems and access to literacy), can be colonised by educational purposes; and, secondly, alternative educational practices with ICT can challenge all those who are interested in education to revise their conceptions about education and learning.

Understanding that social structures (as school) and concepts (as education) can also become *different*, opens the door not to one, but to number of alternatives, because it is a process of de-naturalisation. Concepts are not frozen, "natural" entities. They live their life in the middle of contracts, negotiations, practices, and debates. They have a history, and a context from which they take their meaning. When the context changes, concepts can undergo mutations. If they don't, they become obsolete and are replaced (as it has happened to the notion of phlogiston). That's why examples are important to me, and I have used many in this chapter. It is the old Wittgensteinian methodological rule: see this way, and now see it the other way, but do not stop seeing it other ways. The effect is that we acquire what

Robert Musil called the sense of possibility, as opposed to the sense of reality. So, the fact that new, alternative practices spread does not mean that schools will or should be closed – unless they prove ineffective in relationship with the objectives they assign to themselves; or unless these objectives conflict with wider objectives, which become dominant in the society surrounding and supporting schools (principle of reality). However, the spread of new practices certainly forces us to re-conceptualise what we intend when we talk about schools, education, learning, and knowledge (principle of possibility). For example, peer-teaching practices and self-directed learning induce a mutation in the notion of authority and of education as the transmission of knowledge from someone who possesses information to hollow learners. At the same time, the distinction between formal education and informal learning becomes less important, because learning is no longer bound to official places for transmission.

In the context of information, which is accessible anytime, anywhere, on demand and just in time, even the notion of knowledge as something we possess in our brain undergoes some adjustment. In certain respects, our mind can be considered as an extended structure encompassing the brain under the scalp *and* of the tools in our pockets. In this perspective, the idea that school should transmit all the contents, which *may* be needed in the future becomes redundant.

Thus, it becomes more reasonable to concentrate on learning deep, rather than on learning broad; also because of the possibility of learning from experience in concrete – even if digital – settings that are models of reality with a stress on relevant variables, and relevant variables only. If this re-conceptualisation sounds too extreme let us just side with sceptics, and leave big re-conceptualisations to optimists.

7.4.1 Further Research Directions

As we have seen, the main difference between optimists and sceptics lies in the following opposition: on the one side, the idea that when the infosphere extends to schools, frontiers between schooling and no-schooling are redesigned, as it happens to physical and virtual artefacts in augmented reality (*augmentation*); on the other side, the idea that friction in the circulation of information will always make a difference between places inside the educational system and places outside it, because ICT will be functional to enhance the present state of affairs (*enhancement*).

Evaluation (and the definition of proper systems of evaluation that are apt to monitor the achievement stated objectives) is a crucial condition for asserting that (a certain) technology represents the best tool for *enhancing* education. If we can prove it works, it will become easier to foster the use of new technologies in school, in order to enhance students' performances. Accountability and a systematic use of evaluation and test applied to choose the best teaching strategy is the key for the identification of *good tools*, and for the spreading of ICT tools (that are worth spreading). Some (radicals) might argue that this is not a big gain, and certainly not a revolution in education.

If we shift to the *augmentation* scenario, then measures become more difficult. The difficulty stems from the definition of the objectives of education: do they remain the same as traditional school (learning sciences, and maths), or do they undergo a redefinition? Once schooling and no-schooling have blurred their respective boundaries, one could argue, the aims of education will have changed in a meaningful way. If this is the case, it becomes difficult to prove that ICT represent a gain in comparison with more traditional tools. Accountability and the systematic use of evaluation is no more the key for the spreading of ICT in education.

Philosophical choices about what do we want education to be, in relationship to our description of the world, are now at the heart of the problem. For instance, one can assert, as Mitchell Resnick does, that – in order to cope with twenty-first century issues – soft skills are more important acquisitions than plain, factual knowledge: we live in a world, which has accelerated its own pace of change, because of ICT (Resnick et al. 2009). ICT can help us cope, by making learners more creative, and capable of acquiring the twenty-first century skills: capacity of collaboration, of communication, of management, of directing one's own work. Others could adopt a different image of the twenty-first century, or believe that the principles of education are eternal: citizenship, respect of the other, progress of the learner as a human being, self-empowerment. Evaluations are still possible (dues), but it is very important to recognize that they correspond to specific (and clearly stated) objectives.

It is also important to avoid a major mistake, which consists in believing that ICT *per se* is a sufficient condition for transforming Victorian classrooms into playful spaces, where learners make (real and virtual) experiences and learn to evolve in the infosphere thanks to the tutoring of a variety of infoteachers. What we have seen is that *practices with ICT* exist that can challenge the tenets of traditional education. So, it is the diffusion of practices and not simply of technologies that makes the difference. We should fear that infostudents are asked to use mobile phones and the internet for retrieving and memorizing Napoleon's date of death. I claim that up to a certain point, ICT is neutral to educational models and contents. Hence, that an effort towards a rich and innovative use of ICT must be done, which goes beyond endowing classrooms with computers and interactive whiteboards. This effort consists in imagining different forms of education, in defining their scopes, in diffusing *new practices* (or better: in colonising widespread practices with educational aims), and in *evaluating* their results.

Still, this is only part of the story. Another part concerns the development of a new research field in education, capable of integrating research on technologies and information, and studies on the functioning of mind. This field is still in its infancy, but it promises to bring about serious challenges (and eventually confirmations) to the way we conceive school.

Let us take the case of the idea that education consists in transmitting information from teachers to learners. This idea is strongly connected to the metaphor of learners as hollow boxes, which should be filled with knowledge for a supposed future. This metaphor is at the same time challenged by practices with mobiles and games, and by the research on early knowledge in babies, and on the existence of naïve beliefs that pre-exist (and eventually resist) formal education (Bransford et al. 2000).

From the first years of their life, kids perceive the world as being structured: they use criteria for parcelling the flux of stimuli into separated, consistent, dynamically coherent objects; they distinguish between non-animated and animated entities; they get habituated to regularities, and show surprise when faced with violations of expectations. They also develop beliefs about how the physical, the biological, and the psychological worlds work, and interpret events in terms of these beliefs, which, quite often, can reveal false when compared with scientific theories of the same phenomena. Replacing or updating false beliefs is referred to as "conceptual change", and it is a big challenge for education. This wouldn't be the case had the hollow box been a correct image: hollow boxes do not oppose any resistance at being filled in. But learners, whatever young they are, are not hollow boxes. They are rather complicated interpreting machines that use what they know and their previous experiences to make sense of new events and of the world.

As well as technologies, knowledge from cognitive science is challenging some of the tenets of education and suggesting that education should start from *how we learn* (hence from the observation of good practices and the study of mind) rather than from the consideration of *what is useful to learn* (even in the twenty-first century perspective). How their joint venture will be able to affect education is more a matter of *will*, than of divination.

References

Agrawal, Sandip, et al. 2009. PhonePoint Pen: Using mobile phones to write in air. In *MobiHeld09*. Barcelona, Spain.

Ally, Mohamed. 2009. *Mobile learning. Transforming the delivery of education and learning*. Edmonton: AU Athabasca Press.

Anderson, Craig Alan, Douglas A. Gentile, and Katherine E. Buckley. 2006. *Violent video game effects on children and adolescents. Theory, research, and public policy*. Oxford/New York: University of Oxford Press.

Aviram, Aharon, and Deborah Talmi. 2005. The impact of information and communication technology on education: The missing discourse between three different paradigms. *E-Learning and Digital Media* 2(2): 169–191.

BBC. 2005. Should mobile phones be banned in schools? May 27. http://news.bbc.co.uk/cbbcnews/hi/newsid_4570000/newsid_4579100/4579159.stm

Bloom, Benjamin. 1984. The 2 sigma problem: The search for methods of group instruction as effective as one-to-one tutoring. *Educational Researcher* 13(6): 4–16.

Bransford, John D., et al. 2000. *How people learn: Brain, mind, experience, and school*. Washington, DC: National Academy Press.

Bremner, Charles. 2009. Mobile phones to be bannes in Frenchprimary schools to limit health risks. *The Times online*, May 27. http://www.timesonline.co.uk/tol/news/world/europe/article6366590.ece

Brown, Tom H. 2008. Mlearning in Africa: Doing the unthinkable and reaching the unreachable. In *International handbook of information technology and primary and secondary education,* Springer International Handbooks of Education, vol. 20, no. 9, ed. Jone Voogt and Gerald Knezek, 861–871. New York: Springer.

Bulstrode, Mark. 2008. Half of Cambridge students admit cheating. *The Independent,* October 31. http://www.independent.co.uk/news/education/education-news/half-of-cambridge-students-admit-cheating-980727.html

Butgereit, Laurie. 2007. Math on MXit: the medium is the message. In *Proceedings 13th annual national congress of the association of mathematics education of South Africa,* White River, South Africa.

Byron, Tanya. 2008. *Safer children in a digital world. The report of the Byron review 2008.* http://publications.education.gov.uk/default.aspx?PageFunction=productdetails&PageMode=publications&ProductId=DCSF-00334-2008&

Casati, Roberto. 2009. Learning beyond electrification. Mobile technology offers opportunities for redesigning the teaching process. *Interdisciplines.* http://www.interdisciplines.org/mobilea2k/papers/5

Clark, Andy, and David Chalmers. 1998. The extended mind. *Analysis* 58: 10–23.

CTGV. 1990. Anchored instruction and its relationship to situated cognition. *Educational Researcher* 19(6): 2–10.

Dewey, John. 1997. *Experience and education.* New York: Free Press.

Dukker, Stephen. 2007. Is the OLPC project doomed a failure? *Znet,* August 07. http://www.zdnet.co.uk/news/it-strategy/2007/08/07/is-the-olpc-project-doomed-to-failure-39288450/

FAS. 2006. Harnessing the power of video games for learning. *Summit on educational games.* http://www.fas.org/gamesummit/Resources/Summit%20on%20Educational%20Games.pdf

Felicia, Patrick. 2009. *How are digital games used in schools? Complete results of the study.* European schoolnet. http://games.eun.org/upload/gis-full_report_en.pdf

Floridi, Luciano. 2003. Two approaches to the philosophy of information. *Minds and Machines* 13(4): 459–469.

Floridi, Luciano. 2004. *The Blackwell guide to the philosophy of computing and information.* Malden: Blackwell.

Floridi, Luciano. 2007. A look into the future impact of ICT. *The Information Society* 23(1): 59–64.

Floridi, Luciano. 2010. *The Cambridge handbook of information and computer ethics.* Cambridge/New York: Cambridge University Press.

Gee, James Paul. 2007a. *Good video games + good learning: Collected essays on video games, learning, and literacy.* New York: P. Lang.

Gee, James Paul. 2007b. *What video games have to teach us about learning and literacy.* New York: Palgrave Macmillan.

Gentile, Douglas A. 2009. Pathological video game use among youth 8 to 18: A national study. *Psychological Science* 20: 594–602.

Greenwood, Louise. 2009. Africa's mobile banking revolution. *BBC News,* August 12. http://news.bbc.co.uk/2/hi/8194241.stm

Johnson, Rachel. 2007. A degree in cut and paste. *The Times online,* March 11. http://www.timesonline.co.uk/tol/comment/columnists/rachel_johnson/article1496130.ece

Johnson, Lawrence F., et al. 2010. *The Horizon report.* Austin: The New Media Consortium. http://wp.nmc.org/horizon2010/

Keating, Candes, and Murray Williams. 2006. Schools seek to ban addictive Mxit. *IOLNews,* August 23. http://www.iol.co.za/news/south-africa/schools-seek-to-ban-addictive-mxit-1.290620

Korte, Werner B., and Tobias Hüsing. 2007. Benchmarking access and use of ICT in European Schools 2006. Final Report from Head Teacher and Classroom Teacher Surveys in 27 European Countries. *Elearning Papers,* 2, 1. http://www.elearningeuropa.info/files/media/media11563.pdf

Kozma, Robert. 2008. *International handbook of information technology in primary and secondary education.* Berlin: Springer.

Kukulska-Hulme, A. 2009. Will mobile learning change language learning? *ReCALL* 21(2): 157–165.

Maes, Pattie, and Pranav Mistry. 2009. Unveiling the "Sixth Sense," game-changing wearable tech. In *TED 2009,* Long Beach, CA.

Mitchell, Alice, and Carol Savill-Smith. 2004. *The use of computer and video games for learning. A review of the literature.* London: Learning and Skills Development Agency. http://gmedia.glos.ac.uk/docs/books/computergames4learning.pdf

Mitra, Sugata. 2009. Remote presence: Technologies for 'beaming' teachers where they cannot go. *Journal of Emerging Technologies in Web Intelligence* 1(1): 55–59.

Mitra, Sugata, and Vivek Rana. 2001. Children and the Internet: Experiments with minimally invasive education in India. *British Journal of Educational Technology* 32(2): 221–232.

Mitra, Sugata, et al. 2005. Acquisition of computing literacy on shared public computers: Children and the 'hole in the wall'. *Australasian Journal of Educational Technology* 21(3): 407–426.

Nussbaum, Bruce. 2007. It's time to call One Laptop Per Child a failure. *Businessweek*, September 24. http://www.businessweek.com/innovate/NussbaumOnDesign/archives/2007/09/its_time_to_call_one_laptop_per_child_a_failure.html

Papert, Seymour. 1980. *Mindstorms: Children, computers, and powerful ideas*. New York: Basic Books.

Papert, Seymour. 2004. Entretien avec Seymour Papert. Education et Territoires – Conseil Général des Landes. http://www.dailymotion.com/video/x5zdl4_seymour-papert-2004_webcam/

Prensky, Marc. 2005. What can you learn from a cell phone? Almost anything. *Innovate. Journal of Online Education* 1(5). http://innovateonline.info/pdf/vol1_issue5/What_Can_You_Learn_from_a_Cell_Phone__Almost_Anything!.pdf

Prensky, Marc. 2006. *Don't bother me mom, I'm learning! How computer and video games are preparing your kids for twenty-first century success and how you can help!* St. Paul: Paragon House.

Resnick, Mitchell, et al. 2009. Scratch: Programming for all. *Communications of the ACM* 52(11): 60–67.

Rudd, Peter, et al. 2009. *Harnessing Technology Schools Survey 2009 Analysis report*. Berkshire: National Foundation for Education Research. http://research.becta.org.uk/upload-dir/downloads/page_documents/research/ht_schools_survey08_analysis.pdf

Sharples, Mike. 2005. Learning as conversation: Transforming education in the mobile age. In *Proceedings of the conference on seeing, understanding, learning in the mobile age*, Budapest, Hungary.

The Economist. 2010. Don't shoot the messenger. America's president joins a long (but wrong) tradition of technophobia. May 13. http://the-economist.com/node/16109292/comments

Traxler, John, and Agnes Kukulska-Hulme. 2005. *Mobile learning in developing countries*. Commonwealth of Learning. http://www.col.org/SiteCollectionDocuments/KS2005_mlearn.pdf

Trucano, Michael. 2005. *Knowledge maps: ICTs in education*. Washington, DC: infoDev/World Bank. http://www.infodev.org/en/Publication.8.html

Trucano, Michael. 2009. Mobile phones: Better learning tools that computers? (An EduTech debate). *EduTech*. http://blogs.worldbank.org/edutech/mobile-phones-better-learning-tools-than-computers-an-edutech-debate-0

Van der Klein, Raimo (Thinkmobile). 2008. The box and beyond. *Slideshare*. http://www.slideshare.net/Thinkmobile/the-box-and-beyond-web

Vecchiatto, Paul. 2009. Mxit becomes teachers' pet. *MyDigitalLife*, April 20. http://www.mydigitallife.co.za/index.php?option=com_content&task=view&id=1045673&Itemid=35

Wagner, Daniel, et al. 2005. *Monitoring and evaluation of ICT in education. A handbook for developing countries*. Washington, DC: infoDev/World Bank. http://robertkozma.com/images/ict_ed_ch2_monitoringandeval.pdf

Whitehead, Alfred North. 1929. *The aims of education and other essays*. New York: Free Press.

Chapter 8
Content Net Neutrality – A Critique

Raphael Cohen-Almagor*

8.1 Introduction

In a recent article, Luciano Floridi (2010a, p. 11) argues that we are now experiencing the fourth scientific revolution. The first was that of Nicolaus Copernicus (1473–1543), the first astronomer to formulate a scientifically-based heliocentric cosmology that displaced the Earth and hence humanity from the center of the universe. The second was Charles Darwin (1809–1882), who showed that all species of life have evolved over time from common ancestors through natural selection, thus displacing humanity from the centre of the biological kingdom. The third was Sigmund Freud (1856–1939), who acknowledged that the mind is also unconscious and subject to the defence mechanism of repression, thus we are far from being Cartesian minds entirely transparent to ourselves. And now, in the information revolution, we are in the process of dislocation and reassessment of humanity's fundamental nature and role in the universe. Floridi argues that while technology keeps growing bottom-up, it is high time we start digging deeper, top-down, in order to expand and reinforce our conceptual understanding of our information age, of its nature, less visible implications and its impact on human and environmental welfare, and thus give ourselves a chance to anticipate difficulties, identify opportunities and resolve problems, conflicts and dilemmas (Floridi 2009, 2010a).

---

*All websites were accessed during December 2010. I am most grateful to Jacqueline Lipton and Jack Hayward for their valuable comments.

Raphael Cohen-Almagor (D. Phil., Oxon) is an educator, researcher and human rights activist; Chair in Politics, University of Hull, UK. To date, he has published 15 books, including two books of poetry. http://www.hull.ac.uk/rca; http://almagor.blogspot.com/

R. Cohen-Almagor (✉)
Department of Politics, University of Hull, Cottingham, UK

H. Demir (ed.), *Luciano Floridi's Philosophy of Technology: Critical Reflections*, Philosophy of Engineering and Technology 8, DOI 10.1007/978-94-007-4292-5_8,

Floridi has made many contributions in his attempts to "dig deeper." In this paper I would like to focus on some of Floridi's ideas on information ethics which he describes as the study of the moral issues arising from the availability, accessibility and accuracy of informational resources, independently of their format, type and physical support. He further clarifies that information ethics, understood as information-as-a-product ethics, may cover moral issues arising, for example, in the context of accountability, liability, libel legislation, testimony, plagiarism, advertising, propaganda, and misinformation (Floridi 2008). I wish to add to this list answerability and responsibility and to focus on these two concepts as well as on accountability.

Answerability is closely related to accountability. The former accentuates more the need to respond to external claims, pressures, demands; providing explanation for one's conduct. The accompanying concept of responsibility refers to a person or organization that is able to answer for one's conduct and obligations. When we speak of social responsibility we refer to the responsibility of individuals, groups, corporations and governments to society. The difference between responsibility, on the one hand, and answerability and accountability on the other is that the first connotes a more voluntary and self-directed character. Responsibilities are typically accepted, not imposed by force, although they can be contracted and attributed. In contrast, answerability and accountability have a more external character, although they can also be voluntary. The more voluntary it is, the more conduct is compatible with freedom and even coterminous with responsibility. The accountable person or organization is also answerable.

In other words, responsibility, answerability and accountability complement each other, the one being an extension of the other (McQuail 2003, p. 306; Tavani 2011, pp. 119–123). They are designed to improve the quality of the service or product, promote trust of those who are using the service or product and protect the interests of all parties concerned, including the business at hand. Business known to be responsible, answerable and accountable for its services and/or products enjoys solid reputation and may attract more customers. Responsibility, answerability and accountability are important as sometimes people and organizations seek independence from their responsibilities. Ambrose Bierce (1911) described responsibility as a "detachable burden easily shifted to the shoulders of God, Fate, Fortune, Luck or one's neighbor. In the days of astrology it was customary to unload it upon a star".

In the Internet age, an interesting phenomenon emerged that confuses the concept of moral and social responsibility. In the offline, real world, people know that they are responsible for the consequences of their conduct, speech as well as action. In the online, cyber world, we witness responsibility shake-off. You can assume your dream identity and then anything goes. The Internet has a dis-inhibition effect. The freedom allows language one would dread to use in real life, words one need not abide by, imagination that trumps conventional norms and standards. It is high time to bring to the fore discussion about morality and responsibility. My discussion focuses upon the concept of net neutrality.

In his recent book, *The Philosophy of Information*, Floridi (2010c) addressed the issue of the truthfulness of data which he termed alethic neutrality. I, in turn, wish to speak of different meanings of neutrality: (1) Net neutrality as non-exclusionary business practice, highlighting the economic principle that the Internet should be opened to all business transactions. (2) Net neutrality as an engineering principle, enabling the Internet to carry the traffic uploaded to the platform. (3) Net neutrality as content non-discrimination, accentuating the free speech principle. I call the latter content net neutrality. While endorsing the first two meanings of net neutrality I argue that ISPs should scrutinize content and discriminate against not only illegal content (for instance, terrorism) but also against content that is morally repugnant and hateful. Here the concept of responsibility comes into play. Being cognizant of the possibility that "morally repugnant" might open a wide gate to further restrictions, I emphasise that *hate speech alone* features in this category. Other morally repugnant types of net speech, such as child pornography, terrorism, and criminal activities, are covered by law and are widely considered illegal.

8.2 Responsibility of Internet Service Providers (ISPs) and Web Hosting Services (WHS)

The issue of responsibility of ISPs and host companies is arguably the most intriguing and complex. Their actions and inactions directly affect the information environment. *An Internet Service Provider (ISP)* is a company or other organization that provides a gateway to the Internet, usually for a fee, enabling users to establish contact with the public network. Many ISPs also provide e-mail service, storage capacity, proprietary chat rooms, and information regarding news, weather, banking or travel. Some offer games to their subscribers. *A Web Hosting Service (WHS)* is a service that runs various Internet servers. The host manages the communications protocols and houses the pages and the related software required to create a website on the Internet. The host machine often uses the Unix, Windows, Linux, or Macintosh operating systems, which have the TCP/IP protocols built in (Gralla 2007, p. 173).

It is generally agreed in both the United States and Europe that the access provider should not be held responsible for the contents of messages. In Europe, this has been codified in the E-Commerce Directive of the European Union as well as the German Teleservices Act. In the United States, so-called "common carrier" provisions allow certain carriers of communications to carry all manner of traffic without liability. American courts tend to hold that ISPs are not liable for content posted on their servers, this under Section 230(1) of the Communications Decency Act (1996) (the "Good Samaritan provision" to be discussed *infra*). More recently, Congress granted limited immunity to access providers for violations of copyright law in the Digital Millennium Copyright Act (National Research Council 2001, pp. 119–120).

WHS's, however, are a different story. A host provider may be a portal or a proprietary service that gathers in one place a large amount of third-party content for user access. Being closer to a virtual forum site or bazaar than to a postal system, it provides Web space, helps its subscribers find material more easily, and establishes "bulletin boards" and e-mail services. Generally, the host provider does not have anything to do with the content placed on the server, but a good deal to do with its organization in the "marketplace" (National Research Council 2001, p. 120).

Because the host provider offers more than a connection service, the question of liability is more complicated. Legal systems have to determine when the value added by the host provider's services begins to make it look less like an access provider and more like a content provider. The task is made all the more difficult as new technologies create new business opportunities for inventive entrepreneurs, and the services offered by host providers change. It is unlikely that a simple or permanent resolution to this question will become available soon (National Research Council 2001, p. 120).

Yahoo! has terms of service that prohibit to "upload, post, email or otherwise transmit any Content that is unlawful, harmful, threatening, abusive, harassing, tortious, defamatory, vulgar, obscene, libellous, invasive of another's privacy, hateful, or racially, ethnically or otherwise objectionable" (http://uk.docs.yahoo.com/info/terms.html). However, if such content is not removed by the ISP, neither it nor its partners assume any liability. In the United States, the guiding principle inspired by the First Amendment and the special status that freedom of expression enjoys is of net neutrality. The underlying belief is that the Internet should remain an open platform for innovation, competition, and social discourse, free from unreasonable discriminatory practices by network operators. All content, sites, and platforms should be treated equally, free of any value judgment. In justifying this philosophy, American new media experts explain that the Internet was built and has thrived as an open platform, where individuals and entrepreneurs are able to connect and interact, choose marketplace options, and create new services and content on a level playing field. Richard Whitt, Google's Washington Telecom and Media Counsel, writes that "No one seems to disagree with that fundamental proposition," arguing for the need to "protect that unique environment" and supporting the adoption of "rules of the road" to ensure that the broadband on-ramps to the net remain open and robust (Whitt 2009). Jack Balkin, from Yale Law School, said that the open Internet is crucial to freedom of speech and democracy because it allows people to actively participate in decentralized innovation, form new digital networks, and allows freedom from prior government constraints. People can reach all audiences and find a way around gatekeepers with great new tools and applications (Naoum 2009).

I wish to take issue with these arguments, that of net neutrality and that the Internet environment is unique to the extent that it is a public domain and thus any speech should be freely available on the Internet. I argue that some value screening of content may be valuable and that the implications affording the Internet the widest possible scope can be very harmful. Contra Balkin I think that limitless freedom of speech might undermine democracy and bring about its destruction. Indeed, one of the dangers we face is that the principles that underlie and characterize the Internet might undermine freedom. Because the Internet is a relatively young phenomenon, people who use and regulate it lack experience in dealing with pitfalls involved in its working. Freedom of expression should be respected as long as it does not imperil the basic values that underlie our democracy. Freedom of expression is a fundamental right, an important anchor of democracy; but it should not be used in an uncontrolled manner. Like every young phenomenon, the Internet needs to develop gradually, with caution and care. Since we lack experience, we are uncertain with regard to the appropriate means to be utilized in order to fight explicit anti-democratic and harmful practices. But we should not stand idly by in the face of such phenomena.

Thus, while I accept the concept of net neutrality I also reject the concept of content net neutrality for reasons that I explain below. First we should ensure that no confusion arises between the two.

8.3 Content Net Neutrality

Net neutrality is one of the core principles of the Internet. In October 2009, a group of the world's largest Internet companies wrote a letter of support to the US Federal Communications Commission (FCC). The letter is the latest in an ongoing debate about "network neutrality" – or how data is distributed on the web. The letter, signed *inter alia* by the chief executives of Google, eBay, Skype, Facebook, Amazon, Sony Electronics, Digg, Flickr, LinkedIn and Craigslist, says that maintaining data neutrality helps businesses to compete on the basis of content alone: "An open internet fuels a competitive and efficient marketplace, where consumers make the ultimate choices about which products succeed and which fail… This allows businesses of all sizes, from the smallest start-up to larger corporations, to compete, yielding maximum economic growth and opportunity" (BBC Reporter 2009). This is yet another step in a sustained and till now quite successful effort to grant Internet companies the widest possible freedom and independence to conduct their affairs in a way that best serves their commercial interests. Their responsibility, as these large companies see it, is to provide their customers with efficient service.

Net neutrality is also about the organization of the Internet. No one application (WWW, email, messenger) is preferred to another. All applications should be treated by Internet intermediaries equally. Information providers – which may be websites, online services, etc., and who may be affiliated with traditional commercial enterprises but who also may be individual citizens, libraries, schools, or nonprofit entities – should have essentially the same quality of access to distribute their offerings. "Pipe" owners (carriers) should not be allowed to charge some information providers more money for the same pipes, or establish exclusive deals that relegate everyone else (including small noncommercial or startup entities) to an Internet "slow lane." This principle should hold true even when a broadband provider is providing Internet carriage to a competitor.[1] To this I agree. The public is interested in having a neutral platform that supports innovations and the emergence of the best technological applications.

[1] "Network Neutrality," American Library Association http://www.ala.org/ala/issuesadvocacy/telecom/netneutrality/index.cfm

However, the American Library Association also holds that the principle of net neutrality maintains that consumers/citizens should be free to get access to – or to provide – the Internet content and services they wish, and that consumer access should not be regulated based on the nature or source of that content or service.[2] Similarly, the Norwegian Post and Telecommunications Authority (NPT) holds that netusers are entitled to an Internet connection that enables them to send and receive content of their choice as well as to Internet connection that is free of discrimination with regard not only to type of application and service but also content.[3]

This part of net neutrality that concerns content I find much more complicated, and problematic. It should be separated from the principle of net neutrality. I call it *content net neutrality.*

Content net neutrality holds that we should treat all content that is posted on the Internet equally. ISPs and WHSs should not privilege or in one way or another discriminate between different types of content. Now, it is unclear what the implications of such a view are. One possible implication against which content net neutrality warns is that a specific search engine might pay ISPs fees to ensure that responses from its Web site would be delivered to the user faster than the results from a competing search engine that had not paid special fees. Another possible wrong implication against which we all protest is that an ISP might accord a lower priority to packets transmitting, say, video feeds – unless the customer were to pay a special fee for higher-speed access. The most alarming scenarios involve outright blockage of content by source or by type. An example of blockage *by source* often cited in news stories is that of the Canadian ISP Telus, which blocked subscribers' access to a Web site of the Telecommunications Workers Union, with which it was in conflict (Kabay 2006). Labour disputes should never constitute grounds for content discrimination. The example of *type-based* blocks much mentioned in the debate is that of Madison River telecommunications provider, which blocked voice over IP (VoIP) traffic from Vonage as an anticompetitive move to protect its own long-distance conventional telephony service.[4]

This kind of brute discrimination, motivated by narrow economic interests, is also illegitimate. Such incidents demonstrate the skewed incentives that ISPs might have in controlling content and applications. The present debate is about the extent that ISPs should be allowed to control the size of the pipes: Can ISPs actively control the bandwidth available to certain websites based on the type of content they provide, thus influencing the Internet speed available to netusers?

---

[2] *Ibid.*

[3] *Network Neutrality – Guidelines for Internet neutrality* (Post-og teletilsynet, February 24, 2009).

[4] "FCC Chairman Michael K. Powell Commends Swift Action to Protect Internet Voice Services," *Federal Communications Commission News* (March 3, 2005), at http://tinyurl.com/hscav

8.3.1 *Is the Internet Like the Electric Grid?*

Tim Wu helps us understand the logic behind net neutrality by arguing that a useful way to understand this principle is to look at other networks, like the electric grid, which are implicitly built on a neutrality theory. The general purpose and neutral nature of the electric grid is one of the things that make it extremely useful. The electric grid does not care if you plug in a toaster, an iron, or a computer. Consequently it has survived and supported giant waves of innovation in the appliance market. The electric grid worked for the radios of the 1930s and it works for the flat screen TVs of the 2000s. For that reason the electric grid is a model of a neutral, innovation-driving network.[5]

However, does this mean that as you do not expect to control the content of the electric grid so you should not aim to control the Internet's content? If this is a plausible deduction then this comparison is misleading. The electric grid transmits power that enables the functioning of electric equipment. It does not have content, messages, propaganda, instructions, means to abuse you or to harm you. The Internet, on the other hand, has all this. As Floridi (2010a, p. 13) rightly writes, a digital interface is a gate through which a user can be present in cyberspace. Regarding the electronic grid you cannot develop subjective notions. The Internet which contains the best and worst products of its customers may lead you to develop subjectivity. The Internet contains the power to influence your life in constructive and destructive ways. As thinking people who are able to differentiate between right and wrong, good and evil; as morally responsible beings, we must discriminate between contents. We cannot be neutral about it if we wish to continue leading a free, autonomous lives. The only meaningful aspect in the comparison between the Internet and the electric grid is that in both we insist on some measures to assure our security. These measures do not need to include subjectivity when we consider the electric grid. They do require subjectivity when we consider the Internet. Ethics requires us to care about the consequences of our actions and to take responsibility for them. As Floridi and Sanders (2005, pp. 195–196) rightly note, ethics is about constructing the world, improving its nature, and shaping its development in the right way.

In a House Committee on The Juidiciary, Telecom & Antitrust Task Force, Wu (2006) said that the "instinct" of protecting consumers' rights on network is very simple: Let customers use the network as they please. With due appreciation to instincts which often serve as good guides for conduct, by nature they are not thoughtful. Sometimes, after reflecting and pondering, we act against our instincts, for good reasons. I think there are ample reasons to doubt whether allowing customers to use the network as they please is a good policy to follow. While the majority of netusers appreciate this policy and would not abuse it, some people might opt for abuse. We should respect the users and protect ourselves against the abusers.

---

[5] Tim Wu, "Network Neutrality FAQ," at http://timwu.org/network_neutrality.html

On the other hand, Wu in an earlier article (2003) explains that the basic principle behind a network anti-discrimination regime is to give users the ability to use non-harmful network attachments or applications, and provide innovators the corresponding freedom to supply them. ISPs should have the freedom to reasonably control their network ("Police what they own") and, at the same time, the Internet community should view with suspicion restrictions premised on inter-network criteria (Wu 2003, pp. 142, 145).

What does "reasonably control their network" mean? First, ISPs prohibit netusers from using applications or conduct that could hurt the network or other netusers. For instance, Akamai Acceptable Use Policy states: "Customer shall not use the Akamai Network and Services to transmit, distribute or store material that contains a virus, worm, Trojan horse, or other component harmful to the Akamai Network and Services, any other network or equipment, or other Users."[6]

Blocking denial of service attacks and spam are also within what we perceive as legitimate network management.

Second, some companies market equipment aimed to facilitate application-based screening and control for broadband networks. Companies like Check Point Enterprise[7] and Symantec Gateway Security[8] provide traffic-management features with highly-developed security-management tools. Allot Communications provides facilities to manage traffic and produces a fully integrated, carrier-class platform capable of identifying the traffic flows of individual subscribers.[9]

Packeteer tracks link and provides statistics per application – including peak and average utilization rates (down to 1 min), bytes, availability, utilization, top talkers and listeners, network efficiency and frames. It monitors use and performance through proactive alarming and exception reporting or through comprehensive central reporting tools.[10]

Third, ISPs may prohibit netusers from inflicting harm on others by upholding and promoting crime-facilitating speech designed to uphold harmful conduct. The effort, writes Wu (2003, p. 168) quite rightly, is to strike a balance between prohibiting ISPs, absent a showing of harm, from restricting what netusers do with their Internet connection, while giving them general freedom to manage bandwidth consumption. This non-discrimination principle works by recognizing a distinction between local network restrictions, which are generally allowable, and inter-network restrictions which are suspect. The effort is to develop forbidden and permissible grounds for discrimination in broadband usage restrictions. Wu has in mind illegal activities. I argue that ISPs and WHSs should also consider prohibiting hate speech, which is legal in the USA. While racist Nazi speech is protected under the First Amendment, the same speech is not protected in most European countries. Morally speaking, such speech is repugnant.

---

[6] "Acceptable Use Policy," http://www.akamai.com/html/policies/acceptable_use.html

[7] http://www.checkpoint.com/products/enterprise/

[8] http://www.symantec.com/avcenter/security/Content/Product/Product_SGS.html; http://www.symantec.com/business/products/allproducts.jsp

[9] http://www.allot.com/index.php?option=com_content&task=view&id=2&Itemid=4

[10] http://www.packeteer.com/solutions/visibility.cfm

Fourth, ISPS can block unlawful transfer of unauthorized transfer of copyrighted works. Open Internet principles *prima facie* apply to lawful content, services and applications – not to activities such as unlawful distribution of copyrighted works, which has serious economic consequences. The enforcement of copyright and other laws and the obligations of network openness can and must co-exist. In order for network openness obligations and appropriate enforcement of copyright laws to co-exist, it appears reasonable for an ISP to refuse transmission of copyrighted material if the transfer of that material would violate applicable laws (Sohn 2009).[11] Thus, legitimate network management includes maintaining the technical quality of the network, preventing abuse, and complying with legal dictates. In order to ensure that any network management is legitimate, it is incumbent on providers to disclose network management practices to users and regulators, who should assess these practices against net neutrality principles (Trans Atlantic Consumer Dialogue 2008).

8.3.2 Anti-perfectionism

Conceptually, both net neutrality and content net neutrality emphasize diversity and plurality. Diversity entails openness and more opportunities for living a valuable and richer life. Pluralism is perceived indispensable for having the potential for a good life. Methodologically, the idea of neutrality is placed within the broader concept of anti-perfectionism. The implementation and promotion of conceptions of the good, though worthy in themselves, are not regarded as a legitimate matter for governmental action. The fear of exploitation, of some form of discrimination, leads to the advocacy of plurality and diversity. Consequently, ISPs and WHSs are not to act in a way that might favour some ideas over others. ISPs and WHSs ought to acknowledge that every person has her own interest in acting according to her own beliefs; that everyone should enjoy the possibility of having alternative considerations; that there is no single belief about moral issues and values that should guide all and, therefore, each has to enjoy autonomy and to hold her ideals freely.

The concept of anti-perfectionism comprises the "political neutrality principle" and the "exclusion of ideals" doctrine (Cohen-Almagor 1994). The "political neutrality principle" holds that ISPs and WHSs' policies should seek to be neutral regarding ideals of the good. It requires them to make sure that their actions do not help acceptable ideals more than unacceptable ones; to see to it that their actions will not hinder the cause of false ideals more than they do that of true ones. The "exclusion

[11] Comments of the Motion Picture Association of America, Inc. In response to the Workshop on the Role of Content in the Broadband Ecosystem," Before the Federal Communications Commission, Washington, DC 20554, In the Matter of A National Broadband Plan For Our Future (October 30, 2009).

of ideals" doctrine does not tell ISPs and WHSs what to do. Rather it forbids them to act for certain reasons. The doctrine holds that the fact that some conceptions of the good are true or valid should never serve as justification for any action. Neither should the fact that a conception of the good is false, invalid, unreasonable or unsound be accepted as a reason for a political or other action. The doctrine prescribes that ISPs and WHSs refrain from using one's conception of the good as a reason for state action. They are not to hold partisan (or non-partisan) considerations about human perfection to foster social conditions (Raz 1986, pp. 110–111).

Advocates of content net neutrality, in their striving to convince us of the necessity of the doctrine, are conveying the assumption that the decision regarding the proper policy is crucial because its grave consequences. Content net neutrality entails pluralism, diversity, freedom, public consensus, non-interference, vitality etc. If we do not adhere to neutrality, then we might be left with none of these virtues. This picture leads to the rejection of subjectivity (or perfectionism), while this Essay suggests a rival view that observes conduct of policies on a continuous scale between strict perfectionism, on the one hand, and complete neutrality on the other. The policy to be adopted does not have to be either the one, or the other. It could well take the middle ground, allowing plurality and diversity without resorting to complete neutrality; involving some form of perfectionism without resorting to coercion. For perfectionism does not necessarily imply exercise of force, nor does it impose the values and ideals of one or more segments of society on others, or strive to ensure uniformity, as neutralists fear. On this issue my view comes close to that of Joseph Raz (1986). I call his view the Promotional Approach (PA).

8.4 The Promotional Approach (PA)

My mid-ground position is influenced, even dictated by two principles. I suggest that any liberal society is based on the idea of respect for others, in the sense of treating citizens as equals, and on the idea of not harming others, in the sense that we should address attempts made to harm others, either physically or psychologically. Accordingly, restrictions on liberty may be prescribed when threats of immediate violence are voiced against some individuals or groups and also when the expression in question is intended to inflict psychological offence, morally on a par with physical harm (Cohen-Almagor 2005). Thus I submit that ISPs and WHSs should adhere to PA rather than to neutrality. I reject content net neutrality on important social issues that concern the safeguarding of democracy within which all media operate and flourish.

Let me illustrate with two examples. The first is concerned with clearly illegal speech that should be filtered out the Internet. The second is concerned with speech that is protected under the First Amendment yet morally speaking should be discriminated against in one form or another.

8.4.1 Terror

One of the gravest threats we are facing today is the threat of terrorism. Presently there are more than 40 active terrorist groups, each with an established presence on the Internet. Most active terrorist groups have established their Internet presence with hundreds of websites worldwide. These websites use slogans to catch attention, often offering items for sale (such as T-shirts, badges, flags, and video or audio cassettes). Frequently the websites are designed to draw local supporters, providing information in a local language and giving information about the activities of a local cell as well as those of the larger organization. The website is, thus, a recruiting tool as well as a basic educational link for local sympathizers and supporters (Combs 2006, p. 139).

The Internet is the single most important factor in transforming largely local jihadi concerns and activities into the global network that characterizes al Qaeda today (Atwan 2006, p. 124). The sheer accessibility of cyber warfare capabilities to tens, perhaps hundreds, of millions of people is a development without historical precedent. Thus the ethical dimensions of acts of war and terror conducted by networks of individuals, operating via the virtual realm, might become just as important as the considerations for nation-states (Arquilla 2010).

Unfortunately, many of the terrorist websites are hosted by servers in the western world. In the United States alone, al Qaeda has received funds from numerous social charities based on its own soil. Some of the message boards and the "information hubs" where terrorists post texts, declarations, and recordings are often included in the "communities" sections of popular Western sites such as Yahoo!, Lycos, and others (Wright 2004). The concept of content net neutrality which rejects any responsibility for content facilitates this phenomenon. However, overconfidence, arrogance, dismissiveness, laziness, dogmatism, incuriosity or self-indulgence are no justification or excuse. The Internet is not outside the democratic realm. ISPs and WHSs are a necessary part of it. They also know that democracy and terrorism are mutually exclusive. A zero-sum game exists between them. The victory of one comes at the expense of the other. Therefore, if the spirit and ideas of democracy are dear to ISPs and WHSs and if they wish the democracy that enables their operation to prevail, they cannot shield themselves under the concept of content net neutrality. It is necessary to take sides, distinguishing good from evil, adopting PA.

However, many Internet experts believe that all they need to do is to provide the structure and the rest is up for the public. They preach content net neutrality, which means ignorance. Such ignorant neutrality is aethical at best, and unethical at worse. Let me say something about their belief and conduct. I think all humane people perceive bombing civilian targets – be they buses, trains, airplanes, shopping malls, buildings – as immoral, wrong, wicked, and odious. We also think that these views are true, *i.e.*, in this case we might be sufficiently confident to say that we know they are true, and that people who disagree are making a bad mistake. We think, moreover, that our opinions are not just subjective reactions to the idea of indiscriminate massacre of innocent lives, but opinions about its actual moral character.

We think that it is an objective matter – a matter of how things really are – that terrorism is wrong and wicked. This claim that I am advancing now – that terrorism is objectively wrong – is equivalent to the claim that terrorism would still be wrong even if no one thought it was. That is another way of emphasising that terrorism is plainly wicked, not wicked only because people think it is so (Dworkin 1996, pp. 92–98). Therefore, advancing content net neutrality at the expense of social responsibility serves wicked aims that undermine the platform people wish to protect and the society that promotes the democratic spirit in which they thrive.

8.4.2 Hate Speech

Terrorism, I trust, is not a contested issue. Hate speech, however, is contested and in the United States is protected under the First Amendment. Morally speaking, it is repugnant speech. Hate is a social evil that offends the two most basic principles that underlie any democratic society: Respecting others and not harming others. Generally speaking, hate is derived from one form or another of racism and modern racism has facilitated and caused untold suffering. It is an evil that has taken catastrophic proportions in all parts of the world. Notorious examples include Europe under Nazism, Yugoslavia, Cambodia, South Africa and Rwanda. Elsewhere I argued that in hate messages, members of the targeted group are characterized as devoid of any redeeming qualities and are innately evil. Banishment, segregation and eradication of the targeted group are proposed to save others from the harm being done by this group. By using highly inflammatory and derogatory language, with the tone of extreme hatred and contempt and through comparisons to and associations with animals, vermin, excrement and other noxious substances, hate messages dehumanize the targeted groups (Cohen-Almagor 2010).

Hate messages undermine the dignity and self-worth of the targeted group members and they erode the tolerance and openmindedness that must flourish in democratic societies committed to the ideas of pluralism, justice and equality. Furthermore, hate speech might lead to hate crimes. Benjamin Smith and Richard Baumhammers are two Aryan supremacists who in 1999 and 2000 respectively went on racially motivated shooting sprees after being exposed to Internet racial propaganda. Smith regularly visited the World Church of the Creator website, a notorious racist and hateful organisation.[12] He said: "It wasn't really 'til I got on the Internet, read some literature of these groups that… it really all came together… It's a slow, gradual process to become racially conscious" (Wolf 2004). Rabbi Abraham Cooper (1999) of the Wiesenthal Center argued that the Internet provided the theological

[12] For information on 'World Church of the Creator', see http://www.volksfront-usa.org/creator.shtml; http://www.nizkor.org/hweb/orgs/american/adl/cotc/; http://www.reed.edu/~gronkep/webofpolitics/fall2001/yagern/creator.html; http://www.adl.org/poisoning_web/wcotc.asp; http://www.apologeticsindex.org/c171.html

justification for torching synagogues in Sacramento and the pseudo-intellectual basis for violent hate attacks in Illinois and Indiana (Cohen-Almagor 2006).

On June 10, 2009, James von Brunn entered the U.S. Holocaust Memorial Museum in Washington DC and opened fire, killing Security Guard Stephen Tyrone Johns before he was stopped by other security guards. Von Brunn, a white supremacist anti-Semite, spewed hate online for decades. He ran a hate website called holywesternempire.org and had a long history of associations with prominent neo-Nazis and Holocaust deniers (Cohen-Almagor 2009, pp. 33–42).

ISPs and WHSs should not be neutral regarding such speech. While regarding terrorism they should take active steps to prohibit such speech, regarding hate ISPs and WHSs should take active action to discriminate against such speech by variety of ways. To start with, ISPs and WHSs are under no obligation to provide service to people who spawn hatred. Indeed, Internet providers have terms of service that often include prohibitions against hate messages. When sites cross the bounds of tolerance and violate the terms, providers should enforce their rules and shut down the hate sites. From an ethical perspective, ISPs and WHSs can and should have codes of conduct explicitly stating that they deny service to hate mongers. We humans are capable of discerning between good and evil. PA is in place, not content neutrality. Sometimes, for whatever reasons (laziness, economic considerations, dogmatism, incuriosity, lack of care, contempt), we refrain from doing the right moral thing. But we should. This is not a free speech issue as we are not free to inflict harm on others. It is about taking responsibility for stopping those who abuse the Internet for their vile purposes.

In Canada, Fairview Technology Centre Ltd., an ISP owned by Bernard Klatt whose server was located in Oliver, British Columbia and connected to the Internet via BC Telecom, was identified as a host of a number of websites associated with hate speech and neo-Nazi organizations, including the Toronto-based Heritage Front, WTOTC, and the French Charlemagne Hammerhead Skinheads. The ISP was described as containing the most out-and-out racist, fascist, anti-Semitic, Holocaust-denying Web sites in Canada by a wide margin, and the material on it as the most hateful. About a dozen of the neo-Nazi, white supremacist, and skinhead clients had used Klatt's server to publish material railing against immigration and "the homosexual agenda" while celebrating "Euro-Christianity" and Hitler's accomplishments (Cribb 1998). Klatt explained that Fairview did not control what subscribers choose to upload to their websites. His business was that of selling computers and providing un-censored Internet access services. Klatt added: "If you want to express a controversial viewpoint, I certainly don't have to agree with it, but I believe strongly that you have the right to express it… If you don't agree with what you read or see – switch the topic. You have a choice. You don't have to read or watch any further. It won't come to you" (*Oliver Chronicle,* July 24 1996).

In 1998, Klatt announced that Fairview stopped serving as a provider. The announcement was the result of pressure from anti-racists organization on British Columbia Telephone, the local Internet access provider. The BC Tel demanded, on expiry of their then existing contract with Fairview, that the new contract contain a term indemnifying BC Tel for any damages for which BC Tel might be liable as the result of the material hosted by Fairview. Fairview, rather than sign the contract proposed by BC Tel for renewal, gave up providing internet service (Howard 1998; Matas 2009).

Many other ISPs, WHSs and social networks take a PA responsible stance against hate, barring blatant expressions of bigotry, racism and/or hate.[13] Facebook, the largest social networking site with more than 845 million users,[14] prohibits posting content that is hateful or threatening.[15] XOOM.com of San Francisco, California, bans "hate propaganda" and "hate mongering."[16] Lycos Terms of Service prohibit to "Upload, post, e-mail, otherwise transmit, or post links to any Content, or select any member or user name or e-mail address, that is unlawful, harmful, threatening, abusive, harassing, tortuous, defamatory, vulgar, obscene, pornographic, libelous, invasive of privacy or publicity rights, hateful, or racially, sexually, ethnically or otherwise objectionable."[17] Fortunecity requires its users to agree to "not upload, post, email, transmit or otherwise make available (collectively, 'Transmit') any Content that is unlawful, harmful, threatening, abusive, harassing, tortuous, defamatory, vulgar, obscene, libelous, invasive of another's privacy, hateful, or racially, ethnically or otherwise objectionable."[18]

In this context, let me mention that the American Congress passed the "Good Samaritan provision", included in the 1996 Communication Decency Act (section 230-c-2) which protects ISPs that voluntarily take action to restrict access to problematic material: "No provider or user of an interactive computer service shall be held liable on account of – (A) any action voluntarily taken in good faith to restrict access to or availability of material that the provider or user considers to be obscene, lewd, lascivious, filthy, excessively violent, harassing, or otherwise objectionable, whether or not such material is constitutionally protected."[19]

One may raise the question of how we decide whether something on the Internet is terroristic or hateful? Many cases are quite straightforward. But there might be more obscure or contested cases. One solution is to approach law enforcement agencies or the courts. This surely would be a long and costly process. An alternative is to seek online arbitration. Online arbitration is a private dispute resolution process that involves the intervention of a neutral decision maker, namely the arbitrator, who listens to both parties' arguments and renders a decision that is binding on them. Compared with a court procedure, arbitration is faster, cheaper and also confidential.

Online arbitration is increasingly appreciated by companies active on the Internet because it is more rapid and less costly than legal proceedings or classical arbitration.

[13] For instance, Atlas Systems, http://www.atlas-sys.com/products/aeon/policy.html; Elluminate Online Services, http://www.elluminate.com/license_agreement.jsp; Evehosting.co.uk; Host2Host, http://host2host.com/contract.htm

[14] http://www.facebook.com/press/info.php?statistics

[15] http://www.facebook.com/terms.php?ref=pf

[16] ADL, *Combating Extremism in Cyberspace* (2000): 11.

[17] http://info.lycos.com/tos.php

[18] https://secure.fortunecity.com/order/register/agreement.php?siteid=55527.

[19] CDA 47 U.S.C at http://www4.law.cornell.edu/uscode/47/230.html

Everything is done online: the claimant fills out a form on the Cyber Tribunal site, and the form is then sent to the other party. If the other party agrees to participate in arbitration, he or she is asked to respond to the claim. When they undertake arbitration, the parties agree to comply with the award, no matter what is its decision. In case of non-compliance and in accordance with applicable laws and treaties, the injured party can obtain enforcement of the award (Katsh and Rifkin 2001).[20]

8.5 Conclusions

Luciano Floridi (2001) argues that the ethical use of information and communication technologies and the sustainable development of an equitable information society need a safe and public infosphere for all, where communication and collaboration can flourish coherently with the application of human rights and the fundamental freedoms in the media. Sustainable development means that our interest in the sound construction of the infosphere must be associated with an equally important, ethical concern for the way in which the latter affects and interacts with the physical environment, the biosphere and human life in general, both positively and negatively (Floridi 2001, pp. 18–19). Ethical behavior considers the consequences of one's actions, and it is about being accountable for them. Information professionals cannot be neutral regarding content as this behavior is irresponsible and unprofessional. They have a *prima facie* moral duty to provide stakeholders with certain level of security.

Ethics, Floridi (2010b) rightly notes, is not only a question of dealing morally with a given world. It is also a question of shaping the world for the better. This is a proactive approach which perceives agents as world owners, creators, game designers, producers of moral goods and evils, providers, hosts. Accordingly, ISPs should be able to plan and initiate action responsibly, in anticipation of future events, in an attempt to control their course by making something happen, or by preventing something from happening.

Moreover, I have argued that the Internet is a form of new media but it is still a media. It is not reasonable to prohibit certain expressions in print and allow the same objectionable expression electronically. We cannot be neutral with regard to certain conduct which falls within the parameter of harming others; then the dangers to democracy, to our fellow citizens, to the moral basis of society, to values which we hold dear, might be too grave.

We need to take into account the temper of the time. The level of tolerance is in flux. What is needed is to evoke awareness as to abuse of the Internet for promoting anti-social, criminal activities and the appropriate ways to counter those activities. The discussion, no doubt, will continue for many years to come.

---

[20] See, for instance, CyberTribunal II at http://www.cybertribunal.org/index.en.html; net-ARB at http://www.net-arb.com/; WPO at http://www.wipo.int/amc/en/arbitration/online/index.html

References

Acceptable use policy. http://www.akamai.com/html/policies/acceptable_use.html
Arquilla, J. 2010. Conflict, security and computer ethics. In *Handbook of information and computer ethics*, ed. L. Floridi. Cambridge: Cambridge University Press.
Atwan, A.B. 2006. *The secret history of al Qaeda*. Berkeley: University of California Press.
BBC Reporter. 2009. Big names support net neutrality. *BBC News*, October 20. http://news.bbc.co.uk/1/hi/8315918.stm
Bierce, A. 1911. The devil's dictionary. http://www.alcyone.com/max/lit/devils/
Cohen-Almagor, R. 1994. Between neutrality and perfectionism. *The Canadian Journal of Law and Jurisprudence* VII(2): 217–236.
Cohen-Almagor, R. 2005. *Speech, media, and ethics: The limits of free expression*. Houndmills/New York: Palgrave-Macmillan.
Cohen-Almagor, R. 2006. *The scope of tolerance*. London: Routledge.
Cohen-Almagor, R. 2009. Holocaust denial is a form of hate speech. *Amsterdam Law Forum* 2(1): 33–42.
Cohen-Almagor, R. 2010. In Internet's way. In *Ethics and evil in the public sphere: Media, universal values & global development*, ed. M. Fackler and R.S. Fortner. Cresskill: Hampton Press.
Combs, C.C. 2006. The media as a showcase for terrorism. In *Teaching terror: Strategic and tactical learning in the terrorist world*, ed. J.F. Forest. Lanham: Rowman & Littlefield.
Cooper, A. 1999. Statement, hate crime on the Internet, hearing before the committee on the judiciary, September 14, 1999. United States Senate, Washington, DC.
Cribb, R. 1998. Canadian net hate debate flares. *Wired*, March 25.
Dworkin, R. 1996. Objectivity and truth: You'd better believe it. *Philosophy & Public Affairs* 25(2): 87–139.
FCC Chairman Michael K. Powell commends swift action to protect Internet voice services. *Federal Communications Commission News*, March 3, 2005. http://tinyurl.com/hscav
Floridi, L. 2001. Ethics in the infosphere. *The philosophers' magazine* 6: 18–19.
Floridi, L. 2008. Information ethics: Its nature and scope. In *Moral philosophy and information technology*, ed. J. van den Hoven and J. Weckert, 40–65. Cambridge: Cambridge University Press. http://www.philosophyofinformation.net/publications/pdf/ieinas.pdf.
Floridi, L. 2009. The information society and its philosophy: Introduction to the special issue on 'The philosophy of information, its nature and future developments'. *The Information Society* 25(3). http://www.philosophyofinformation.net/publications/pdf/tisip.pdf.
Floridi, L. 2010a. *Information – A very short introduction*. Oxford: Oxford University Press.
Floridi, L. 2010b. Ethics after the information revolution. In *Handbook of information and computer ethics*, ed. L. Floridi. Cambridge: Cambridge University Press.
Floridi, L. 2010c. *The philosophy of information*. Oxford: Oxford University Press.
Floridi, L., and J.W. Sanders. 2005. Internet ethics: The constructionist values of homo poietcus. In *The impact of the Internet on our moral lives*, ed. R.J. Cavalier. Albany: State University of New York Press.
Gralla, P. 2007. *How the Internet works*, Eighth ed. Indianapolis: Que Publishing.
Howard, R. 1998. Notorious Internet service closes. *The Globe and Mail*, April 28.
Kabay, M.E. 2006. The net neutrality debate. *Ubiquity* 7(20), May 23–29. http://delivery.acm.org/10.1145/1140000/1138694/v7i20_neutrality.html?key1=1138694&key2=7250511921&coll=DL&dl=ACM&CFID=113263028&CFTOKEN86603576
Katsh, E., and J. Rifkin. 2001. *Online dispute resolution: Resolving conflicts in cyberspace*. San Francisco: Jossey-Bass.
Matas, M. 2009. Combating hate on the internet without recourse to law. In *Freedom of speech versus hate speech. Panel contribution for the INACH Conference 2009,* Amsterdam, November 9, 2009.
McQuail, D. 2003. *Media accountability and freedom of publication*. New York: Oxford University Press.

Motion Picture Association of America Inc. 2009. *Comments in response to the workshop on the role of content in the broadband ecosystem*. Before the Federal Communications Commission, Washington, DC 20554. In the Matter of a National Broadband Plan for Our Future, October 30, 2009.

Naoum, C. 2009. Web content producers favor net neutrality. *Reject regulation of search engines*, December 16. BroadbandBreakfast.com

National Research Council. 2001. *Global networks and local values: A comparative look at Germany and the United States*. Washington, DC: National Academy Press.

Network neutrality – Guidelines for Internet neutrality. 2009. Post-og teletilsynet, February 24, 2009.

Network neutrality. American Library Association. http://www.ala.org/ala/issuesadvocacy/telecom/netneutrality/index.cfm

Oliver Chronicle, July 24, 1996.

Raz, R. 1986. *The morality of freedom*. Oxford: Clarendon.

Sohn, G.B. 2009. Content and its discontents: What net neutrality does and doesn't mean for copyright. Yale Information Society Project, Yale Law School, New Haven, October 27, 2009. http://www.publicknowledge.org/node/2740

Tavani, Herman T. 2011. *Ethics and technology: Controversies, questions, and strategies for ethical computing*. Hoboken: Wiley.

Trans Atlantic Consumer Dialogue. 2008. Resolution on net neutrality. DOC No. INFOSOC 3608.

Whitt, R. 2009. Time to let the process unfold. *Google Public Policy Blog*, October 22. http://googlepublicpolicy.blogspot.com/2009/10/time-to-let-process-unfold.html

Wolf, C. 2004. Regulating hate speech qua speech is not the solution to the epidemic of hate on the Internet. In *OSCE meeting on the relationship between Racist, Xenophobic and Anti-Semitic Propaganda on the Internet and hate crimes*, Paris, June 16–17, 2004.

Wright, L. 2004. The terror web. *The New Yorker*, August 2.

Wu, T. 2003. Network neutrality, broadband discrimination. *Journal of Telecommunications and High Technology Law* 2: 141–179.

Wu, T. 2006. Testimony, hearing on "Network neutrality: Competition, innovation, and nondiscriminatory access." House Committee on the Juidiciary, Telecom & Antitrust Task Force.

Wu, T. Network neutrality FAQ, at http://timwu.org/network_neutrality.html

Chapter 9
Information Science and Philosophy of Information: Approaches and Differences

Armando Malheiro da Silva and Fernanda Ribeiro

9.1 Information Science: A Theoretical Overview

It is usually accepted that Information Science (IS) has its early origins in Documentation, conceived and implemented by Paul Otlet and Henri La Fontaine at the end of nineteenth century. However, the designation of "Information Science" only appeared at the end of the 1950s in close connection to scientific and technical information, which was growing strongly at that time (Rayward 1997; Saracevic 1996; Shera and Cleveland 1977; Silva and Ribeiro 2002; Williams et al. 1997).

This new field of study and work developed alongside the traditional areas – Archivistics and Librarianship – which emerged as scientific disciplines in the mid-nineteenth century, in the framework of Historicism and Positivism, but with an "auxiliary" status to History and characterized by high-level erudition.

A gradual technological revolution, initiated with the telegraph, the telephone, the typewriter, the wireless set, cinema and photography, was at the origin of new forms of communication and new information media, different from the traditional paper format. Thus, new documents such as the graphic, sound and audiovisual were produced and joined books, journals and manuscripts, giving rise to thought and reflection which differed from the norm. Paul Otlet and Henri La Fontaine shared such concerns and searched for the foundations for a new area that they called "Documentation".[1] This field of work did not mean a break with ways of

[1]The major expression of their work appears in Paul Otlet's *Traité de Documentation*, published in 1934. A translation of this book has been published some years ago by Universidad de Múrcia (Otlet 1996). On Paul Otlet's work, see for instance: (Day 1997).

A.M. da Silva (✉) • F. Ribeiro
Faculty of Arts and Humanities, University of Porto/CETAC.MEDIA, Porto, Portugal
e-mail: malheiro@letras.up.pt

H. Demir (ed.), *Luciano Floridi's Philosophy of Technology: Critical Reflections*, Philosophy of Engineering and Technology 8, DOI 10.1007/978-94-007-4292-5_9,

viewing and doing of the traditional disciplines, but put the emphasis on the technical aspects of processing documents and the organization of services, in order to improve access to and the use of information.

The growth of IS, in the continuity of Documentation and the expansion of its technical aspects, accompanied technological development and took place in close connection to scientific and technical information since 1958.[2] Some concerns with its definition and theoretical foundations quickly arose. In fact, over the last half century, the evolution of IS has been quite significant in what concerns its scientific consolidation, above all in the academic sphere. As testimony to this growth we can mention the proliferation of undergraduate programmes and advanced studies (masters and doctorates) all over the world, but with great emphasis in Europe and the USA, as well as the emergence of several journals connected to universities and research groups that involve teachers and researchers from academic institutions in the majority of the countries.

The technological revolution of the last decades and society's involvement in the information phenomenon, today completely linked to digital media, provoked profound changes in the IS field, because of the urgency to provide answers to new problems and challenges, whose solutions demanded increasingly more consistent theoretical and methodological groundings, able to support applied research and intervention in the diverse organizational and social contexts. But, in spite of the quick growth of IS, the scientific *consensus* as to its nature and identity are, still today, a problem, because its disciplinary construction did not occur at the same time and in the same way across all countries and contexts and, consequently, its degree of development varies significantly and makes unitary thinking on the disciplinary field itself quite difficult.

These constraints however do not hinder us from taking a clear position as to the scientific nature and identity of IS, which may be understood as a contribution to the epistemological, theoretical and methodological foundation of this field of knowledge.

The perspective we defend and have attempted to consolidate over the last decade at the University of Porto, assumes IS as a unitary yet transdisciplinary field of knowledge, included in the overarching area of the human and social sciences, which gives theoretical support to some applied disciplines such as Librarianship, Archivistics, Documentation and some aspects of Technological Information Systems. The way in which we see the cartography of the IS scientific field at the University of Porto has been explained in an epistemological paper, edited in 2002, and represented in a diagram, that gave support to the education model developed in the undergraduate and master's curricula lectured at the University of Porto

[2] Anthony Debons states that, before 1958, the term *information science* rarely appears in specialized literature (Debons 1986); and according to Shera and Cleveland, the happening that marked the transformation of *documentation* into IS was the International Conference on Scientific Information, that took place in Washington, in 1958, resulting from cooperation between ADI, FID, National Academy of Sciences and National Research Council. This meeting jointed the greatest names in *documentation* at world level (Shera and Cleveland 1977).

(Silva and Ribeiro 2002). Later on, this diagram was redesigned and improved in the context of another theoretical paper, (Silva 2006), and is presented below:

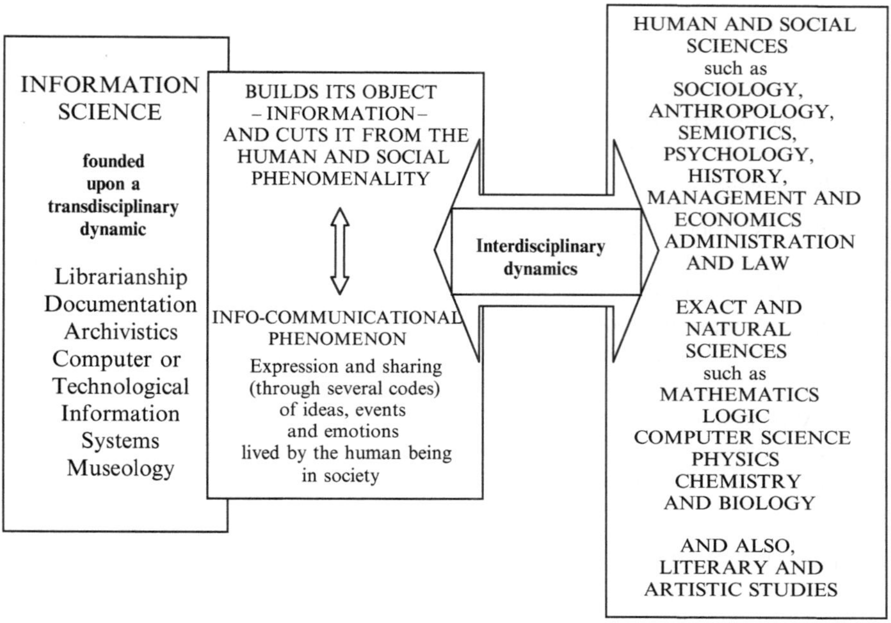

Diagram of the trans- and interdisciplinary construction of information science

In the perspective we defend, besides establishing the boundaries of IS, it is also crucial to define its object of study and to assume a research method adapted to the characteristics of Information as a social phenomenon, emphasizing its qualitative component, as is appropriate in the scope of the social sciences.

When it comes to IS's object of study and work – Information – it is essential to have a definition as a starting point, because it acts as an operative and foundational concept. The definition we propose is as follows:

> **Information**: Structured set of codified mental and emotional representations (signs and symbols), modelled with/by social interaction, and capable of being recorded on any material medium and, therefore, communicated in an asynchronous and multidirectional way (Silva 2006; DeltCI 2007).

Complementing the definition, the characterization of the informational phenomenon is broadened by the enunciation of its properties. In his book *A Ciência da Informação* [Information Science], Yves-François Le Coadic (2004) attempted to formulate the properties of information, but, in our opinion, in a way that is rather unclear. So, we attempt to complete the above definition by listing the *properties* of information, formalized as general axioms. Information is:

1. *structured by an action* (human and social) – the individual or societal act structurally establishes and models information

2. *integrated dynamically* – the informational act is involved with, and results from, conditions and circumstances both internal and external to that action
3. *has potentiality* – a statement (to a greater or lesser extent) of the act which founded and modelled the information is possible
4. *quantifiable* – linguistic, numeric or graphic codification is capable of quantification
5. *reproducible* – information can be reproduced without limit, enabling, therefore, its subsequent recording/memorization
6. *transmissible* – informational (re)production is potentially transmissible or communicable.

These six properties, and especially the last two, characterize information, not only as a *phenomenon* but also as a *process*. In this second dimension we include the idea of information behaviour, as well as all the activities related to the creation, organization, representation, storage, retrieval and use of information. Thus, information comprises the core (single and cross-disciplinary) of an academic field, which is itself dynamic and closely interrelated with other disciplines, as the diagram in the Appendix demonstrates.

The assumption of social information as the object of knowledge has wide-ranging and unexpected implications. The main one is the emergence of a *scientific-informational* paradigm, shaped by the following factors:

(a) the value of information (and not the medium on which it is recorded) as a human and social phenomenon/process, with its own historicity (organic and contextual) and its cultural importance;
(b) the statement of the natural and continuous dynamism of information in opposition to documental immobility;
(c) the impossibility of keeping the traditional divisions of information according to the institutional or technological space where it is preserved (archival service, library or computer package) because such a criterion does not embrace the dynamic context of its production, of its recording and of its use/access (functionality);
(d) the need to know (to understand and to explain) social information through theoretical-scientific models, increasingly more effectively, instead of an empirical practice reduced to a set of technical procedures such as arrangement, description and retrieval;
(e) the replacement of the process-oriented perspective evident in the terms 'records management' or 'information management' by a new scientific view that tries to understand the information involved in the management process of any organization; this means that the informational practices/procedures are aligned with managers' conceptions and practices and with the organizational culture.

These characterizing elements, together with the definition of Information, can be considered the minimum and fundamental basis of a scientific approach to that which we consider to be the object of study and work of IS, understood as a theoretical and practical field in consolidation that supports multifaceted professional competencies, in accordance with the contexts and demands of professional activities.

In what concerns the methodological component of IS, we can sum up the ideas largely explored in the book mentioned previously (Silva and Ribeiro 2002). According to the topological model proposed by Paul de Bruyne, J. Herman and M. de Schoutheete for research in the social sciences (De Bruyne et al. 1974; Lessard-Hébert et al. 1994), the method of information science is achieving greater acceptance and tends to find consolidation through quadripolar research dynamics, which are operated and continuously repeated within the field of knowledge itself. This action combines quantitative approaches (there are aspects of the object which can be observed, experimented on and measured) and qualitative approaches, in which the subject's interpretative/explanatory ability necessarily has modelling implications. The research dynamics mentioned thus imply permanent interaction on four poles, that is, the epistemological, theoretical, technical and morphological.

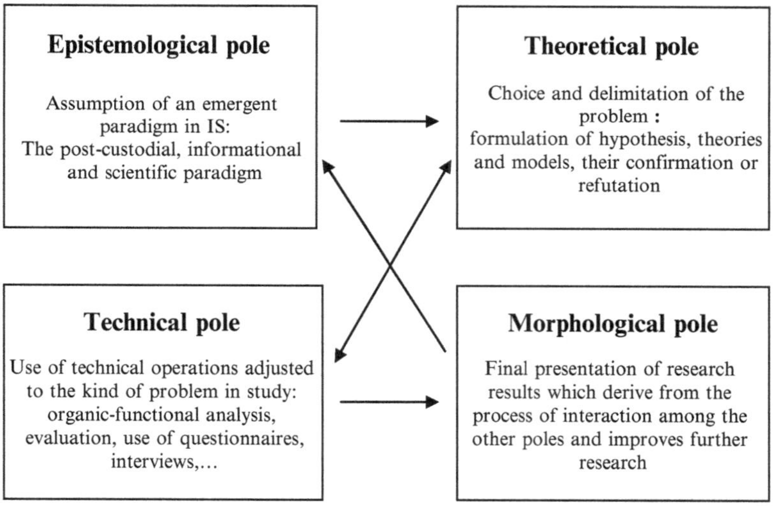

Quadripolar method of research: interactions between the four poles

The *epistemological pole* – the scientific community of information professionals, their schools, institutes, working places, with their own political, ideological and cultural references – operates the permanent construction of the scientific object and the definition of the boundaries of the research problems. The discursive parameters are constantly reformulated, as are the paradigms and scientific criteria (objectivity, reliability and evaluation) which guide the whole research process. Empirical procedures and archival knowledge gradually substantiate this pole, which is by no means static but, on the contrary, must be subject to periodic reflection on the occurrence, or otherwise, of epistemological continuity or gaps.

The *theoretical pole* operates the rationality of the subject (who knows and approaches) over the object, as well as the postulation of laws, the formulation of hypotheses, theories and operational concepts and the consequent validation or refutation of the "theoretical context" elaborated.

On the *technical pole*, contact with objectified reality is operated through instrumental application, thus verifying the validation capacity of the methodological mechanism. It is here that crucial operations are developed, such as the study of cases and variables and retrospective and prospective evaluation, always keeping in mind the confirmation or refutation of the postulated laws or principles, the theories elaborated and the operational concepts formulated.

On the *morphological pole*, the results of the research carried out are formalized through the representation of the object of study and the description of the whole research process which enabled the scientific construction around it. It deals with the organization and presentation of data, objectively checked on the theoretical and the epistemological poles, what shows the interactive character of the quadripolar method of research.

In this quadripolar dynamic, the theoretical pole assumes particular relevance, because it supports the technical and instrumental component and gives meaning to the explanation of the results in the morphological pole. There are, naturally, different theories and models applied to the interpretation of the informational phenomenon/process, but we prefer the Systemic Theory, whose origins derive from Ludwig von Bertalanffy's studies, developed since the 1920s. This preference is based on the fact that Systemic Theory enables a holistic view and adjusts quite well to the complex and diffuse universe of Information (Mella 1997).

The epistemological, theoretical and methodological foundation of IS, here briefly reviewed, is mirrored, obviously, in research projects, in educational and training models and in professional activities, developed in the most diverse organizational contexts. Only in this way does the theoretical and practical *corpus* that sustains IS as a scientific field gain meaning and a reason to exist.

9.2 Philosophical Implications of a Scientific Project

The epistemological concept we have been working on and developing at the University of Porto is rooted, as described above, in a (re)constructive process of an applied social science which needs to delimit its object of study as clearly as possible, built upon a complex phenomenon which has to be brought to the "surface" of reality, on which research activities are anchored. Apparently, our efforts seem to be restricted to the rationalization of practices, converted into professional procedures, which when passing from the sphere of common sense to systematic insights and explanations, call upon the status of scientificity, as legitimate as it is subject to lively debate.

However, what is called to question in the process of 'scientification' of practical disciplines such as Archivistics and Librarianship, through a transdisciplinary dynamic that can generate a new and consistent IS, is not of a strictly and instrumentally epistemological order. Although it is true that what is at hand is the need to confirm the respective epistemic validity of this transdisciplinary field, it is above all necessary to understand the deeper philosophical implications resulting from the alternative to the epistemological impasse which has afflicted the debate on the status of IS – the impasse of reducing this "science" to an inter-discipline, which seems to be no more than a "non place" ....

Clearly, in this exploratory paper, we do not intend to overview all the implications and much less go into them in detail. But there is a need to at least identify and highlight those which arise as central points in structural and analytical reflection on the epistemological project in which we are deeply involved. So as to understand this intention, we have to retake here the operational definition of information put forward previously, and attempt a "deconstruction" which clearly reveals the underlying epistemological and, particularly, philosophical assumptions.

Since having begun and broadened the epistemological debate on IS, we feel that the need for an operational definition of information and another of communication is strategic, so as to clearly understand the "texture" and contours of the object of study of our scientific discipline. Since it is apparent that the object is a discursive and social construct by a group or a community of practicians/researchers, it also seems clear that this constructed object has to point to a phenomenal reality or to phenomena which arise from an external reality independent of the subject-researcher. This positioning can be seen in philosophical terms, or in terms of theory of knowledge, as a mitigated realism which aggregated reconciles the representational subjectivity of the themes and related problems as a set to be explored by IS, with the objective rooting of those themes and problems in a reality to which the concepts of information and communication report. Specifically, the human mind and body, socially embedded, ultimate instance which we intend to know and which is beyond the palpable materiality of the documental, seen as epiphenomenon of *semiosis*, that is, of the signifying and symbolic capacity of the (human and social) production of meaning. This view can and should be completed with that of Robert Escarpit in *L'Information et la communication: théorie générale*, in which document is defined as a visible or touchable informational object endowed with a dual independence in relation to time: synchrony or internal independence of the message which is no longer a linear sequence of events, but a multidimensional juxtaposition of traits; and the stability or global independence of the informational object which is no longer an event registered in the course of time, but a material medium of the trait which can be preserved, transported, reproduced (Escarpit 1991:123). Traits or codified representations? There is at the core of this question a certain divergence with Escarpit and the idea that information does not dematerialize itself, even when it is produced in the mind and can be absorbed by another by means of phonetic and direct communication, or by means of a record on a physical medium (document).

The concepts of information and communication, within the historical specificity of IS, emerge and are adopted and used less by influence of the Mathematical Theory of Information of Claude Shannon and Warren Weaver (1949), and more by the reflexive deconstruction of the old notion of document. The importance of documental action as a potentially communicational practice and the natural criticism of Shannon and Weaver's mathematical theory mark the specificity of the info-communicational object in the epistemological conception defended here.

Let us retake the operational definition of:

Information: Structured set of codified mental and emotional representations (signs and symbols), modelled with/by social interaction, and capable of being recorded on any material medium and, therefore, communicated in an asynchronous and multidirectional way (Silva 2006; DeltCI 2007).

This definition already establishes the bridge with human and social interaction, which the concept of communication substantiates and to which it is intrinsically complementary, although not to be confused with information, despite some authors who have accepted this mistaken overlap.

Communication: Process of transmitting information among agents who share a set of signs and semiotic rules (syntactic, pragmatic and semantic), whose objective is the construction of meaning. Synonymous of human and social interaction and necessarily assuming information in the form of messages or contents which are transmitted, shared, in sum, communicated (Silva 2006; DeltCI 2007).

Information and communication are two operational concepts which serve to name and understand a human and social phenomenon, that of the innate and acquired ability to "give form" (ideas, sensations, emotions, etc.) and interact with other(s) or to "make common" that to which form has been given (Silva 2006:81–109). Information is thus synonymous of *knowledge* (explicit) and *data* (any codified representation no matter how small), and is counter to *cognition* (implicit or tacit knowledge definable as a function which realizes knowledge (material), since it is physiologically determined by the brain's structures and modes of operation – Tiberghien 2002:71) and, also, to *data*, understood as physical or natural impulse. However, the info-communicational phenomenon is rooted in the psyche; which is why Raymond Ruyer (1902–1987), a French philosopher who is rather unknown among us, highlighted the psychological information which prevails over the physical.

Having started with his PhD thesis (1930) by elaborating a vast philosophy of the world, Ruyer became interested a short time later, with the publication of *La conscience et le corps* in 1937, in the relationship between conscience and organism, focusing particularly on visual sensation. This is considered a turning point in his work, operating a radical distinction between mechanical (physical) structures and that which he came to call "true forms". A turning point and the embryo of the *Ruyerian* philosophy of information, as he explained it in 1950, following the works of Claude Shannon/Warren Weaver and Norbert Wiener: the "true form" of 1937 is naturally converted into psychological information or quasi-information, described in 1954 in his book, *La cybernétique et l'origine de l'information*. And the "mechanical structures" come to be called physical information. Ruyer thus operated a profound, well-articulated review/rereading of the mechanistic assumptions of the mathematical theory of communication (Shannon and Weaver) and cybernetics (Wiener), which soon had a perverse influence on the conceptions and perceptions around information, together with the development of computers and the advent of the internet. But this rereading did not, unfortunately, have sufficient force to take hold, and it should thus be recovered and emphasized, given its value and relevance.

Sylvie Leclerc-Reynaud (2006:67) provided a clear overview of Ruyer's thought, countering under three categories psychological information or quasi-information[3] (PsyI) and physical information (PhyI):

Nature
PsyI is theme, meaning; PhyI corresponds in Cybernetics to a negentropy or improbable structure.

Place
PsyI is on the subject's side, in the brain, in the areas of the occipital lobe for visible information, of the temporal lobe for audible information, etc.; PhyI is on the object's side (external to me), on a page for a text, a magnetic camp for a text codified into 0 and 1, in the air for photons or in electrical wires for electrical impulses (telephone), etc.

Properties
PsyI is not measurable, not visible, a source of negentropy, has meaning (vertical information), is dynamic (directive information) and amorphous; and PhyI is measurable, visible, tends towards entropy, has no meaning (horizontal information), is not dynamic and has precise form.

In action, I inform//Quasi-information is initiator and directive//Passing from meaning to structure.

PsyI is meaning in the form of intention, need, motivation, that is, of tendency oriented towards an end; PhyI consists of the means to accomplish and accomplished action (mail and material message; chisel and statue; software and equipped machine).

In acquiring knowledge, I inform myself//Quasi-information is receiver//Passing from structure to meaning.

PsyI is read message, inclusion, meaning under the form of signification, of idea or expressiveness, terminal and nutritious information; PhyI is material message (sequence of letters, photon pattern, etc.), circulating and horizontal information (without meaning).

In communication//Peter informs John who, by listening to him, becomes informed// From meaning to structure (expression) and from structure to meaning (understanding).

Framing information: at the beginning, intention as envisioned ideal (meaning to be communicated) and at the end, understanding, meaning in the form of signification, of idea or expressiveness. Transmission of data: at the beginning, message as framed phenomenon, circulating and horizontal information (without meaning), and at the end, message as framed phenomenon, circulating and horizontal information (without meaning).

Ruyer's philosophical proposal, thus summarized, inspires and supports the distinction we have been making between information and document, but above all, allows us to come stimulatingly closer to the bold and synthetic proposal of the Danish semiotics professor Soren Brier, an approach we will not be able to explore in this article. Departing from biosemiotics, as a theory of cognition and communication unifying life with the cultural world, Brier goes further and contemplates the ongoing technological revolution, formulating a Cybersemiotics as a global interpretative framework (Brier 2008). The author's intention is immediately put forward in the book's Introduction.

[3]Psychological Information is quasi- or proto-information by reference to the mathematical and physical conception of Shannon and Weaver.

The present book goes further: it is an inter- and trans-disciplinary project in the philosophy of science that analyses modern efforts to arrive at a unified conceptual framework, one that encompasses the complex fields of information, cognition and communication science, and semiotic scholarly studies – fields that together are often referred to as information science.

This book offers an interpretation of those "information science" research programs of the sort which unified information science can offer; it also discusses what is needed to supplement present approaches. As such, it is part of the Foundation of Information Science (FIS) research program, in that it asks whether there can be a transdisciplinary information science that encompasses the technical, natural, and social sciences, as well as the humanities, in its under-standing of understanding and communication, a vision that originally came from Norbert Wiener in his book *Cybernetics; or, Control and Communication in the Animal and the Machine (1961)* (…)

This book aims to formulate a new transdisciplinary framework based on Peirce's semiotics, second-order cybernetics, Luhmann's systems theory, cognitive semantics, and language game theory. I apply concepts found in second-order cybernetics and the semiotics of Charles Sanders Peirce to solve various transdisciplinary conceptual problems at the heart of cognitive science, since cybernetics was among the original contributors to modern information and communication science. I will refer to this transdisciplinary framework as 'Cybersemiotics' (Brier 2008:3–4).

Bearing in mind these citations, used here for merely illustrative purposes, there is undoubtedly an intense dialogue to be developed with Brier's Cybersemiotics, as with the Luciano Floridi's Philosophy of Information, although the latter presumes at the outset a less linear and more complex dialogic process. It is precisely for this reason that we seek to begin this dialogue here in this article, indicating points of convergence and deviation, more even than points of divergence.

The very recent book by Floridi, *Information: A Very Short Introduction* (2010), may prove a good starting point. The author summarizes the brief and recent "history" of a concept which has been appropriated by and adjusted to different areas of activity and scientific knowledge, much as Anthony Wilden did in the Information entry of the *Enciclopedia Einaudi* (Wilden 2001:11–72). And among the several meanings collected, we will see how Floridi presents semantic information, since this "conceptual variant" comes quite close to the psychological information of the operational definition we used in our epistemological approach to IS. However, there is a prior aspect to consider in the way Floridi conceives semantic information and which has to do with the relationship that he establishes between data and information in Chap. 2 – *The Language of Information* – of the above-mentioned book:

Over the past decades, it has become common to adopt a *General Definition of Information* (GDI) in terms of *data + meaning*. GDI has become an operational standard, especially in fields that treat data and information as reified entities, that is, stuff that can be manipulated (consider, for example, the now common expressions "data mining" and 'information management'). A straightforward way of formulating GDI is as a tripartite definition (Table 9.1):

According to (GDI.1), information is made of data. In (GDI.2), 'well formed' means that the data are rightly put together, according to the rules (*syntax*) that govern the chosen system, code, or language being used. Syntax here must be understood broadly, not just linguistically, as what determines the form, construction, composition, or structuring of something (…).

Regarding (GDI.3), this is where semantics finally occurs. 'Meaningful' means that the data must comply with the meanings (*semantics*) of the chosen system, code, or language

Table 9.1 The general definition of information (*GDI*)

(GDI) *a* is an instance of information, understood as semantic content, if and only if:
GDI.1) *a* consists of n data, for $n \geq 1$;
GDI.2) the data are *well formed*;
GDI.3) the well-formed data *are meaningful.*

in question. Once again, semantic information is not necessarily linguistic. For example, in the case of the car's operation manual, the illustrations are supposed to be visually meaningful to the reader (Floridi 2010:20–21).

We retain and highlight the statement contained in this extract that information is made of data that are rightly put together, according to the rules (syntax) that govern the system, code, or language being used. It should also be noted that, for Floridi, syntax here is meant broadly and is not restricted to the linguistic dimension. It covers other codes and systems. This is a highly relevant aspect implied in the initial part of our operational definition of information – *codified (mental and emotional) representations* address a plurality of codes, from spoken and written word to Braille code, Morse or "programming languages", including musical notation, mathematical codification (digits, propositions, equations, algorithms, etc.), geometry and chromatic code. The intention is to identify, through the concept of information, a broad and unified object of study, which aggregates types of representation which are still persistently classified and "arranged" into different and even incompatible "categories". In this perspective, there is the assumption of complex thought as founding matrix of the info-communicational field and that an "IS approach" to people's informational behaviour contemplates everything. To maintain Floridi's example, this means the illustrations included in a car manual intersected in the same support with text, but we can go further, and connect the manual to publicity spots about that car model, the audiovisual pieces with that car's test-drives, etc. Floridi seems to converge on our perspective, by highlighting the amplitude of the notion of semantics and by emphasizing the importance of syntax (code, system and languages), in the correct structuring of the data that compose information. This defining strategy does not clarify a distinction we have come to make between data.1 and data.2. Data.1 in Computer Science is the conventional representation, through codification, of a piece of information which enables its electronic processing. This allows us to say that there in absolutely no difference between data and information and they address the same phenomenon (cerebral mental and psychological activity). Data.2 means the physical, electromagnetic, seismic, etc., impulse or vibration which, through specific technological devices, is converted into graphic representations (information) – in this sense, data and information differ, in that they address distinct phenomena (Silva 2006:145).

Floridi seems to address this distinction, which we believe is enlightening, when he says that *data can be lacks of uniformity in the real world. There is no specific name for such 'data in the wild'. One may refer to them as* dedomena, *that is, 'data' in Greek (…). They are pure data, that is, data before they are interpreted or subject to cognitive processing* (Floridi 2010:23); and when a little later, still in Chap. 2, he talks of *environmental information*, he provides the following explanation: *One of the most often cited examples of environmental information is the series of concentric rings visible in the wood of a cut tree trunk, which may be used to estimate its age. Viewers of* CSI: Crime Scene Investigation, *the crime television series, will also be well acquainted with bullet trajectories, blood spray patterns, organ damages, fingerprints, and other similar evidence* (Floridi 2010:32).

The "father" of Philosophy of Information again comes closer to the symbiosis which we believe exists between the concept of information and that of data.1, by listing several types of data/information:

Information can consist of different types of data. Five classifications are quite common, although the terminology is not yet standard or fixed. They are not mutually exclusive, and one should not understand them as rigid: depending ont circumstances, on the sort of analysis conducted, and on the perspective adopted, the same data may fit different classifications.

Primary data
These are the principal data stored in a database, for example a simple array of numbers in a spreadsheet, or a string of zeroes and ones. (…)

Secondary data
These are the converse of primary data, constituted by their absence. Recall how John first suspected that the battery was flat: the engine failed to make any noise, thus providing secondary information about the flat battery. (…)

Metadata
These are indications about the nature of some other (usually primary) data. They describe properties such as location, format, updating, availability, usage restrictions, and so forth. Correspondingly, *metainformation* is information about the nature of information. The copyright note on the car's operation manual is a simple example.

Operational data
These are data regarding the operations of the whole data system and the system's performance. Correspondingly, *operational information* is information about the dynamics of an information system. (…)

Derivative data
These are data that can be extracted from some data whenever the later are used as indirect sources in search of patterns, clues, or inferential evidence about other things than those directly addressed by the data themselves, e.g. for comparative and quantitative analyses. (…) Credit cards notoriously leave a trail of derivative information (Floridi 2010:29–31).

However Floridi's typology does not separate informational data (data.1) from the non-informational (data.2), which can be interpreted and converted into semantic information. This lack of distinction is not very enlightening. We put forward that a more precise classification of informational data may be convenient, but Floridi raises reservations and doubts because it places data proceeding from nature or from mechanical and artificial systems, and not from human and social cognitive activity, as secondary data. Furthermore, we feel that the distinction between metadata and metainformation is redundant, since the indication of copyright considered as metainformation is a codified mental representation, much as the indication of place of edition, considered as metadata.

Continuing with this comparison of perspectives, we will now focus on Chap. 4, *Semantic Information*, which begins with a highly relevant warning from Floridi: the *MTC* [mathematical theory of communication] *is not interested in the meaning, reference, relevance, reliability, usefulness or interpretation of the information exchanged, but only in the level of detail and frequency in the uninterpreted data that constitute it* (Floridi 2010:48). This is a fitting insight, particularly bearing in mind that, for Floridi, the difference between MTC and semantic information is to the same order, different yet related, as the Newtonian description of the physical laws that govern the dynamic of a tennis match and the narration of that same match by a sports commentator – *The two are certainly related, the question is how close* (Floridi 2010:48). In the perspective of IS, both descriptions are information in codes which obey different rules (syntaxes), thus configuring the respective object of study which includes the way in which a certain type of information is produced, in which context and with which aims, how it is organized, stored, made accessible, used and reproduced. For PI, on the other hand, what needs to be discussed is what type of relationship exists between apparently different phenomena, such as the physics of movement and the description with mental signs and symbols of a specific game, taking place in time and space. Through this and other specifications, it is possible to show where IS and PI clearly function on differentiated planes and how they do not intersect or do so very contingently.

Throughout his instructive book, Floridi provides several diagrams, some of which are reproduced with certain indications and signs for the readers, such as "*You are here*", letting them know where they are or the matter being discussed. This image thus reproduced several times emerges as a "tree" scheme through which we come to perceive a precise conceptual sequence: the (structured) data are divided into *environmental (information)* and *semantic (content)*; these, in turn, are subdivided into *instructional* (with a linking trait to *environmental*) and *factual*, which are further subdivided into *untrue* and *true (information)*; *true information* generates knowledge, whereas *untrue* information is *unintentional (misinformation) intentional (disinformation)*. Here, we are interested in the figure in which Floridi focuses on the conceptual point of *factual information*, from which he guides us. The most relevant distinction under *factual* is between "*semantic content* and *semantic information: the latter needs to be true, whereas the former can also be false*" (Floridi 2010:50). At the basis of this distinction lies the definition (DEF) of *factual semantic information* thus formulated: ***p*** *qualifies as factual semantic information if and only if* ***p*** *is (constituted by)* ***well-formed, meaningful*** *and* ***veridical data*** (Floridi 2010:50). And there are at least three advantages in this DEF: the first is that it clarifies that false information is not a genuine type of information (*when semantic* *content is false, this is a case of* ***misinformation****. If the source of misinformation is aware of its nature, as when John intentionally lied to the mechanic, one speaks of* ***disinformation*** – Floridi 2010:50); the second is that it establishes a robust and intuitive link between *factual semantic information* e *knowledge* (*Knowledge encapsulates truth because it encapsulates semantic information, which, in turn, encapsulates truth, as in a three dolls matryoshka. Knowledge and information are members of the same conceptual family* – Floridi 2010:51); and the third is that it may enable a solution to the so-named Bar-Hillel-Carnap paradox, mentioned by Floridi at the end of the chapter on *semantic information*. But before that, he focuses on the *scandal of deduction* and analyzes *informativeness*, highlighting that, considering this topic, *semantic information* differs from *MTC* because the former aims to answer questions such as *'how can something count as information? and why?', 'how can semantic information be generated and flow?', 'how is information related to error, truth and knowledge?'* etc., and because it seeks to obtain *more complex forms of epistemic and mental phenomena, in order to understand what it means for something, such as a message, to be informative* (Floridi 2010:52). However, it is possible to detect two important connections between *semantic information* and *MTC*: the model of communication, which Floridi explains in Chap. 3, and the so-named *Inverse Relationship Principle (IRP)*, which *refers to the inverse relation between the probability of* ***p*** *– where* ***p*** *may be a proposition, a sentence of a given language, an event, a situation, or a possible world – and the amount of semantic information carried by* ***p****. IRP states that information goes hand in hand with unpredictability (Shannon's surprise factor*) – Floridi 2010:53).

The *IRP* appears in the item in which Floridi presents and elaborates on the Bar-Hillel-Carnap paradox, reminding us that *the less probable or possible* ***p*** *is, the more informative it is* (Floridi 2010:58). But, the author proceeds, if we continue to make ***p*** gradually less probable, there will be a point at which the probability of ***p*** is almost zero, that is, it is impossible or equivalent to a contradiction. But, according to *IRP*, it is in that state that ***p*** achieves its maximum *informativeness*. That is, using the example Floridi employs throughout the book, *John would be receiving the highest amount of semantic information if he were told that the car's battery is and is not flat (at the same time and in the same sense). This other counterintuitive conclusion has been called Bar-Hillel-Carnap paradox (because the two philosophers were among the first to make explicit the counterintuitive idea that contradictions are highly informative)* – Floridi 2010:58).

Since its formulation, the problem was considered unfortunate, although the perfectly correct and logically inevitable consequence of any *quantitative* ***theory of weakly semantic information,*** *'Weakly' because truth values play no role in it* (Floridi 2010:58). Although often ignored or merely tolerated, the paradox is retaken by Floridi, having as its resolving-key the DEF of semantic information: *It is now easy to see why: if something qualifies as factual semantic information only when it satisfies the truthfulness condition, contradictions and indeed falsehoods are excluded* ***a priori****. The quantity of semantic information in* ***p*** *can then be calculated in terms of distance of* ***p*** *from the situation* ***w*** *that* ***p*** *is supposed to address* (Floridi 2010:59). The rigorous application of this DEF leads to the consideration of four enunciations of a concrete situation – *A. there will or will not be some guests for* *dinner tonight; or B. there will be some guests tonight; or C. there will be three guests tonight; or D. there will and will not be some guests tonight* – only (c) has a maximum degree of *informativeness* because it fully corresponds to the truth of situation ***w***.

The solution proposed by Floridi for the Bar-Hillel-Carnap paradox has naturally direct implications on the scientific study of human and social communication: the interaction between a sender A and a receiver B will be all the more complete and perfect the higher the degree of *informativeness* transmitted, that is, correct communication has to be based on the assimilation of this assumption. A good journalist, for example, depends entirely on this assumption. However, this formulation may have fault by reducing the complexity of a well-conditioned "logical environment", that is, built on the basis of respect for prior good conditions, which may not translate or capture the incoherence, the usury and irrationality of daily life.

This imbalance is aggravated if we take into account the concept of IS which we put forward, which means that we are clearly positioned within the emergent post-custodial, informational and scientific paradigm. Within this paradigm, IS faces the complexity of the real world more clearly: to study information scientifically does not exclude the recourse to hermeneutics, as has been suggested by Rafael Capurro (2002), and in this sense, it is important to bear in mind the meaning of words, images, drawings, colours, sounds, etc., but the internal or inherent meaning(s) of each text do(es) not comprise IS's main object of study. IS does not preferentially search for truth in each unit of information or meaning produced and which can be communicated, but rather, it seeks the (possible) truth in the info-communicational cycle or process developed through a number of stages and significant moments, such as the production of information in a certain context, the respective organization, arrangement and storage in that context or another, its use according to the specific needs of the user acting in situation and in context, its reproduction … Seen from this perspective, information cannot be reduced only to the notion of *semantic information*, a notion we accept without any restrictions in particular. However, the phenomenon underlying IS's object of study intertwines individual psychology and social dynamic, thus making the study situations much more complex. Thus, taking as a critical example the video games produced according to an inferential narrative logic involving the linear resolution of problems and obstacles, naturally distinct from, for example, the classical literary narrative. There is no concern with the truth in this process, but rather with building the plausible and the degree of *informativeness* cannot be measured by the correspondence to the truth of situation *w* to which it reports. In the perspective of IS, it is not the degree of *informativeness* which is studied, but who produces the video games, in which context and to what end, how are they accessed (and this implies knowing how are they organized, accumulated and disseminated), what informational needs do they satisfy (how are they generated, reproduced, modified and contextualized), and what impact they have on the personal and professional lives of their "consumers".

We have here undoubtedly a topic for further analysis and debate which we hope to retake in other contributions. Before concluding, there are two more topics we would like to mention.

By considering *semantic information* and *knowledge* as members of the same conceptual family, Luciano Floridi shines a different light on the growing distinction between data, information and knowledge in the literature on computerized management, information and knowledge management and competitive intelligence. In this "universe", a number of interpretative proposals have received some attention, such as that of Nonaka and Takeuchi (1998), who distinguish between explicit knowledge and tacit knowledge: the former is codified knowledge, being hence easy to observe, since it is transmitted through conventional external languages; the latter has a personal nature, which makes it much more difficult to formalize and communicate, since it is deeply rooted in action, in commitment and development in a specific context. If we accept this "reading", it becomes impossible to distinguish information from explicit knowledge and the expression "knowledge management" can only report to tacit knowledge. From the perspective of IS we defend, the operational distinction to take account of is between information and document: the first corresponds to the intellectual content it is possible to register on a material support external to the subject (document). Based on this, the proximity between information and knowledge is very high, as Floridi has highlighted: the distinction between one and the other has to do with the sieve of truth. Knowledge is true factual information. Confronted with this logical formalism, tacit knowledge is remitted to an obscure zone which is not worth considering. From the perspective of IS, information forms in the human mind, where emotion and cognition operate and blend, being converted there into a signifying and symbolic unit which through voice, writing or gestures, is externalized and able to be communicated. In this view, it is possible to distinguish information from cognition (cerebral biochemical faculty), but how can we distinguish it from knowledge? The distinction which is used recurrently in the management literature is rejected by authors like Capurro (2002), for whom, at the end of Modernity, it is not possible to recognize the difference between knowledge and information, since at this time we have abandoned the Platonic idea of human knowledge as separate from the knowing subject or cognoscente, from where the relevance of concepts derives, such as that of mediation (information technology, for example, disseminates all kinds of knowledge in a form prefigured by the press).

A second and final topic with which we will conclude our contribution to this collective work of analysis and reflection on Luciano Floridi's Philosophy of Information, has to do with a more general and profound question: in what way and at what level can the mission-aims of Philosophy of Information directly guide and influence research in IS?

In a broader study it would be interesting to analyze the question using more of Floridi's texts and, of course, the indispensible *The Blackwell guide to the philosophy of computing and information* (2004). For our purposes here, the article published in the journal *The Information Society* (2009) suffices, in which Floridi defines Philosophy of Information as a field in which two dimensions emerge: "*(a) the critical investigation of the conceptual nature and basic principles of information, including its dynamics, utilization, and sciences, and (b) the elaboration and application of information-theoretic and computational methodologies to philosophical problems*" (Floridi 2009:154). PI is clearly positioned in a space of research and of epistemological reflection in which the problems, strategies and methodologies of the computational and information sciences have a place: *It is therefore essential to stress that PI critically evaluates, shapes, and sharpens the conceptual, methodological, and theoretical basis of ICS – in short, that it also provides a philosophy of ICS, as has been obvious since early work in the area of philosophy of artificial intelligence (IA)* (Floridi 2009:155). In what concerns IS directly, and without regarding PI as merely converted into one of its epistemological extensions, it seems clear that PI's questionings and reflections should be incorporated in IS's epistemological programme: conceptual, methodological questions which are clearly centred on the nature and characteristics of knowledge, which have been produced through research in IS within the broader framework of the Information and Knowledge Sciences, and at the intersection with the inter-science of Information Systems and Computational Sciences, receive enriching insights and readings from PI.

To specify a bit further how PI can be decisive and positive for the development of the sciences, particularly of IS, we can take one of PI's main research topics, that of *whether nature (physis) and technology (techne) may be reconcilable.* The path that can be opened by PI in this domain is certainly useful to IS researchers who, evidently, by positioning themselves in the new post-custodial, informational and scientific paradigm, explore their object from a holistic and integrating perspective, overcoming reductionist dichotomies which cannot block the progress of research: for example, the paper-digital (support) antinomy naturally enters as a variable in the scientific study of information, but the focus is on the latter and its properties. Through this focus, it is possible to explore different phenomena which intersect and overlap – psychological information (biological and social phenomenon) is converted in document, that is, it is materialized, passing to another phenomenal order (it is fixed on any type of material – physical phenomenon – it is possible to inscribe/record signs and symbols which express cognitive and emotional representations – information).

IS can thus make its way through challenges and technological novelties which appear at a blinding rate in this Information Age (or in the *Fourth Revolution*, according to Floridi), concerned with understanding problems, exploring cases and adjusting applications (new or existing) to the info-communicational processes of people and, increasingly, of *informational organisms* – the *inforgs* announced by Floridi.[4] Meanwhile, PI will keep providing the bases of a macro-reflection on the nature and evolution of the info-sphere, proposing even a metaphysics whose master traits are the following:

> Within the information society, it seems that we are modifying our ontological perspective, from a materialistic one, in which physical objects and processes still play a key role, to an informational one, in which (a) objects and processes are dephysicalized, typified, and perfectly clonable; (b) the right of usage is perceived to be at least as important as the right

[4] Floridi believes that "*we are now slowly accepting the idea that we might be information organisms among many others, significantly but not dramatically different from natural entities and agents and smart, engineered artifacts*" (Floridi 2009:156).

to ownership; and (c) the criterion for existence is no longer being immutable (Greek metaphysiks) or being potentially subject to perception (modern metaphysics) but being interactable (Floridi 2009:156).

Together with a Metaphysics, Floridi, in line with other authors such as Capurro, has come to raise the foundations of an Ethics, both forming the fundamental components of PI and a space to reformulate the classical, crucial philosophical problems according to the current state of the World and Mankind. A project of this type does not override or substitute the specific path of Science and, particularly, of the *information and computational sciences-ICS*, on which Floridi explicitly focuses in his article, and among which IS stands in its own right, but can and should accompany them in regular, intense debate, bringing clear benefits to all.

References

Brier, Soren. 2008. *Cybersemiotics: Why information is not enough!* Toronto: University of Toronto Press. ISBN 978-0-8020-9220-5.

Capurro, Rafael. 2002. La hermeneutica y el fenómeno de la información. Available in: http://www.capurro.de/herminf.html. Accessed 14 Apr 2010.

Day, Ron. 1997. Paul Otlet's book and the writing of social space. *JASIS – Journal of the American Society for Information Science.* 48(4): 310–317. New York. ISSN 0002-8231.

De Bruyne, P., et al. 1974. *Dynamique de la recherche en sciences sociales de pôles de la pratique méthodologique*. Paris: PUF.

Debons, A. 1986. Information science. In *ALA world encyclopedia of library and information services*, 2nd ed, 354–358. Chicago: American Library Association. ISBN 0-8389-0427-0.

DeltCI – Dicionário Eletrônico de Terminologia em Ciência da Informação. 2007. http://www.ccje.ufes.br/dci/deltci/index.htm. Accessed on 14 Apr 2010.

dir Tiberghien, G. 2002. *Dictionnaire des sciences cognitives*. Paris: Armand Colin. ISBN 2-200-26247-7.

Escarpit, R. 1991. *L'Information et la communication: théorie générale*. Paris: Librairie Hachette.

Floridi, L. (ed.). 2004. *The Blackwell guide to the philosophy of computing and information.* Malden: Blackwell Publishing. ISBN 0-0631-22918-3.

Floridi, L. 2009. The Information society and its philosophy: introduction to special issue on "The philosophy of Information, its nature, and future developments. *The Information Society* 25: 153–158. London: Routledge. ISSN 0197–2243.

Floridi, L. 2010. *Information: A very short introduction.* Oxford: Oxford University Press. ISBN 978-0-19-955137-8.

Le Coadic, Y.-F. 2004. *A Ciência da Informação*. Trad. de Maria Yêda F. S. de Filgueiras Gomes, 2.ª ed. Brasília: Briquet de Lemos – Livros. ISBN 85-85637-23-4.

Leclerc-Reynaud, S. 2006. *Pour une documentation créative: l'apport de la philosophie de Raymond Ruyer.* Paris: ADBS-Association des Professionnels de l'Information et de la Documentation. ISBN 2-84365-2.

Lessard-Hébert, M., et al. 1994. *Investigação qualitativa: fundamentos e práticas*. Lisboa: Instituto Piaget. ISBN 972-9295-75-1.

Mella, P. 1997. *Dai Sistemi al pensiero sistémico: per capire i sistemi e pensare com i sistemi.* Milano: Franco Angeli. ISBN 88-464-0336-3.

Nonaka, I. and H. cop. Takeuchi. 1998. A Theory of the firm's knowledge-creation dynamics. In *The Dynamic firm: the role of technology, strategy, organization and regions*, ed. A. Chandler, P. Hagström, and Ö. Sölvell, 214–241. New York: Oxford University Press. ISBN 019-829604-5.

Otlet, P. 1996. *El Tratado de Documentación : el libro sobre el libro: teoria y práctica*. Trad. Maria Dolores Ayuso Garcia. Múrcia: Universidad. ISBN 84-7684-766-1.

Rayward, W.B. 1997. The origins of information science and the International Institute of Bibliography//International Federation for Information and Documentation (FID). *JASIS – Journal of the American Society for Information Science* 48(4): 289–300. New York. ISSN 0002-8231.

Saracevic, T. 1996. Ciência da informação: origem, evolução e relações. *Perspectivas em Ciência da Informação* 1(1): 41–62. Belo Horizonte. ISSN 1413–9936.

Shera, J.H., and D.B. Cleveland. 1977. History and foundations of information science. *Annual Review of Information Science and Technology, Washington* 12: 249–275.

Silva, A.M. 2006. *A Informação: da compreensão do fenómeno e construção do objecto científico*. Porto: Edições Afrontamento; CETAC.COM. ISBN 972-36-0859-3.

Silva, A.M., and F. Ribeiro. 2002. *Das "Ciências" Documentais à Ciência da Informação: ensaio epistemológico para um novo modelo curricular*. Porto: Edições Afrontamento. ISBN 972-36-0622-4.

Wilden, A. 2001. Informação. In *Enciclopédia Einaudi. Vol. 34. Comunicação-Cognição*. Lisboa: Imprensa Nacional-Casa da Moeda. ISBN 972-27-0923-2.

Williams, R.V., L. Whitmire, and C. Bradley. 1997. Bibliography of the history of information science in North America, 1900–1995. *JASIS – Journal of the American Society for Information Science* 48(4): 373–379. New York. ISSN 0002-8231.

Part IV
Epistemic and Ontic Aspects of the Philosophy of Information

Chapter 10
Skepticism and Information

Eric T. Kerr and Duncan Pritchard

Philosophers of information, according to Luciano Floridi (2010, 32), study how information should be "adequately created, processed, managed, and used." It is unlikely that we can do this without linking that study to the epistemic purposes of creating, processing, managing, and using information. Doing so, we claim with Floridi, requires attention to the epistemic value of information. In particular, our interest in information has a number of purposes, one of which is to acquire knowledge.[1]

If I visit the doctor to be told that I am suffering from Creutzfeldt–Jakob disease he will likely bombard me with information, either through leaflets and documents or through informing me verbally. Our joint purpose is that I will know more about the disease (so as to better arm myself against its effects as well as to make me more comfortable with my condition). I will be informed that CJD is a rare but fatal brain disorder; it affects about one person in every one million people per year worldwide; symptoms typically occur at about age 60; about 90% of patients die within 1 year; and so on. We are rarely occupied with collecting information for information's sake. I want this information because it is relevant to knowledge I wish to acquire about CJD. The information is epistemically valuable to me in the situation

[1]Notes: It should be noted that gathering, creating, processing, managing and using information is not always done for the acquisition of knowledge or other epistemic standings. Sometimes, for example, information is collected for the sake of collecting more information or for justifying policy decisions. Nevertheless, the kind of information-based inquiry we explore here is that which is pursued with the final purpose of gaining knowledge about the matter at hand. This is the kind of inquiry pursued in Dretske (1981) and Floridi (2010), among others. These scholars accordingly view information as, in their own distinctive ways, an important component of epistemology.

E.T. Kerr (✉) • D. Pritchard
School of Philosophy, Psychology and Language Sciences, University of Edinburgh,
Dugald Steward Building, Charles Street, Edinburgh EH8 9AD, UK
e-mail: E.T.Kerr@sms.ed.ac.uk; duncan.pritchard@ed.ac.uk

H. Demir (ed.), *Luciano Floridi's Philosophy of Technology: Critical Reflections*, Philosophy of Engineering and Technology 8, DOI 10.1007/978-94-007-4292-5_10,

I am in (Fallis and Whitcomb 2009). As has been noted elsewhere (e.g., Himma 2007), we are overloaded with information in the modern age. In this paper we examine these paths from information to knowledge and how constricting the range of relevant information is critical to information management.

With the development of fairly recent technology, information has become a ubiquitous cultural buzz-word: the Information Age; information overload; the Information Superhighway; freedom of information; information technology; information science; and so on. Information and knowledge appear together frequently both in popular writing and scientific disciplines either as conflated terms for the same phenomena or related terms in some way involved in practices of inquiry, discovery, knowledge acquisition, and so on. The job of relating these concepts more precisely has tended to be undertaken by various academic disciplines that take information as a key theoretical concept. Although it is other disciplines such as information technology, knowledge management and library science that have devoted sustained analysis to information, such growing cultural awareness of information has provoked some philosophers to comment on its societal, epistemological, ontological, or axiological significance and sometimes to use it as a component in their philosophical work.[2]

Two philosophers of particular note in this regard are Fred Dretske and Floridi. Both have developed technically complex epistemologies with information playing a central role (See especially Dretske 1981, 1983, 2000, 2006; Floridi 2005, 2010). Dretske connects information to knowledge via an ordinary dictionary definition of the former:

> [*By information*] I mean nothing very technical or abstract. In fact, I mean pretty much what (I think) we all mean in talking of some event, signal or structure carrying (or embodying) information about another state of affairs. A message (i.e., some event, stimulus or signal) carries information about *X* to the extent to which one could learn (come to know) something about *X* from the message. (Dretske 1983, 10)

By relating information to knowledge in this way, Dretske's information-based epistemology becomes allied to the relevant alternatives theory of knowledge (or 'RAT', for short) that he puts forward. According to this view, possessing knowledge depends on an agent's capacity to rule out a certain range of alternatives which varies according to what kind of alternatives are relevant.

The notion of relevancy in play here has been notoriously difficult to pin down (Floridi 2010, 300–24; Shope 2002, 37). Duncan Pritchard states the RAT view as applied to perceptual knowledge as follows:

> *S* has perceptual knowledge that *p* only if *S* can discriminate the target object at issue in *p* from the objects at issue in relevant alternative (not-*p*) propositions, where a relevant alternative is an alternative that obtains in a near-by possible world (Pritchard 2010, 3).

According to this rendering of the RAT view, our capacity to possess perceptual knowledge is heavily affected by our environment. Pritchard (2009, 5) makes the

[2] See, for example, Fallis (2004), Harms (1998), and Goldman (1999, 161–182).

distinction between epistemically friendly and unfriendly environments. Most of the time, it will be very easy for us to make the necessary discriminations between, for example, hands and stubs, or canaries and crows. But epistemology is replete with thought experiments which arrange our environment such that it will not be so easy. For instance, the 'barn façade' case describes one such environment where we no longer have the easy capacity to discriminate between barns and other things which may be in the environment (on account of how many of the items in this environment which look like barns are in fact barn façades). On this rendering of the RAT view, then, one consequently fails to know that the object before one is a barn.[3]

It is because of the possibility of deceptive environments like this that Dretske denies that information alone could ever answer a skeptical doubt. The argument for this is as follows: I have many defeasible reasons for thinking that I am writing these words in Edinburgh, Scotland just now (memory, testimony, observation, etc.). This gives me an informational basis for believing that I am writing these words in Edinburgh. However, I do not have an informational basis for believing that I am not a brain-in-a-vat (BIV) on Alpha Centauri who is being fed the illusion that he is writing these words in Edinburgh, Scotland. Even if the standards for knowledge are very low, and even if I know that were I in Edinburgh then I would not be a BIV on Alpha Centauri, this would not give me an informational basis for denying the skeptical hypothesis. The reason for this is my inability to discriminate between the scenario in which I am in Edinburgh and the skeptical BIV scenario in which I am on Alpha Centauri. Accordingly, argues Dretske, it follows that I receive exactly the same information in either scenario, and hence that I can have no informational basis to reject the alternative skeptical scenario.

In general, Dretske argues that no signal can carry the information that a skeptical hypothesis—an hypothesis explicitly designed such that it is indiscriminable from normal circumstances, and yet involves a high degree of error—is false. In his *Knowledge and the Flow of Information*, for example, he writes: "*No* signal can rule out *all* possibilities if possibilities are identified with what is consistently imaginable. No signal, for instance, can eliminate the possibility that it was generated, not by the normal means, but by some freak cosmic accident, by a deceptive demon, or by supernatural intervention" (Dretske 1981, 130). And, later: "This is true of all indicators, all sources of information. That is why there is nothing in the world [...] that indicates that there is a material world" (Dretske 2005b, 22).

So on Dretske's view I can have an informational basis for believing that I am in Edinburgh but I can have no informational basis for believing that I am not a BIV on Alpha Centauri (a skeptical hypothesis which entails that I am not in Edinburgh), even whilst I know that if I am a BIV on Alpha Centauri then I am not in Edinburgh. It is for this reason that Dretske denies epistemic closure.[4] In its crudest form,

[3]The barn-façade case was first put forward in print by Goldman (1976), who credits the example to Carl Ginet.

[4]For Dretske's initial rejection of epistemic closure, see Dretske (1970, 1971). See also his recent exchange with Hawthorne (Dretske 2005a, c; Hawthorne 2005). For a critical discussion of the implications of Dretske's informational epistemology on epistemic closure see Jäger (2004) and Shackel (2006).

epistemic closure is the principle that if an agent knows one proposition, and knows that it entails a second proposition, then that agent also knows the second proposition. So, for example, if one knows that one is presently in Edinburgh, and one knows that this entails that one is not a BIV on Alpha Centauri, then one knows that one is not a BIV on Alpha Centauri. Although this principle has broad intuitive support, Dretske rejects it.[5] But why is it that on Dretske's view I can acquire knowledge about a proposition but not about a proposition which I know full well is entailed by it? Dretske is led into this position through two closely related commitments: (i) that perceptual information is never relevant to skeptical hypotheses, and (ii) that information is essentially non-factive evidence.

We noted the first commitment above. Since, *ex hypothesi*, agents cannot discriminate between normal scenarios and skeptical alternatives, so it follows, according to Dretske, that agents lack an informational basis for dismissing skeptical alternatives. The second commitment becomes clear once we reflect that if information could be factive evidence for what it is evidence for—if, that is, it could entail the truth of what it is evidence for—then it would follow that the information we have to support our beliefs in normal circumstances might well suffice to entail the denial of the target skeptical scenario. Clearly, however, Dretske does not think that we ever have evidence of this sort, and hence a non-factive view of the evidence provided by information is clearly implicit here.

In order to more closely examine these commitments, consider the following local skeptical hypothesis, which we will call 'Zebra':

> *Zebra*
> Fred is at the zoo. If he perceives what he takes to be a zebra, Fred can have no informational basis for believing that what he perceives is not, in fact, a cleverly-disguised mule. In other words, the signal carrying this information does not allow him to discriminate between 'a zebra in my perceptual field' and 'a cleverly disguised mule in my perceptual field'.

Fred may interpret the signal as evidence that there is a zebra in front of him as a matter of habit, or perhaps relying on other evidence such as the sign on the fence or assumptions about what kinds of animals are in a zoo. However, his information is, it seems, non-factive. Just because I receive a signal such as this does not entail that there is in fact a zebra in the pen. More generally, as Dretske claims, it appears that none of the information that the subject possesses which indicates that he is perceiving a zebra is information which offers him an adequate epistemic basis on which he can dismiss the 'cleverly disguised mule' skeptical scenario.

This way of thinking about our evidential position with regard to skeptical challenges has, however, been challenged. Ram Neta (2002, 2003), for example, has argued that the scope of your evidence is affected by context. Under this account, there is a range of contexts in which evidence (read: information) is factive. Neta argues that the skeptic only appears to succeed by restricting what counts as evidence. In normal contexts my evidence typically is factive, and it only becomes

[5]Although there are few philosophers these days who deny this principle, it was also famously denied by Nozick (1981), for reasons very similar to the reasons offered by Dretske.

non-factive in skeptical contexts in which very demanding standards for what counts as evidence are in play. Hence, in the zebra case, my evidence for believing that there is a zebra before me could well be factive in normal contexts. For example, if my evidential state in normal contexts is that of *seeing that* there is a zebra before me, then, since seeing that *p* entails *p*, my evidential state actually entails that there is a zebra before me, and which hence entails that I am not currently being presented with a cleverly disguised mule. Relatedly, if my evidence, in normal contexts, for believing that I have two hands is that I can see them before me, then I have evidence which entails not only that I have two hands, but also that I'm not a handless BIV on Alpha Centauri.

According to Neta, however, the context can change in such a way as to restrict the scope of one's evidence. If I were to gain evidence that cast doubt upon my belief that I have hands—for example, if I were to witness a room of BIVs—then this would make the possibility that I am a BIV a relevant alternative. This is effectively what the skeptic does: to describe such a scenario and cast doubt upon what was previously undoubted. There are two ways in which this may be done.

On the first, the skeptic may simply suggest the possibility of a skeptical hypothesis that had previously been ignored or unexamined by the subject. This may place an onus on the subject to now eliminate that possibility in order to be correctly said to know the proposition. This intuition suggests that we cannot know a proposition until we have ruled out all relevant alternatives and that the range of relevant alternatives is determined by the conversational context (Pritchard 2010, 19). In other words, being made aware of an alternative, however implausible or absurd, can make that alternative relevant.

The second way in which the skeptic can make the alternative relevant is by actually offering *evidence* for thinking that a skeptical scenario has obtained. For example, consider an extension to the case of Zebra:

> *Zebra**
> Fred's friend and skeptic, Frank, mentions to Fred that he once read a science-fiction story in which all the world's zebras are replaced by hologram zebras and the real zebras are taken to a neighbouring planet. A little while later, Frank notices a pot of paint lying beside the animal and brings this to Fred's attention by gesturing towards it. He also tells Fred that the sign on the outside of the pen appears to have been written over an older sign, suggesting that a different message was once written there.

In this example, Frank initially merely presents Fred with a radical skeptical hypothesis. In the view of some epistemologists such pronouncements can change the conversational context in which evidence requirements and relevant alternatives are set.[6] Frank's story may thus rob Fred of his knowledge that there is a zebra in the pen before him. In the subsequent details of the story, however, Frank presents Fred with perceptual *information* and testimonial *evidence* for calling into doubt Fred's knowledge of what is in the pen.

[6] See, for example, DeRose (1995) and Lewis (1996).

According to Neta, in these skeptical contexts Fred's evidence is no longer factive. In particular, it is now no longer the case that one's evidence can entail the denials of skeptical hypotheses, given that they are in play and problematising our epistemic position. So although in normal contexts my evidence that I am seeing two hands could be that I see that I have two hands, in skeptical contexts where the skeptical hypothesis is at issue my evidence can at most be that I *seem to see* that I have two hands, where this evidential standing clearly does not entail the target proposition. It is, on the other hand, possible to gain evidence supporting local skeptical claims. In the case of Zebra, if I were to notice a pot of paint next to the animal or its flaking 'skin', then this may provide an evidential or informational basis for believing that the animal is a cleverly disguised mule. If one subscribes to Dretske's relevant alternatives theory or Neta's contextualism, then the absence of such signals means, respectively, that either we are not required to rule out this possibility or that we are in an ordinary context in which the denials of skeptical hypotheses are known.

Consider O. K. Bouwsma's (1965) 'adventures': when Tom peels away part of his face he receives a signal carrying the information that he is in a world made of paper (i.e., that a skeptical hypothesis—*viz.*, that the world he perceives is not what 'real'—is true). Of course, one could take this a stage further and ask if the perception of a paper world is also subject to a skeptical trick but there the same test will apply. Whilst Tom is in the paper environment he has the capacity to discriminate and can come to know. Information, in these local skeptical scenarios, is relevant to what Tom knows. In Zebra, it appears, we have perceived signals that carry the information that the animal may be a painted mule. What is relevant information is constrained by skeptical or non-skeptical environments. Just as the victim of CJD does not need to know about the controversies over the aetiology of CJD (because he is a sufferer not a specialist doctor), he does not need to know the denials of skeptical hypotheses which may cast doubt upon what knowledge he possesses about CJD (because he is an epistemic agent and not an epistemologist).

The upshot of this is that information not only has the function of providing a basis for knowledge but also an alternatives or context-defining function. This gives pluralist epistemologies such as relevant alternatives theory and contextualism practical application as epistemic sorting-machines for information managers: in what contexts can we know what we want to know, what information is relevant, what information changes the contexts for knowledge, what are the epistemic limits of information? To return to our original scenario, the knowledgeable specialist is one who can inform me of relevant information about CJD and also point me in the direction of reliable information sources elsewhere (and steer me away from dodgy websites and quack medical treatments). In most cases, these sources will not be denials of skeptical hypotheses but they will be sources of information which will increase the likelihood of my acquiring knowledge about my condition and how to cope with it. It would be an odd special case if local skepticism were the only epistemological problem that can be affected by informational signals in a context. Information services such as libraries, databases and internet search engines can also make use of relevant alternatives in order to organize and structure their resources and content.

Here are two apparent truisms. First, that our interest as inquirers in information is often motivated by our desire to gain knowledge about something.[7] Second, that we are almost always faced with limited information about the target issue. At the very least, one can always think that it would be better if one had more information about this subject matter. What falls out of these two statements? One might think that, as Aristotle claimed of knowledge (*De Anima*, 402a1), more information is always better than less and so we should endeavor to collect as much information as possible on the matter in question with the hope of, at some point, turning it into knowledge. Cursory reflection reveals that this is evidently false (Himma 2007). Internet search engines are a good example. Type in a random search string and it will probably return hundreds of thousands of results. No human could sort through that amount of information and so the search engine is designed to return those results that are likely to be most beneficial to the user first. A great problem of the Information Age is our inability to keep the technology for sorting and filtering relevant information apace with the rapidly developing technology for collecting information. This is a familiar problem for anyone tasked with making use of any of the many web search engines out there. Access is almost always there, but relevancy is sporadic and limited. Thus, in order to deal with problems as they arise one needs to put constraints on what evidence and information is relevant. According to Neta, the skeptic unduly restricts evidence in certain contexts. What information management effectively does is make the same judgments about appropriate restrictions.

Dretske's account is primarily an account of perceptual knowledge and information. He therefore feels entitled to conclude that, since the mere appearance of an object cannot communicate its non-skeptical status, any signal which carries information about appearance cannot answer a skeptical doubt. However, we have provided examples (such as Bouwsma's adventures and Zebra) where perceptual information does justify a skeptical hypothesis or a non-skeptical proposition. It would seem that Dretske is wrong to think that information is irrelevant to combat local skeptical scenarios. Agents can receive information (even if we think of information as non-factive) for dismissing such scenarios (once we do not limit their information to the bare visual scene) (Pritchard 2010). Whether Dretske is right about radical skeptical scenarios depends on whether information is ever factive. If it is always factive then Dretske has no need to deny closure. Even if information is only sometimes factive (i.e., in ordinary contexts, *à la* Neta) then Dretske is still wrong.

---

[7]For an extended discussion of the goal of information collection and dissemination see Fallis (2002). Note that even those who deny that the goal of information services is for users to acquire knowledge grant that in a large range of contexts our goal in collection and disseminating information *is* to acquire knowledge. For example, the information management scholar Chun Wei Choo expresses, albeit in different terms, a widely held view when he states that the primary goal of information management is to 'harness the information resources and information capabilities of the organization in order to enable the organization to learn and adapt to its changing environment' (Choo 2002, xv). Later, Choo writes that the 'transfiguration of information into knowledge is the goal of information management' (Choo 2002, xiv).

Let us consider an argument that reasons (under which heading we may include perceptual evidence) are factive, which is from John McDowell (1995). Earlier in the Chapter, we discussed Neta's comment that external world skepticism is not meant to cast doubt upon certain 'inner' reasons such as 'that I am not having a visual experience of a white expanse before me'. McDowell argues against a tacit assumption throughout epistemology that these inner reflections can encompass factive empirical reasons (Pritchard 2008, 10).

However, McDowell does not think that *no* empirical reasons are factive. In the case of veridical perception, we have a kind of perceptual evidence which is not present in cases of non-veridical perception such as illusion or hallucination. McDowellian epistemological disjunctivism presents an option for Dretske which has so far been left unexplored but which may undermine his case against epistemic closure, with concomitant implications for his theory of information. In brief, if perceptual evidence is (sometimes) factive, then Dretske is wrong to say that there is no perceptual evidence which can serve as evidence against skeptical hypotheses. Dretske's view is that all perceptual evidence is defeasible when it comes to radical skeptical hypotheses. No matter how competently one receives and judges the information one is presented with, these processes never amount to something which entails the denial of the target skeptical hypothesis. The view is intuitive and persuasive but the McDowellian view offers one alternative: that there is a disjunct between cases of factive and non-factive reasons. That is, there is some reason or warrant or a kind of support missing in cases of radical skepticism that is present in so-called 'ordinary' cases.

Dretske takes it for granted that any given knowledge claim can be subject to a skeptical rebuttal. Such rebuttals challenge the upgrading of an information-based belief (that something appears to be the case) to information-based knowledge (knowledge that something is the case). In the case of Zebra* there is information that carries the signal to Fred that what is in the pen is a painted mule. Dretske might insist that this does not undermine his thesis as these pieces of information may themselves be subject to skeptical hypotheses and are providing only non-factive evidence. However, if one follows McDowell down his disjunctivist path then it is not inevitable that Dretske takes such a position and consequently not inevitable that he is lead to reject the principle of epistemic closure.

Neta presents a contextualist account of evidence or reasons in which the evidential requirements for knowledge are affected by context. Dretske closely links information to non-factive evidence but under the contextualist account there are cases of factive evidence which would provide information-based knowledge of the denials of skeptical hypotheses in some cases. Additionally, McDowell provides a non-contextualist account of evidence or reasons in which there is an epistemic component present in some cases, not present in others (such as cases of hallucination or illusion—the hallmark of skeptical hypothesizing), and in which factive evidence warrants the denial of skeptical hypotheses (Gomes 2011). As a consequence, these distinctions between skeptical and ordinary contexts or between factive and non-factive evidence present alternatives to Dretske's inference that perceptual information can never give us evidence or reasons to refute skeptical hypotheses.

At the beginning of this paper we described a scenario in which a patient may seek information as a means to gaining knowledge about a medical matter. If information such as this were always susceptible to skeptical challenges then this susceptibility would be uncomfortably passed on to the knowledge claims based upon the evidence it carries. Such worries caused Dretske to abandon a key principle explaining how we reliably expand our knowledge: epistemic closure. We have presented an alternative epistemological picture here which does not have such drastic consequences.

References

Bouwsma, O.K. 1965. Descartes' evil genius. In *Meta-meditations: Studies in Descartes*, ed. A. Sesonske and N.Fleming Belmont. Belmont: Wadsworth.

Choo, C.W. 2002. *Information management for the intelligent organization: The art of scanning the environment*, 3rd ed. Medford: Information Today.

DeRose, K. 1995. Solving the skeptical problem. *Philosophical Review* 104: 1–52.

Dretske, F. 1970. Epistemic operators. *Journal of Philosophy* 67: 1007–1023.

Dretske, F. 1971. Conclusive reasons. *Australasian Journal of Philosophy* 49: 1–22.

Dretske, F. 1981. *Knowledge and the flow of information*. Cambridge, MA: MIT Press.

Dretske, F. 1983. The epistemology of belief. *Synthese* 55(1): 3–19.

Dretske, F. 2000. The pragmatic dimension of knowledge. In *Perception, knowledge and belief: Selected essays*, ed. F. Dretske. Cambridge: Cambridge University Press.

Dretske, F. 2005a. The case against closure. In *Contemporary debates in epistemology*, ed. E. Sosa and M. Steup, 13–26. Oxford: Blackwell.

Dretske, F. 2005b. Is knowledge closed under known entailment? In *Contemporary debates in epistemology*, ed. E. Sosa and M. Steup, 13–26. Oxford: Blackwell.

Dretske, F. 2005c. Reply to Hawthorne. In *Contemporary debates in epistemology*, ed. E. Sosa and M. Steup, 43–46. Oxford: Blackwell.

Dretske, F. 2006. Information and closure. *Erkenntnis* 64: 409–413.

Fallis, D. 2002. Introduction. *Social Epistemology and Information Science*, special issue of *Social Epistemology* 16(1): 1–4.

Fallis, D. 2004. Epistemic value theory and information ethics. *Mind and Machines* 14(1): 101–117.

Fallis, D., and D. Whitcomb. 2009. Epistemic values and information management. *The Information Society* 25(3): 175–189.

Floridi, L. 2005. Is semantic information meaningful data? *Philosophy and Phenomenological Research* 70(2): 351–370.

Floridi, L. 2010. *The philosophy of information*. Oxford: Oxford University Press.

Goldman, A. 1976. Discrimination and perceptual knowledge. *The Journal of Philosophy* 73: 771–791.

Goldman, A. 1999. *Knowledge in a social world.* Oxford: Oxford University Press.

Gomes, A. 2011. McDowell's disjunctivism and other minds. *Inquiry* 54(3): 277–292.

Harms, W.F. 1998. The use of information theory in epistemology. *Philosophy of Science* 65(3): 472–501.

Hawthorne, J. 2005. The case for closure. In *Contemporary debates in epistemology*, ed. E. Sosa and M. Steup, 26–43. Oxford: Blackwell.

Himma, K.E. 2007. The concept of information overload: A preliminary step in understanding the nature of a harmful information-related condition. *Ethics and Information Technology* 9: 259–272.

Jäger, C. 2004. Skepticism, information, and closure: Dretske's theory of knowledge. *Erkenntnis* 61(2–3): 187–201.

Lewis, D. 1996. Elusive knowledge. *Australasian Journal of Philosophy* 74: 549–567.

McDowell, J. 1995. Knowledge and the internal. *Philosophy and Phenomenological Research* 55: 877–893.

Neta, R. 2002. S knows that P. *Noûs* 36: 663–681.

Neta, R. 2003. Contextualism and the problem of the external world. *Philosophy and Phenomenological Research* 66: 1–31.

Nozick, R. 1981. *Philosophical explanations*. Cambridge, MA: Harvard University Press.

Pritchard, D.H. 2008. McDowellian Neo-Mooreanism. In *Disjunctivism: Perception, action, knowledge*, ed. A. Haddock and F. Macpherson, 283–310. Oxford: Oxford University Press.

Pritchard, D.H. 2009. Wright *Contra* McDowell on perceptual knowledge and scepticism. *Synthese* 171: 467–479.

Pritchard, D.H. 2010. Relevant alternatives, perceptual knowledge and discrimination. *Noûs* 44: 245–268.

Shackel, N. 2006. Shutting Dretske's door. *Erkenntnis* 64: 393–401.

Shope, R.K. 2002. Conditions and analyses of knowing. In *The oxford handbook of epistemology*, ed. P.K. Moser, 25–70. Oxford: Oxford University Press.

Chapter 11
Levels of Abstraction; Levels of Reality

Joseph E. Brenner

11.1 Introduction

11.1.1 Philosophies of Technology and Information

As stated by Luciano Floridi in the Introduction to his *Philosophy of Information* (2010), Information and Communication Technologies (ICTs) have achieved the status of the characteristic technology of our time. The computer and its related devices constitute a "culturally defining technology", and Information and Communications Systems (ICSs) and ICT applications are among the most strategic factors governing science, the life of society and its future directions of development. The concept of levels enters inevitably into the philosophy of information, involving their nature, content, and the relations between them, starting from the 'lowest' levels of information constituted by physical electronic data themselves.

In parallel with what Floridi has called the Informational Fourth Revolution, in the context of his Philosophy of Information, the *Philosophy of Technology* has emerged as a separate field of study, as summarized by Franssen et al. (2010). Despite differences in perspective and detail, many issues, including those of levels of analysis and the ethical impact of all new technologies are common to both. I have discussed

J.E. Brenner, Ph.D. (✉)
International Center for Transdisciplinary Research, Paris, France

Chemin du Collège1, P.O. Box 235. CH-1865 Les Diablerets, Switzerland
e-mail: joe.brenner@bluewin.ch

fiAmerican Association for the Advancement of Science; New York Academy of Sciences; Swiss Society for Logic and the Philosophy of Science; International Center for Transdisciplinary Research, Paris. Associate Director, International Center for the Philosophy of Information, Xi'An, Jiaotong University of Social Sciences, China.

H. Demir (ed.), *Luciano Floridi's Philosophy of Technology: Critical Reflections*, Philosophy of Engineering and Technology 8, DOI 10.1007/978-94-007-4292-5_11,

Floridi's conception of an Information Ethics (Floridi 2008b) in a previous paper (Brenner 2010a). However, since this Volume will contain other articles dealing specifically with the Philosophy of Technology, I will focus here on issues in the Philosophy of Information.

11.1.2 Rationale and Objective: Information and Levels

Among the key issues in what van Benthem and van Rooy (2003) called the "lively present stage of investigations of information" is the integration of its qualitative, content-oriented and quantitative aspects. Theories of information as a process or an operator, changing the states of receivers and embodying meaning on the one hand, and on approaches that concentrate on how much information is communicated by a message coexist, somewhat incoherently.

Hofkirchner (2009) among others has argued for the desirability of a unified theory of information (UTI) that would encompass the different manifestations of information processes. Such a UTI should be capable of balancing the apparently contradictory properties of information – physical and non-physical, universal and particular – without reduction. Its underlying principle should be "as abstract as necessary but as concrete as possible at the same time."

As an integral part of his Philosophy of Information (PI), in fact as the core strategy for analysis of informational issues and solving information-related problems, Luciano Floridi has made a critical construction of epistemological Levels of Abstraction (LoAs), a notion from computer science. In applying LoAs in various fields, Floridi correctly critiques other uses of 'levels' in philosophy (levelism), especially, the lack of a satisfactory concept of ontological levels.

This chapter approaches the problem of levels in the philosophy of information from a novel perspective, namely, that of an extension of logic to complex real processes, including those of information production and transfer. The proposed non-propositional, non-truth-functional logic – Logic in Reality; LIR (Brenner 2008) – is grounded in the fundamental dualism (dynamic opposition) inherent in energy and accordingly present in all real phenomena. The picture of the world that is used is one of different, physical levels of reality, to all of which LIR applies. As Capurro (Capurro 1996) notes, technology is "non-neutral", and hence LIR is appropriate to it, rather than standard logics that are virtually required to be topic-neutral and context-independent.

11.1.3 The Method of Levels of Abstraction

In his Philosophy of Information (PI), Floridi defines a Level of Abstraction, LoA, as a finite but non-empty set of observables, where an *observable* is an *interpreted typed variable*, that is, a typed variable together with a statement of what feature of the system under consideration it stands for. For completeness, a *typed variable* is a uniquely-named conceptual entity (the *variable*) and a set, called its *type*, consisting of all the values that the entity may take. An *observable* is an interpreted typed variable, that is, a typed variable together with a statement of what feature of the system under consideration it represents. The additional key notion is that of *behavior* of a system that defines the relationships holding between observables. Behavior at a given LoA is a predicate whose free variables are those observables.

Being an abstraction, an observable does not necessarily result from quantitative measurement or empirical perception. The feature of the system under consideration may be a physical magnitude, but or an artifact of a conceptual model, constructed for the purpose of analysis. Roughly, a LoA can account for the behavior of a discrete system, describing the latter in a formalism that corresponds functionally to that of differential calculus in analog systems. The output of a LoA is a model of the system, comprising information, whose amount is lower at higher levels.

For Floridi, the finality of introducing Levels of Abstraction and their combination into Gradients of Abstraction (see Sect. 11.4) as a *method* is to bring additional rigor into theories of information and the systems, models and structures that can be constructed from experiential data. Floridi limits his discussion, however, to LoAs as epistemological and avoids the question of whether the method of abstraction used may be exported to, especially, ontological contexts. Rather, he defends a version of epistemological levelism that is compatible with criticisms of other forms of levelism. This position leaves open the option, however, that Floridi's constructionist view of information might be *supported* by an interpretation of ontological levels that does not suffer from the weaknesses of the levelism he correctly critiques (cf. Sect. 11.3). I will show that in fact application of Floridi's Levels of Abstraction (LoAs) to informational issues can be supported by a concept of ontological levels of reality (LoRs) based on LIR, defined in terms of the different but isomorphic laws applicable to them.

The concept of Levels of Abstraction can then be seen as a component of a broader theory of information and information technology in which LoAs coexist and interact with Levels of Reality. Such a joint theory might provide additional explications of the properties of informational entities and of the behavior of the informational component present in all phenomena. In this chapter, I claim that Logic in Reality provides an interpretation of the ontological content and properties of Levels of Reality that accomplishes this objective.

11.1.4 Outline

Since Logic in Reality is both relatively unfamiliar and is the framework in which all the subjects in this chapter will be discussed, I give first a brief outline of it in Sect. 11.2, as a complete but non-standard logic, including its approach to information. In Sect. 11.3, I return to Floridi's critique of ontological levels and discuss the LIR categorial ontology and conception of ontological levels of reality. These are contrasted with some different conceptions of the concept of Levels of

Reality and Levels of Complexity in the categorical interpretation of Poli (2006). With the preceding as background, Sect. 11.4 returns to Floridi's conceptions of Levels of Abstraction and Organization and suggests both the existence and value of key areas of convergence with LIR. Section 11.5 shows the relation between Salthe's hierarchies and Floridi's Gradients of Abstraction (GoAs). The concept of Levels of Logical Openness of Minati et al. (1998) and Licata (2008), which are applied in a systems context, is also discussed briefly and compared with Floridi's LoAs, and the chapter concludes with a discussion of emergence from the LIR and Floridi perspectives.

11.2 Logic in Reality

11.2.1 Fundamental Postulate and Components

Floridi's development (2006) of a *logic* of and for information (Information Logic; The Logic of Being Informed) fills a major gap in the current effort to characterize information, since standard epistemic and doxastic logics fail to capture some of its essential characteristics. In my previous chapter in this Series on the work of Floridi (Brenner 2010a), I referred to several of the remaining open problems in information to which Floridi has called attention (2004) and proposed an even more radical change in logical approach for their solution, as follows.

Logic in Reality (LIR) is a new, non-propositional kind of logic, based on the work of Stéphane Lupasco (1947), that extends the domain of logic to real processes. LIR is grounded in a particle/field view of the universe, and its axioms and rules provide a framework for analyzing and making inferences about complex real world entities and interactive processes at biological, cognitive and social levels of reality or complexity.

The term Logic in Reality (LIR) is intended to imply both (1) that the principle of change according to which reality operates is a *logic* embedded in it, *the* logic in reality; and (2) that what logic really *is* or should be involves this same real physical-metaphysical but also logical principle. The major components of this logic are the following:

- The foundation in the physical and metaphysical dualities of nature
- Its axioms and calculus intended to reflect real change
- The categorial structure of its related ontology
- A two-level framework of relational analysis

Details of LIR are provided in Brenner (2008). Stated rapidly, its most important concepts are that (1) every real complex process is accompanied, logically and functionally, by its opposite or contradiction (Principle of Dynamic Opposition; PDO), but only in the sense that when one element is (predominantly) present or actualized, the other is (predominantly) absent or potentialized, alternately and reciprocally, without either ever going to zero; and (2) the emergence of a new entity at a higher level of reality or complexity can take place at the point of equilibrium or maximum interaction between the two.

LIR should be seen as a logic applying to processes, in a process-ontological view of reality (Seibt 2009), to trends and tendencies, rather than to 'objects' or the steps in a state-transition picture of change. Processes are described formally as transfinite chains of chains of chains, etc. of alternating actualizations and potentializations of implications, considered with the other logical operators, conjunction and disjunction as real processes themselves. The directions of change are either (1) toward stable macrophysical objects and simple situations, the result of processes of processes, etc. going in the direction of a "non-contradictory" identity or diversity: or (2) toward a state of maximum contradiction (T-state for included third term) from which new entities can emerge. LIR is, therefore, a logic of emergence, a new non-propositional, non-truth-functional logic of change. There is an interesting connection to be explored between the LIR conception of potential and Floridi's use of 'virtual' information to by-pass (my term) standard deduction (Floridi 2011, p. 171).

Standard logic underlies, rather, the construction of simplified models which fail to capture the essential dynamics of biological and cognitive processes, such as reasoning (Magnani 2002). LIR does not replace classical binary or multi-valued logics but reduces to them for simple systems and situations. The interactive relationships within or between levels of reality to which LIR applies are characteristic of entities with some form of internal representation, biological or cognitive.

In contrast to standard logics, LIR has no difficulty in accepting inconsistency, interpreting it as a natural consequence of the underlying oppositions in physical reality. Many if not most of the problems in the (endless) debate about the nature of change, as pointed out by Mortensen (2008), seem to require a fundamental inconsistency in the world, which LIR naturalizes. Logic in Reality, then, is an information system that is not "brittle, like a classical logic system" (Floridi 2011, p. 161) in the presence of an inconsistency. Inconsistency is in the former is not only not as destructive as in the latter, but is accepted as an essential part of its ontology.

11.2.2 Information in LIR

Logic in Reality does not pretend to offer or to constitute an independent theory of information that would supersede any or all existing approaches. LIR provides a new interpretation of the concept of qualitative information or information-as-process (Brenner 2010b) as contrasted with quantitative information. Given its contradictorial approach to all complex real phenomena, LIR can be seen as a method that complements Levels of Abstraction as a method, a logical methodology that would encourage the retention and use of partially conflicting notions and theories of information, among others.

Among the key open problems in the philosophy of information, Floridi (2004) includes several concerning the relation between information and the actual world. Thus, information can be viewed from three perspectives: information *as* reality (e.g. as patterns of physical signals, which are neither true nor false), also known as *environmental* information; information *about* reality (semantic information, alethically qualifiable); and information *for* reality (instructions, like genetic information, algorithms, orders, or recipes).

Many extensionalist approaches to the definition of information *as* reality or *about* reality provide different starting points for answering the question of what information *is*, but the broad theory of information proposed by Floridi requires an understanding of the properties and role of information at all levels of reality, in all entities. Whatever contributes to this understanding must accordingly be valuable for philosophy in general, and I propose this chapter as a clarification of the relevant *ontological* properties of information.

The definition of information that is most congenial to LIR was made by Kolmogorov (Mindell and Gerovitch 2003) to the effect that information is any operator which changes the distribution of probabilities in a given set of events. This is quite different from his well-known contribution to algorithmic information theory, but fits the process conceptions of LIR. In LIR, logical elements of real processes resemble (non-Kolmogorovian) probabilities, and the logical operators are also processes, such that a predominantly actualized positive implication, for example, is always accompanied by a predominantly potentialized negative implication. It is possible to analyze both information and meaning (higher level information, cf. Brenner 2010a) as having the potential or being a mechanism to change the informational context.

LIR thus can provide bridging concepts or 'glue' between the concept of semantic information that Floridi defines at the lowest data level and the broader applications that he looks forward to. It is also Floridi's view that higher LoAs subsume aspects of semantic information. LIR places this concept, and thus the "superconcept" (Hofkirchner 2009) of information, in a naturalized physical, metaphysical and logical context. Information is both a means to model the world and part of the world that is modeled (by LoAs), and LIR describes the dialectic relation between them. Floridi finds the concept that semantic information is true if it points to the actual state of the world somewhat equivocal, but I believe it fits the LIR processual logic, in that logical (in the LIR sense) information *is* the actual state of the world.

11.3 The Problem of Levels

11.3.1 Levelism

The idea that reality is divided into levels that are more or less distinct and involve different degrees of complexity has been proposed, in various forms, since antiquity, but it has received more rigorous attention since the advent of quantum mechanics and insight into brain functioning. Floridi uses the term levelism to reflect a tendency toward the end of the last century to make philosophical descriptions in terms of ontological levels of reality and epistemological levels of observation or interpretation.

While accepting the concept of epistemological levels of organization, description and explanation, he criticizes the former as untenable, pointing to recent debates in the literature about multiple realizability and problems with ontological and methodological levelism (Heil 2005).

This position, however, reflects a standard concept of ontology that depends on *standard* category theory; a superannuated cosmological concept of a non-apparent background space-time; and absolute dichotomies between abstract and concrete and particular and universal. The possibility of a theory of levels of reality based on other principles, *e.g.*, those outlined here[1] is excluded *a priori*. It is an ontology that has been, effectively, *abstracted* from reality.

The counterargument I make for the existence and utility of ontological Levels of Reality does not undermine Floridi's theory of epistemological Levels of Abstraction. As we will see, it provides an opportunity for strengthening and extending its purport. My first task, therefore, is to show that a principled ontological theory of Levels of Reality can be formulated that avoids the difficulties pointed to by Floridi.

11.3.2 The Categorial Ontology of LIR

Many theoretical arguments depend on some form of absolute separability of dichotomous terms *via* the importation, explicit or implicit, of abstract principles of propositional binary logic exemplified in standard notions of time, space and causality. LIR discusses philosophical problems in physical, dynamical terms that do not require abstract categorial structures that separate aspects of reality. In the categorial ontology of LIR, the sole material category is Energy, and the most important formal categories are Process, including the sub-category of Emergence, and Dynamic Opposition, including the crucial sub-categories of Separability and Non-Separability.

From the LIR metaphysical standpoint, for real systems or phenomena or processes in which real dualities are instantiated, their terms are *not* separated or separable! Real complex phenomena display a contradictional relation to or interaction between themselves and their opposites or contradictions. Note that the requirements in (1) classical category theory of exclusivity and exhaustivity and in (2) of absolute separation of sets and their elements do not apply: they are bivalent logic in another form.

LIR approaches in a new way the inevitable problems resulting from the classical philosophical dichotomies as well as such concepts as space and time, simultaneity and succession and indeed levels as categories with *totally independent categorial features*. Non-Separability underlies all other metaphysical and phenomenal dualities, such as cause and effect, determinism and indeterminism, subject and object, continuity and discontinuity, and so on. I thus claim that non-separability

[1] In a later paper, Heil modified his identity theory to permit some interaction between his key notions of dispositions and qualities.

at the macroscopic level, like that being explored at the quantum level, provides a principle of organization or structure in macroscopic phenomena that has been neglected in science and philosophy.

The formal ontology that I propose is a theory that provides *non*-mathematical formulations of the properties and relations of certain categories of phenomena at different levels of reality or complexity. It is intended to be systematic in the sense of stating formally at least some aspects of what all entities are, as well as relating all entities of a certain kind to one another. The approach I have taken is that of Hartmann (Werkmeister 1990) who developed the categories of *his* new ontology "step-by-step from an observation of existing realities". The fundamental assertions of an ontology are about being and have the character of universal constitutive principles. In my analysis, the realities are the manifold dualities of physics, biological science and the dialectics of human thought and behavior. I define a constitutive principle[2] here as one that establishes the relation to an object of experience, while at the same time incorporating the even more fundamental LIR Principle of Dynamic Opposition (PDO) that obtains throughout Nature.

The philosophy of LIR can thus be characterized as a non-naïve dualistic realism that postulates a real, interactive, oppositional relation between all the classic dualities when they are instantiated in reality. It is part of the new *ontological* turn in philosophy. The LIR view, critical for any discussion of ethics and the origin of moral responsibility, is that the world is ontologically deterministic and epistemologically indeterministic, in the contradictorial relation suggested above.

11.3.3 The LIR Notion of Levels

Two kinds of levels physically exist in reality: those determined by (a) simple macrophysical differences in a gravitational field (height) and (b) energy differences in quantum entities more complex than the quark or lepton. The notions of levels of anything else, be it reality, complexity, abstraction or information are intellectual constructs, closely related to that of the emergence of the concepts, phenomena or properties that are designated as "inhabiting" that Level of Reality (O'Connor and Wong 2002). If the event, process or property is new, it must also have an origin and/or be different in some fundamental way from that origin, again, and/from other entities designated as being in other levels. Since the original discussions of the British emergentists, much debate has taken place as whether the entities at a new level have anything in common with the old ones, posing the question of determinism.

The key issue in the discussion of Levels of Reality is not their number, but the existence of ontological "intermediate" or "sub-levels" with real, significantly different properties that are nonetheless tied together by intra- and inter-level interactions.

[2] Below and in Brenner (2008), I discuss the regulative aspects of the PDO.

The major problems in the notion of levels are to characterize (a) the relationships that hold between the entities in a given level and between entities at different levels; and (b) the theories proposed to account for such relationships.

I claim that the fundamental LIR ontology of energy enables a new, useful interpretation of levels that cuts through much of the debate. I have formalized these ideas further (Brenner 2008) in my Logic in Reality (LIR) as a Two-Level Framework for Functional Analysis. In LIR, there are two types of tools for dealing with complex interactive phenomena at the object- and meta-levels. For the structure of theories and their inter-relations, in particular reduction, the PDO is used as a metatheoretical methodological principle for looking at the relations between entities in a domain of dualities or dichotomies, between either classes of entities or two individual terms. For the structure of reality as revealed by physical and biological science, PDO can be used as a quasi-natural law within the language of the scientific theory itself.

Critical examples of interacting object level and meta-level entities, to which Non-Separability applies in the LIR process ontology, are syntax and semantics; types and tokens; data of theories and theories; theories and metatheories; and individuals and groups. All are contradictorially related by the LIR axiom of the functional association of any entity with its opposite or contradiction. Another relational structure is that between processes or events and the explanations of those events.

According to LIR, any total separation between theoretic (epistemological) entities and those of science is arbitrary, since the same object-level and meta-level relations are involved in both. LIR refers to the non-separability of some pairs of those entities, and their alternating actuality and potentiality, and states that both horizontal and vertical part-whole relations are instantiated that follow this dialectics. LIR avoids the difficulties resulting from classical mereology that closely mirrors classical binary logic for the same reason as above: it is a restatement of the standard theory of classes or sets as wholes and their elements as totally separated parts of those wholes.

LIR states that the relation of parts to wholes may be dynamic, that is, that parts and wholes can share one another's properties, in the sense that aspects of the whole are potentialized in the parts, and aspects of the parts are potentialized in the whole. The parts that constitute the content of the object level share properties of the meta-level as a whole. At the level of physical individuals and groups, the situation is the same: the group has some of the characteristics of the individuals that comprise it and the latter have or have internalized aspects of the group.

11.3.4 Other Conceptions of Levels of Reality

11.3.4.1 Basarab Nicolescu

The above discussion is based on the notion of a *logic* as instantiating the dynamic opposition in energy and following the law of a logical included middle. It was first proposed by Stéphane Lupasco (1987) and subsequently extended by Basarab

Nicolescu by his concept of logical Levels of Reality (2002). According to Nicolescu, there are six major levels of Reality:

- the microphysical or quantum mechanical, characterized by non-thermodynamic (timeless) included middle states;
- the next levels are all in thermodynamic world;
- the macrophysical, characterized energetically by global entropy and gradual homogenization of its components;
- the biological, characterized by local negentropy and the emergence of new forms (heterogenization);
- the psychological-social, this level would appear to contain sub-levels or nested levels to account for the ontological aspects of mental phenomena;
- the cosmological, in which both interactions and emergent states of extreme complexity may be present (Nicolescu 1998).

To these, Nicolescu added that of "Cyber-Space-Time", describing it as "both natural and artificial, potentially pre-existent but actualized at present, this "entity" at the interface between man and computer, with its source in the quantum world, involving a dimensionality higher than four, as well as a non-linear causality can be considered as a level separate from both the microphysical, the macrophysical and the (ordinary) psychological/conceptual." Cyber-Space-Time can also, according to Nicolescu, make evident new levels of *perception*.

This concept of levels of reality, in *all* of which at least *some* of the same basic principles are instantiated, is based on an isomorphism of the underlying laws of nature. The detailed laws are different at each level, but they all instantiate the Principle of Dynamic Opposition. For a discussion of the operation of laws, I refer the reader to Poli (2010).

Such a division is an idealization, and reality is a coherent whole. Thus, independently of the (lawlike) properties that are proposed as the basis for the location of the cuts between levels, an additional principle is necessary, namely, to explain the transition from one level to the next. This is, in other words, the problem of emergence, and the Axiom of The Included Middle in LIR suggests a mechanism[3] for emergence that 'emerges' naturally from its dynamics.

11.3.4.2 Roberto Poli

The above attempt to answer the question: What is a level of reality? constitutes what Poli terms an 'objectual' approach (2006) and to which he offers his own categorical approach as an alternative (Floridi includes Poli's description of levels of reality and analysis of the complex relations that obtain both between and within levels (Poli 2001) in his compendium of ontological levelism). The following

[3] I use the term 'mechanism' here in an informal descriptive sense without implying that computable models exist for all the transitions between levels. Indeed, my position is that such models for living organisms *cannot* be constructed.

methodological steps summarize Poli's approach, which takes the work of Hartmann (Werkmeister 1990) as its starting point[4]:

1. Distinguish three strata, rather than levels, of reality: the material, the psychological and the social (the latter encompassing all phenomena of history, language, science, morals, in fact, the entire body of human knowledge and ideation).
2. Define the hierarchical relations of dependence between strata.
3. Define the hierarchical relations within strata, organized into levels (or layers). The layers within strata correspond to "levels of organization", different structurings of the same fundamental laws (Nicolescu 2002).

Each stratum has its own principles, laws and ontological categories, and there are clear discontinuities between strata. This approach is also realistic in that it seeks to extract the relevant categories directly from objects. Levels of reality are radically different from levels of organization; the latter do not presuppose a rupture of fundamental concepts. Several levels of organization or hierarchies can belong to one and the same level of reality, that is, sets of different structures governed by the same fundamental laws. On this point Nicolescu, Poli and I are in agreement. (Poli (2006) also suggests an index of complexity based on the relations between levels and sub-levels of reality defined by Hartmann which I will not develop here.)

Poli makes the further important distinction between ontological levels of reality and epistemological levels of interpretation. In his view, only some of the latter can be taken as levels of reality, namely those that are grounded on ontological categories. Levels of reality constrain the 'items' (JEB: real entities) of the universe as to which types of causation and agency are admissible. A level of reality can be taken to be a level of interpretation endowed with an appropriate web of causes or an appropriate type of agency. One might say that this concept offers a relation between Levels of Abstraction and Levels of Reality, but it remains too abstract, and Poli admits that in his approach, "the links connecting together the various levels of reality are still unknown". I have suggested above the LIR view of those 'links'.

In a subsequent dialogue with Nicolescu (Poli 2010), Poli further emphasized the categorical aspects of his approach, stating that the main reason for distinguishing different levels is to identify "the entirely different new categorical series" that may be needed for their respective analysis. In criticizing the grounding of Nicolescu's theory in the logic of energy (that of LIR), Poli stated that the logic appropriate for his view of levels of reality was an *intuitionist* logic, which maintains an unmodified principle of non-contradiction. This logic is adequate for the entities of classical ontologies and their categories, but it does not describe real ontological levels in the LIR sense, that is, involving contradictorial interactions. For example, the tendencies in and between levels toward physical homogeneity or biological heterogeneity are not independent but are related as discussed above.

[4] Hartmann's "fourth law" of categorical relationships states that "each individual category implies all the others in the same stratum, where 'implication' does not mean standard logical implication, but is an ontic relationship basic to that stratum." This is close to the LIR view of implication as a real process.

In summary, Logic in Reality offers a principled way of using some of the insights of several approaches to levels, without conflating them. Let us now return to the Floridi approach and discuss both his Levels of Abstraction and his Levels of Organization in relation to the conception of ontological Levels of Reality (LoRs) I have outlined.

11.4 Levels and Gradients of Abstraction. Convergence with LIR

11.4.1 Definitions

Floridi proposes the *method of levels of abstraction* (LoAs) "as a more inter-subjective, socially constructible, dynamic and flexible way to further an approach to the knowledge of reality that is still Kantian". In Floridi's terms, this is a step away from *internal realism* (the kinds, categories and structures of the world are only a function of our conceptual schemes), but not yet a step into *external* or *metaphysical realism* (the kinds, categories and structures of the world belong to the world and are not a function of our conceptual schemes, either causally or ontologically). If necessary, it might be called *liminal realism.*

Going beyond the overview in Sect. 11.1.3 above, I note that Floridi further defines the input of a LoA as consisting of the *system* under analysis, comprising a set of *data*; and its output is a *model* of the system, comprising *information*. The quantity of information in a model varies with the LoA: a lower LoA, of greater resolution or finer granularity, produces a model that contains more information than a model produced at a higher, or more abstract, LoA. Thus, a given LoA provides a quantified commitment to the kind and amount of relevant semantic information that can be extracted from the system. The choice of a LoA pre-determines the type and quantity of data that can be considered and hence the information that can be contained in the model. Knowing at which LoA any system is being analyzed means knowing the scope and limits of the *model* being developed.

In the method of Levels of Abstraction, Floridi notes the following as important ways to speak about the levels of analysis of a system:

1. *Levels of explanation* (LoEs) support an *epistemological* approach and do not really pertain to the system or its model, but provide a way to distinguish between different epistemic approaches and goals. In the LIR Two-Level Framework for Analysis, it is not necessary to maintain an absolute dichotomy between *explanandum* and *explanans*, as real processual entities (Brenner 2008), but this issue will not be discussed further here.
2. *Levels of Organization* (LoOs) support an *ontological* approach, according to which the system under analysis is supposed to have a (usually hierarchical) structure in itself, or *de re* – its 'Organization' – which is allegedly captured and

uncovered by its description. It is important to note, however, that ontological is being used here in a standard categorical sense in which the dynamics of the system still do not appear. These structures are discussed below, Sect. 11.4.4.

In Floridi's words, however, the further concept of Gradients of Abstraction (GoAs) is critical, inspired by the concept of simulation used in computer science to relate Levels of Abstraction correctly. A GoA models a system and its observables in terms of LoAs and is a method of insuring the conformity of behavior between them. For a given (empirical or conceptual) system or feature, different LoAs correspond to different representations or views. A *Gradient of Abstractions* (GoA) is a formalism defined to facilitate discussion of discrete systems over a range of LoAs. While a LoA formalizes the scope or granularity of a single model, a GoA provides a way of varying the LoA in order to make observations at differing levels of abstraction.

The functional organization is the net of LoAs (GoA) constructed by epistemic agents It is the relational structure produced by various realizations at various LoAs and by the simulation relation that connects them. However, as Floridi warns us, "GoAs ultimately construct models of systems. They do not describe, portray, or uncover the intrinsic nature of the systems they analyze. We understand systems derivatively, only insofar as we understand their models." No direct translation of GoAs into LIR terms seems possible.

11.4.2 Distinctions and Kinds

In all descriptions of levels, not excluding those in this chapter, it is often difficult to say what property or scalar or vector quantity distinguishes them. Floridi indicates that LoAs can be discrete or analog, more or less abstract or concrete, or can have can have a higher or lower behavioral structure, depending on the complexity of the relations involved, or differ in granularity. One GoA may include a different number of LoAs, one or many. The concept of Gradients[5] of Abstraction implies that different LoAs are of different *kinds* that differ by some parameter or value which may be, but does not have to be their complexity, *e.g.*, disjoint GoAs (whose views are complementary) and nested GoAs (whose views provide successively more information).

As noted above, however, what is essential is not the number of Levels of Abstraction or Levels of Reality or Complexity that it is convenient to designate, but their fundamental characteristics of properties. LIR thus supports Floridi's statement that the assumption that reality must be digital/discrete (grainy) or continuous/analog (smooth) is not justified. "Digital and analog are features of the LoA modeling the system, not of the modeled system in itself." This statement clears the ground for

[5] The concept of 'gradient' itself is suggestive. I feel that we are dealing with an epistemological 'field' that is something like a physical energy gradient. Albeit only metaphorically, the GoA points toward the non-separability I have proposed between the epistemology and the ontology of LIR.

Informational Structural Realism (ISR) that treats the ultimate nature of reality as relational. In the same way, the fundamental principle of LIR leads to a contradictorial concept of reality which is both continuous and discrete, and in which the relations between entities are as important as the entities (relata) themselves.

11.4.3 Levels of Abstraction and Levels of Reality

The fundamental principle of LIR leads to the conclusion that Levels of Abstraction and Levels of Reality are not totally distinguishable or separable. There is always something real about a level of abstraction, as a perspective, a method or a stance, and always something abstract about a level of reality, even if one or the other aspect predominates at a particular time. This can be seen also in the tension in Floridi between the conceptualization of the method of Levels of Abstraction and the experiential use of that method 'in reality'.

In Floridi's critique, metaphysics, when used as a negative label, is what is done by 'sloppy reasoning' without taking into consideration, at least implicitly, the level of abstraction at, and hence the purpose for which a theory is being developed. "Metaphysics is that LoA-free zone where anyone can say anything without fear of ever being proved wrong, as long as the basic law of non-contradiction is respected."

There is no place in the LIR picture of reality for a "basic law of non-contradiction". Rather, there is a principled theoretical relation possible between experience, in which the evolution of contradictorial components are inferred, and the model which is constructed in the process of employing the method of Levels of Abstraction (MLA). Clearly, the model is not the experience, but LIR defines rules for the evolution of that experience. The model can be seen as the result or consequence of the MLA being a selection process not unlike a Husserlian bracketing, in which important elements are (temporarily) set aside, without disappearing totally. Model and reality constitute a dialectically related pair, with one or the other predominating at any time, according to the Principle of Dynamic Opposition (PDO) as a scientific principle. From this perspective, the MLA functions as a Kantian *regulative* principle for LIR, in the sense of Cassirer (Brenner 2008): "A scientific principle fulfills a regulative task of systematizing and conferring order on empirical knowledge, while being an integral part of that knowledge". The resulting metaphysics is no longer a domain where "anything goes".

In the definition of a Gradient of Abstraction, a surjective function guarantees that a relation exists and can be described between the observables at the LoAs the GoA "contains". LIR postulates, on the other hand, that a given complex process entity has a contradictory counterpart. A joint method would start as Floridi proposes, by first stating explicitly the Levels of Abstraction of interest and their grouping into a Gradient of Abstraction, as a rigorous method of limiting the domain of analysis, and then making inferences about the behavior (evolution) of the elements in a contradictorial, interactive process.

The following concepts are examples of where and how Floridi's theory of LoAs can be placed into correspondence with and supported by Logic in Reality:

11.4.3.1 Liminal Realism; Interfaces

Internal realism as defined in Sect. 11.4.1 is a basically anti-realist position. The realism of LIR is external or metaphysical, but it can accept the existence of an intermediate epistemological domain. This intermediate domain, which is that of Levels of Abstraction that Floridi designates as liminal, can be considered to overlap or interact dialectically with the 'external' domain.

Liminal realism is thus related to the informal description of LoAs as interfaces. The description of Levels of Abstraction in Sect. 11.4 above is the formal description, but the informal description is to look at LoAs as being conceptually positioned between data sources and the information spaces of an agent, a 'place' where independent systems meet, act on or communicate with each other. In this domain, the LIR description of the dynamics of the processes involved in the 'movement' across the interface would seem appropriate.

11.4.3.2 Direct and Indirect Knowledge

The second example is in the specification of the meaning of indirect knowledge. Direct knowledge is understood by Floridi as typically knowledge of one's mental states, which is apparently not mediated. Indirect knowledge is usually taken to be knowledge that is obtained inferentially or through some other form of mediated communication with the world, in fact defined in terms of knowledge mediated by a LoA. Note, however, that the fact that data may count as *resources* for (i.e. inputs an agent can use to construct) information, and hence for knowledge, rather than *sources*, leads to constructionist arguments against mimetic theories that interpret knowledge as some sort of picture of the world.

At the heart of the distinction between direct and indirect knowledge is an absolute dichotomy between epistemology and ontology, between knowing and knowledge as interactive real processes and as cognitive 'immaterial' and causally inefficient entities, theories concepts, etc. In the epistemology/ontology of LIR, however, it is neither desirable nor necessary to postulate and maintain a total separation between the disciplines. Knowledge, in my mimetic (?) theory is not a static picture of the world, but a series of cognitive processes interacting dialectically in and between conscious and unconscious domains.

Any cognitive Level of Abstraction will be in a dialectic relationship with the knowledge, or better knowing process, mediated by it. Direct knowledge then is not, in my view, knowledge "of" one's mental states, it *is* one's mental states,[6]

[6] LIR is thus clearly an anti-representationalist theory.

but the operation of the Principle of Dynamic Opposition avoids the problems associated with standard Identity Theories of Mind.

Further points of convergence, adapted from Brenner (2010a), are outlined in the next two sections.

11.4.4 Applying LoAs (1): Informational Structural Realism

Floridi's position is that the ultimate nature of reality is informational. It thus makes sense to select Levels of Abstraction (LoAs) that commit our theories to a view of reality as mind-independent and constituted by structural objects that are neither substantial nor material but *at least* informational. The 'at least' is my suggestion that, without arguing the entire case, would allow for a unified theory in which theories of LoAs and LIR interact, both characterizing the structures of reality seen as dynamic processes.

Floridi's Informational Structural Realism (Floridi 2008a) is a version of Ontic Structural Realism (OSR) (cf. Ladyman and Ross 2007) that supports the ontological commitment to a view of the world as a totality of informational objects dynamically interacting with each other. I refer the reader to the historical development of OSR as a response to the problems of naïve Scientific Realism (SR), the anti-realist empirical critique of SR, and the limitations of simple Structural Realism consequent on its primarily mathematical orientation in Floridi (2011).

ISR provides an ontology applicable to both sub-observable and to observable structural objects by translating them into *informational objects*, defined as cohering clusters of *data*, not in the alphanumeric sense of the word, but in an equally common sense of *differentiae de re*, i.e. mind-independent, concrete points of lack of uniformity. These cohering clusters of data as relational entities are the elementary *relata* required in by Floridi's modified version of OSR. Thus, the structuralism in question here is based on relational entities (understood structurally) that are *particular*, not on patterns that are abstract and universal.

Another area of convergence, then, as noted in Brenner (2010a), is that, as Floridi makes clear, the interpretation of structural objects as informational objects is not meant to replace an ontology of concrete things (better, processes) with one of virtual entities. By conceptualizing concrete *differentiae de re* as data structures and hence as informational objects, he defends a version of structural realism that supports 'at least' an irreducible, fundamental dualism as a more correct description of the ultimate nature of reality.

My claim is that the *epistemological* point to which Floridi has arrived is the *ontological* foundation of Logic in Reality as a dualist metaphysics, grounded, as noted, in the self-dualities of quantum entities and the thermodynamic dualities of our world of experience. LIR is thus compatible with the informational portion of Floridi's approach, and is available, so to speak, to offer insights into the dynamics of processes at higher Levels of Abstraction and Organization.

The method of LoA is an efficient way of making explicit and managing the ontological commitment of a theory. As stated by Floridi, ISR supports the adoption of LoAs that carry a minimal ontological commitment in favor of the structural properties of reality and a reflective, equally minimal, ontological commitment in favor of structural objects. However, unlike other versions of structural realism, ISR supports an informational interpretation of these structural objects. Floridi is concerned, however, that "the adoption of any LoA supporting a degree of ontological commitment less minimal than the epistemic-structuralist one endorsed by ESR (Epistemic Structural Realism) seems metaphysically risky and suspicious. This is the point usually made by supporters of ESR: it is better to limit one's ontological commitment to the existence and knowability of (properties of) relational structures. Any other commitment to the knowability of (properties of) the relata in themselves seems unnecessary, and not backed up by a general conception of knowledge, understood as an indirect relation with the world, even if this were not explicitly interpreted in terms of a LoA methodology".

I quote this passage *in extenso* since it justifies my proposal: LIR offers a (1) a general conception of knowledge as being, in part, a *direct* relation with the world, and (2) knowledge of at least some of the specific dynamic properties of the relata, which is backed up by it, namely the operation of the Principle of Dynamic Opposition. LIR postulates the primacy of relations in a processual framework. Elements and events are not the 'material' terms of a relation, but are themselves always relations. That relata constructed as abstractions from relations doesn't imply that there are no relata, rather the opposite. A core aspect of the claim that relations are logically prior to relata is that the relata of a given relation always turn out to be relational structures themselves on further analysis.

I agree with Floridi that ontological commitments are initially negotiated through the choice and shaping of LoAs, which therefore cannot presuppose a metaphysical omniscience. But LIR offers a way of making an ontological commitment that is clearly greater than minimal. Accordingly, the scope and domain of application of the method of LoAs in the ISR informational interpretation is correspondingly greater as well.

11.4.5 Applying LoAs (2): The Intrinsic Value of Informational Objects

Information Ethics (IE) (Floridi 2008b) also focuses on entities as informational objects, constituted by information at a fundamental Level of Abstraction (LoA). The most important consequence of this strategy that it generalizes the concept of moral agents, as IE is ontologically committed to an informational modeling of *being* as the whole infosphere. The result is that no aspect of reality is extraneous to IE and the whole environment is taken into consideration.

In its "environmental" approach, Information Ethics looks at information from an object-oriented perspective (OOP) and treats it as an entity, moving from an epistemological conception to one which is ontological, albeit from the LIR perspective still 'liminal' (see above). Informational systems as such, rather than just living systems in general, are raised to the role of agents and patients of any action, with environmental processes, changes and interactions are equally described informationally.

Floridi says that he is not "limiting the analysis to (veridical) semantic contents – as any narrower interpretation of IE, as a microethics inevitably does". Floridi goes beyond, here, a definition of information as *solely* meaningful, truthful and well-formed data. This statement justifies, in my opinion, the use of Floridi's conceptions as a basis for a discussion of ethical issues at higher Levels of Abstraction, to which the application of Logic in Reality and its concept of Ethical Information (Brenner 2010a) may be useful (see also Marijuan 2009).

For LIR, the respect due to informational entities is a logical consequence of our general dialectic relationships to "external" objects, and to ourselves as patients as well as agents who have internalized these relationships. As Floridi assures us, the minimalism advocated by IE is only methodological. Its intent is to support the view that entities can be analyzed by focusing on their lowest common denominator, represented by an informational ontology.

Logic in Reality operates at such a higher LoA[7] since it uses a definition of a moral agent from a process standpoint rather that as a transition system that, like Floridi's, sees change as discrete steps from one state to the next. All the entities considered by LIR are *interactive* and *adaptable* in Floridi's sense, but they are not *autonomous*, in any case not completely so. LIR accepts that higher level entities, in particular human beings, share the basic informational aspects of their existence with all entities, through their minimal common ontology. Other Levels of Abstraction can then be adduced to deal with more human-centered values.

11.5 Other Epistemological and Ontological Concepts of Levels

11.5.1 Hierarchies

As summarized by Salthe 2009, hierarchies are examples of partial ordering in standard logical terms: hierarchies order entities or processes into levels. Hierarchy theory is a method of analysis for the observation of natural systems and attempting to discover "Nature's joints". In my terminology, this means much the same as discovering where another set of laws applies in reality.

There is no general ontological commitment in hierarchy theory as to the existence of levels in reality. Like Levels of Abstraction, a hierarchy is a conceptual construction, and use of it does not imply that the world itself is actually

[7] By Floridi's definition, the level should be *more* abstract and involve *less* semantic information, but I would argue that this is offset by the increased functionality of the information (information-as-operator).

hierarchically organized. Hierarchy theory is neither more nor less than another epistemological method of modeling events and interactions in the material world.

Salthe defines a compositional hierarchy and a subsumption hierarchy which differ in the following ways: (1) the level considered the 'focal' level; (2) the kinds of complexity, intensional or extensional, which they embody; and (3) the categorization of the elements in each (part-whole, classes-sub-classes) and their conceptual evolution, that is, how new levels can be seen as appearing in the construction, by interpolation and emergence respectively.

Floridi's view of the nesting of LoAs would seem to place them in the category of synchronic compositional hierarchies. LIR has the properties of a subsumption hierarchy since its informational relations are definitely transitive, and in fact Salthe also uses the term operator to refer to the causal properties of information. Salthe talks about "interpolation of a new level" when one goes from the real world to the conceptual world ("a hierarchy is a conceptual construction"). In the real world, it is new entities that emerge at another level of reality or complexity. Salthe's 'included level' is the epistemological equivalent or projection of a new ontological entity.

According to Floridi, LoAs can be connected together to form broader, structures of abstraction, going from linear hierarchies of abstractions to nets of abstraction. Similar non-linear hierarchies of Gradients of Abstraction are possible, where the relation is more complex than nesting. These constructions, however, remain in the epistemological domain.

I conclude that hierarchical concepts and Levels of Abstraction can both be considered as 'pointers' or 'meta-pointers' to reality that answer the question *where.* 'Where' means: where in the evolving real world the LIR Principle of Dynamic Opposition is in operation. There is no problem in using standard set and category theory (*e.g.*, in defining nesting, classes and sub-classes, etc.) with regard to hierarchies and LoAs because they do not have *their own* dynamics; they 'point to' where the dynamics are.

11.5.2 Levels of Logical Openness

Another concept of epistemological levels has been applied to the modeling of thermodynamically open systems by Minati et al. (1998) and Licata (2008). For such systems, it is assumed that (they) are observer-dependent such that different levels appear as consequences of the observer-system interaction. These authors use the term "levels of logical openness" to describe a hierarchy of kinds of conceptual systems, eventually providing better and better models of complex biological and social processes. Similar pictures of the 'evolution' of models can be found in the approaches of Floridi and Salthe.

Applications of the Minati approach have been focused, however, on the development of simulations and computable models of conceptual collective entities, in which emergent behavior, while real, does not necessarily correspond to what is found empirically. There is an expression here, in another form, of the Gödel principle: if a system is real, it cannot be modeled completely, and if it is modeled completely, it cannot be real. As pointed out by Minati for collections of birds or other creatures (swarms) neither the behavior of the group nor of any individual in it can be predicted in reality.

From the LIR standpoint, the difference between Levels of Abstraction and Levels of Logical Openness is in their degree of approach to reality. Although the Minati approach requires the participation of an external observer, exercising his competence to effect classifications and analyses at different levels, this participation falls short of an actual interaction, as in a process of information exchange, in which the observer is physically involved.

11.5.3 Emergence and Process

Floridi's method of abstraction is suited to the stepwise study of complex systems, by their gradual disclosure at increasingly fine or alternative Levels of Abstraction. A key concept in such an approach to complex systems is that of emergent behavior, that is, behavior that arises in the move from one LoA to a finer level. Emergence, unsurprisingly for such an important concept, can take various forms. It derives from the idea that properties at higher levels are not necessarily predictable from properties at lower levels. (I recall that higher levels are more abstract and lower levels more detailed or concrete.) Floridi captures the idea of emergence using a GoA containing two nested LoAs, where the conceptual 'movement' is clear.

I agree with Floridi's position that without the notion of Levels of Abstraction, the various levels at which an emergent system is discussed cannot be formalized. In his language, emergence is a relational concept: a property is emergent not in a model but in a comparison between models. It arises typically because if a *more concrete* LoA embodies a mechanism or rule for determining an observable, it may have been overlooked at a *more abstract* LoA. Frequently, the breakthrough in understanding some complex phenomenon has come by accounting for emergent behavior which has resulted in turn from considering the process by which it occurs, rather than taking a static view of the ingredients involved.

Logic in Reality is *a priori* designed for discussion of the non-static observables, to use Floridi's term, of real processes (with whose importance Floridi agrees). New properties emerge from the contradictorial relations in their underlying dynamics. Indeed, to take Floridi's example, the most fundamental process of emergence (which provides for our existence) is from the timeless non-thermodynamic domain of quantum mechanics, where each action (excluding its observation) is reversible, to the non-quantum world. Thus when observations are made of macroscopic physical systems, despite the components of those systems obeying the laws of quantum mechanics, the epistemic laws of thermodynamics *emerge* at the same time as ontological process (change) becomes possible.

The LIR view of real-world processes as emergent is not an epistemological but an ontological one (cf. Minati 2009). As Hofkirchner stresses in his discussion of information and computation (2005), "only if computation is meant as a self-organizing process involving emergence in a *non*-epistemological sense can it do justice to the generation of information." I propose LIR as a candidate for a "theory able to model emergent processes embedded in reality", in which ontological and epistemological approaches inform and complement each other, with either ontological or epistemological aspects of the processes studied predominating at any time.

11.6 Conclusion and Outlook

Luciano Floridi has developed the concept of epistemological Levels of Abstraction (LoAs) as a method for the analysis of issues in the theory and philosophy of information. In this chapter, I have argued for the compatibility of Floridi's LoAs with the principles of the new extension of logic to real phenomena I have proposed (Logic in Reality, LIR).

LIR provides an ontological theory of Levels of Reality (LoRs) that applies to complex processes including a concept of information-as-process, and I argue that LIR supports and complements Floridi's approach. The Floridi theory and LIR converge in that the lowest levels of data in both are physical, have information content and value which percolate up through higher levels. It is the value that leads to the foundation of Floridi's Information Ethics and to Ethical Information. LIR also enables an increase in the ontological commitments of the LoAs in Floridi's Informational Structural Realism. I have shown how some of the LIR considerations clarify other essentially epistemological conceptions of levels, namely, conceptions of hierarchies and levels of logical openness.

My discussion has focused on ontological Levels of Reality as seen from the LIR perspective. Future research may determine if other ontological theories of levels can also be related to Floridi's higher or more complex Levels of Abstraction. I believe they can, in view of the ubiquity of information in the relevant interactions, at the social level where the existence of interactions is taken for granted, and also, its current status, throughout cyberspace.

References

Brenner, J. 2008. *Logic in reality*. Dordrecht: Springer.

Brenner, J. 2010a. The logic of ethical information. *Knowledge, Technology and Policy, Luciano Floridi's Philosophy of Technology: Critical Reflections* 23: 109–133, ed. H. Demir. Springer.

Brenner, J. 2010b. Information in reality. Paper for presentation at the fourth international conference on the foundations of information science, Beijing, August, 2010.

Capurro, R. 1996. Information technologies and technology of the self. *Journal of Information Ethics* 5(2): 19–28.

Floridi, L. 2004. Open problems in the philosophy of information. *Metaphilosophy* 35(4): 554–582.

Floridi, L. 2006. The logic of being informed. *Logique et Analyse* 49(196): 433–460.

Floridi, L. 2008a. A defence of informational structural realism. *Synthese* 161(2): 219–253.

Floridi, L. 2008b. Information ethics: Its nature and scope. In *Moral philosophy and information technology*, ed. J. van den Hoven and J. Weckert, 40–65. Cambridge: Cambridge University Press.

Floridi, L. 2011. *The philosophy of information*. Oxford: Oxford University Press.

Franssen, M., G.-J. Lokhorst, and I. van de Poel. 2010. The philosophy of technology. In *The Stanford encyclopedia of philosophy*, Spring 2010 edn, ed. Edward N. Zalta. http://plato.stanford.edu/archives/spr2010/entries/technology/

Heil, J. 2005. Dispositions. *Synthese* 144: 343–356.

Hofkirchner, W. 2009. How to achieve a unified theory of Information. *triple-C* 7(2): 357–358. http://www.triple-c.at/index.php/tripleC/article/viewFile/114/138/

Hofkirchner, Wolfgang. 2005. Does computing embrace self-organisation? In *Information & computation*, ed. G. Dodig-Crnkovic and M. Burgin. Singapore: World Scientific Publishing. http://www.idt.mdh.se/ECAP-2005/INFOCOMPBOOK/

Ladyman, J., and D. Ross. 2007. *Every thing must go. Metaphysics naturalized*. Oxford: Oxford University Press.

Licata, I. 2008. *La logica aperta della mente*. Turin: Codice edizioni.

Lupasco, S. 1947. *Logique et contradiction*. Paris: Presses Universitaires de France.

Lupasco, S. 1987. *Le principe d'antagonisme et la logique de l'énergie*. Paris: Editions du Rocher. (Originally published in Paris: Éditions Hermann, 1951).

Magnani, L. 2002. Preface. In *Model based reasoning: Science, technology, values*, ed. L. Magnani and N. Nersessian. Dordrecht: Kluwer.

Marijuan, P. 2009. The advancement of information science: Is a new way of thinking necessary? *triple-C* 7(2): 369–375. http://www.triple-c.at

Minati, G. 2009. General theory of emergence. Beyond systemic generalization. In *Processes of emergence of systems and systemic properties*, 241–256. Singapore: World Scientific.

Minati, G., M.P. Penna, and E. Pessa. 1998. Thermodynamic and logical openness in general systems. *Systems Research and Behavioral Science* 15(3): 131–145.

Mindell, D., and S. Gerovitch. 2003. Cybernetics and information theory in the United States, France and the Soviet Union. In *Science and ideology: A comparative history*, ed. M. Walker, 66–95. London: Routledge.

Mortensen, C. 2008. Change. In *The Stanford encyclopedia of philosophy*, Fall 2008 edn, ed. Edward N. Zalta. http://plato.stanford.edu/archives/fall2008/entries/change/

Nicolescu, B. 1998. Relativité et physique quantique. In *Dictionnaire de l'ignorance*, ed. Michel Cazenave, 118. Paris: Albin Michel.

Nicolescu, B. 2002. *Manifesto of transdisciplinarity*. Albany: State University of New York Press.

O'Connor, Timothy, and Hong Yu Wong. 2002. Emergent properties. In *The Stanford encyclopedia of philosophy*, Spring 2009 edn, ed. Edward N. Zalta. http://plato.stanford.edu/archives/spr2009/entries/properties-emergent/

Poli, R. 2001. The basic problems of the theory of levels of reality. *Axiomathes* 12(3–4): 261–283.

Poli, R. 2006. Levels of reality and the psychological stratum. *Revue Internationale de Philosophie* 2006(2): 163–180.

Poli, R. 2010. Two theories of levels of reality. In *Dialogue with Basarab Nicolescu* (in press).

Salthe, S.N. 2009. Summary of the principles of hierarchy theory. Pre-print for publication.

Seibt, J. 2009. Forms of emergent interaction in general process theory. *Synthese* 166: 479–512.

Van Benthem, J., and R. van Rooy. 2003. Connecting the different faces of information. *Journal of Logic, Language and Information* 12(4): 375–379.

Werkmeister, W.H. 1990. *Nicolai Hartmann's new ontology*. Tallahassee: Florida State University Press.

Chapter 12
The Floridian Notion of the Information Object

Steve T. McKinlay

The world is the totality of facts, not of things.

(Wittgenstein, 1961, [1.1])

12.1 Introduction

Ontological questions are questions about the nature, existence or reality of objects. And whilst there is a deceptive air of simplicity about the most basic ontological question,[1] "What is there?" the equally simple and somewhat obvious answer, "Everything" leaves us somewhat unsatisfied. Obvious controversies arise when a scientist or philosopher argues that there is something or other which she purports exists, to which I or another scientist or philosopher would not agree. Thus with regard to questions of ontology Quine reminds us "there remains room for disagreement over cases" (1953a, p. 1).

It's perhaps no coincidence that the Object Oriented (OO) programming community has adopted a similar maxim to their own end. To the question, "What is an object?" the OO analyst would also answer, "Everything". Yet just what exactly an "Object" is, is still by and large up for grabs, not only to ontologists, and information theorists, but perhaps surprisingly to OO programmers themselves who, one would have thought, had a mortgage on such terminology. Thus like information,

[1] This question was famously coined by Quine in his 1953a article "On What There Is".

S.T. McKinlay (✉)
School of Information Technology, Wellington Institute of Technology,
Buick Street, Petone, New Zealand

Faculty of Arts, Charles Sturt University, Wagga Wagga, NSW, Australia
e-mail: steve.mckinlay@weltec.ac.nz

H. Demir (ed.), *Luciano Floridi's Philosophy of Technology: Critical Reflections*,
Philosophy of Engineering and Technology 8, DOI 10.1007/978-94-007-4292-5_12,

meaning, truth or belief our concept of "object" is multifarious, capturing both abstract ideas as well as concrete ones. Correspondence between what computer programmers often call "real world objects" and the more or less analogous abstract predicates they like to call "object classes" has always been a tenuous one. The ultimate aim of the OO designer is to capture as much semantic information as is possible about the *real world* objects she hopes to model. However, there is always the recognition that any OO model, whether at conceptual design level or logically implemented as a working OO application, is merely a representation of some (usually tightly defined) subset of external reality.

Further, whilst it is probably true that most OO designers are out and out *realists* about the *real world* objects they seek to model, there remains considerable debate both within the OO fraternity and in philosophy in general about just what an object is. The fact is all of our computational modelling efforts, whether they are OO by nature or that they might employ some other methodology *du jour* really only represent a flimsy link with reality tainted thoroughly with all the human-made ideas, theory and language associated with that particular conceptual scheme. This is similar to the problem which, at least from the philosophical perspective, led Quine to maintain that, "the very notion of *object*, or of one and many, is indeed as parochially human as the parts of speech; to ask what reality is *really* like, however, apart from human categories, is self-stultifying" (1992, p. 9).

Drawing heavily upon OO theory and terminology Luciano Floridi utilises these human-made constructs and language in order to clarify his *information object* notion. The notion is not only a critical element of his Information Ethics (IE) but plays a central role in the service of his wider theory of Informational Realism. The Floridian information object however seems be an unusual ontological case.[2] This is not just because an information object is a non-spatiotemporal type of thing; we might find and (perhaps) agree upon the existence of all manner of abstract objects, including information objects. And whilst there is plenty of debate surrounding the ontological status of abstract objects, in general that is not the primary focus here, instead our attention will converge upon the nature of Floridi's information objects and his use of OO theory and ideas to clarify this concept.

The information object is unusual because the information contained by any information object seems to be *about* something else – that is, it seems to be *about* the actual object or state of affairs that is described or referenced by the information object. Of course this is also the case for OO objects in that any object class, following Quine (1992, p. 6), is just an abstract entity to which each real world object bears a cryptic epsilon relation, in other words, my object class for "cat" for example may well be quite different from yours even though we generally agree upon every cat. This is because cats are almost without exception learnt about via ostension whereas structural abstract objects such as object classes are known only with regard to their role in cognitive discourse and never by ostension (*ibid.*).

[2] We are obliged to point out that Floridi does limit the scope of his adoption of OO concepts and theory by saying "OOP is not a viable way of doing philosophical ontology, but a valuable methodology to clarify the nature of our ontological components" (2004a, p. 5).

Thus there are significant differences in the way Floridi talks about and wants to utilise information objects, and the way in which an OO designer or application uses OO objects and in the way their progenitors use object classes. Accordingly the *information objects'* unusualness is amplified by the way Floridi seems to want to treat information objects, that is, as independent and external[3] objects of themselves, almost as if they were something more than abstract and worthy of genuine ontological status. It may be that this talk is merely a convenience or some kind of metaphor about information objects; if indeed this turns out to be the case then our job will be to clarify such talk.

Consequently this paper is about Floridi's conception of the *information object* and whether we can rightly confer ontological status upon such objects. During this investigation I will consider the validity of using OO theory/terminology as a means of "clarifying" the information object concept. As part of this investigation I will argue that there appears to be a fundamental distinction between OO objects and from a wider perspective the concept of the information object as discussed by Floridi. It may be noted before long that I am something of a gentle nominalist with regard to conceptual objects such as OO objects, their corresponding classes, Floridian informational objects and the like. As such I will continue to draw upon Quine's ideas (as well as others) to support my arguments and will explain my nominalist position shortly. I do see value, particularly with regard to Floridi's Information Ethics, in the notion of the information object, thus we shall see if we can salvage the idea in the face of this critical analysis.

On the one hand, Floridi acknowledges questions surrounding the nature of information as legitimate threads of enquiry (2004a, 2008b). If information is not an independent ontological category then to which category could it be reducible? On the other hand, if it (information) indeed does constitute a valid ontological category then another problem emerges, just how does it relate to the objects to which it usually refers? Such questions lead to enquiry *vis-à-vis* the nature of information *per se*, its relationship with meaning and its status as a natural human independent phenomenon or entity.

Although Floridi relies upon the terminology and the conceptual framework that is representative of the OO programming and design paradigm, his literal application of the concept of the information object differs considerably from the service the OO object[4] is put to within OO computing. That the concept warrants any ontological status seems, on the face of it, at odds with the OO conception of the *object*

---

[3] By "independent and external" I mean something whose existence is independent to human thinking or perceiving, and therefore would exist whether or not (for example) humans existed, in other words "observer-independent".

[4] When I use the phrase "OO Object" I am talking about the structure and function of objects in the service of some OO application, design or model. I want to distinguish this from the phrase "Information Object" which embodies the meaning explicit in Floridi's IE and Informational Realism and while Floridi uses OO terminology and method to explain his conception of the information object I want to show how, even if we do accept the "information object" concept, they cannot really be like OO objects.

which considers objects to be *referents*. To explain, just like Floridi's Information Objects, the object concept in OO programming is inextricably linked to the concept of abstraction yet there are some significant differences. Firstly the sense in which an OO object is more or less representative of some *real world object*. By the term "real world object" I mean to refer to any material or non-material (abstract) object, concept, state of affairs or idea that may be represented (or modelled) using the usual constructs offered by OO-theoretic means. Indeed this tenuous relationship, between OO objects and their real world counterparts, constitutes a necessary condition of all OO objects – such objects are always modeller (observer) dependant.

Secondly, the OO object is a logical and abstract structure that acts as a *referent* to data elements physically stored on disk. The OO object does not and cannot refer directly to any real world object insofar as it is a computationally implemented model-theoretic interpretation represented by a particular class of structures – those structures being an OO domain class.[5] Just as a landscape painting is a more-or-less accurate representation of the landscape it seeks to represent, so too is an OO domain class a representation of some business, engineering or scientific problem.

Finally, the work that such an object performs is delimited by its duty and scope within the OO application within which it resides. Such objects cannot exist meaningfully in isolation, either outside their particular domain class or without some supporting technology framework. Nor can such models come into existence without prior careful analysis of the laws and relationships governing the behaviour and the nature of the entities, and attributes themselves of the extant system that the object model seeks to represent.

Thus there is no doubt whatsoever that all object structures within the OO paradigm, modelled or instantiated within an OO application, are through and through human-made entities or artifacts – they could never be mind-independent or external by nature. And so to suggest that reality could be something like OO class structures or objects seems to me to be like saying that the extant reality represented by a landscape painting could be something like the painting. Of course whilst appearances are identifiable with the model (or the painting) – and this suggests some level of empirical adequacy of the model – our structural representations applied to either concrete or abstract objects within our vicinity are "…rooted in innate predisposition and cultural tradition. The very notion of object at all, concrete or abstract, is a human contribution, a feature of our inherited apparatus for organizing the amorphous welter of neural input" (Quine 1992, p. 6).

[5] Development of an *object oriented domain class* begins with the systematic identification and modelling (and diagramming) of all the entities or objects, attributes, operations and relationships that an OO designer perceives to be important about a particular problem domain, be it business-oriented problems such as invoicing or accounts receivable or scientific problems such as the modelling of biological or genetic systems and so on. Individuals within the domain class are then generalised and represented as object classes which characterise the structure and behaviour common to all objects in that class.

At this point we flag an important distinction. My position so far is not directly in conflict with Floridi's claim that information *qua* information is prime, prior or observer independent, however we ought to be wary of accusations of theory-ladeness[6] associated with directly comparing seemingly naturally occurring phenomena, in this case information *per se*, presuming such a thing does exist independently to human perception, with information *qua* OO objects. We shall have more to say on this specific distinction later.

To summarise, where the OO paradigm considers the object concept to be structural, abstract, referential and artifactual by its very nature, Floridi's informational realism attempts to apply this idea to the age-old philosophical dilemma, "What is the nature of reality?", to which he answers the "Totality of informational objects dynamically interacting with each other" (2004b, p. 1). This informational ontology is then explicated with extensive and explicit reference to Object Oriented Programming (OOP) methodology, "OOP provides an excellent example of a flexible and powerful methodology with which to clarify and make precise the concept of the informational object" (*ibid.*, p. 5).

The point is this, OO objects are explicit about their referential relationships, OO objects are indeed *referents*, but informational objects, if they are to be considered as basic ontological units surely could not refer to any other object. There seems to be a fundamental contradiction going on here, and it is to this that this paper directly speaks.

12.2 Object Oriented Objects

Let us begin with a brief examination of the OO programming and design conception of the object. Later we will look at how Floridi's notion squares with this construal.

Any review of the OO literature reveals a key notion; *abstraction*. The *abstraction* concept is the underlying idea from which all other OO terminology emerges. Common concepts (terminology) used in OO circles include the concepts of *class*, *inheritance*, *encapsulation*, *polymorphism* as well as the *object* concept itself and these are all implemented via an abstraction layer realised by the *method*.

Although the concept of class and object are interwoven there is an important conceptual distinction. Class structures are considered abstract in that they are defined and realised as a generalisation of the concrete, instantiated individual

[6] The idea of theory-ladeness comes from philosophy of science whereby scientific observations are said to be theory-laden when the language and terminology used to describe such observations in question is largely derived from the theory itself. Thus, discussions about the nature of *information* using OO terminology could be accused of being non-theory neutral. Having said that, it is difficult to see how any discussion regarding information could not be influenced by various aspects of culture and language.

objects of that type. Of course instantiated objects are not *really* concrete; instead they represent individual members of the particular abstract class which defines them. Thus the class *reptile* represents the properties common to all reptiles, cold blooded, scaly skin and the like. While we typically identify (real world) individuals in any class via ostension, this particular lizard or that particular crocodile, OO objects are always an abstract expression representing an instance of the defining class. The distinction between abstract classes and concrete instantiated objects is relative. By analogy an object class is a predicate and an object a proposition.

Inheritance is a function built into an object's structure whereby objects in a hierarchy inherit the data elements and behaviours of their parent object class. Thus subtypes in the *reptile* class, such as *crocodile*, inherit all the reptilian attributes and behaviours and then add a few properties specific to crocodiles. *Encapsulation* is an OO specific mechanism whereby an object's components are restricted from being directly accessed. That is, the internal representation of the object is hidden from the outsider's view – just how the attributes and behaviours of the class *reptile* are implemented within the OO application is not available for scrutiny by users of the system.

The *method* is the mechanism by which all object interactions and manipulations are performed. Note here that under all self-respecting OO development environments objects are instantiated via what is often termed a class constructor method, they certainly do not pop into existence spontaneously just because some corresponding real-world object needs to be represented. It is directly due to the abstraction approach that one cannot have direct access to data that might exist inside an object; instead the *method*, sometimes utilising a *message* (often called *parameters* in programming), must be invoked. The message may also contain some identity condition – that is a way of identifying which object you wish to refer to. Furthermore object creation within OO is explicit. Objects are created as per the needs of the application or the application user.

Polymorphism refers to the ability of an object's methods (which might also be implemented as operators), as designed by an OO designer or programmer, to be utilised in more than one way depending upon the context within which it is used. Thus, consider the operator "+", appropriately defined we might use to add up number data types or alternatively if presented with text strings the operator may concatenate or append them to a list depending upon the usage desired by the designer.

Persistent objects usually refer to real world objects, or states of affairs – something that the OO designer wishes the OO application to represent. Objects may persist, in which case they would be correspondingly represented in a database somewhere after the application closes or they may be temporary, existing only while the OO application is running. Temporary objects are often associated with the general operation or running of the application. For example in a Windows application scroll bars, dialog boxes, menus and the like are all instantiated objects which exist and are represented in the memory of the machine (or server) only whilst the application is running.

There is no doubt, much more to say about the OO object concept, however this brief overview should provide a starting point for our discussion and comparison.

It should be clear by now that OO object classes and their instantiated objects are nothing much like their real world counterparts. The OO class *reptile* is an abstract representation, something more akin to "reptileness" than any individual reptile and any instantiated OO *reptile* object is a highly-stylised, conceptual and extremely simplified model of a reptile, nothing like an actual reptile. Furthermore each instantiation of an OO *reptile* object, provided it carries the same attribute values, is logically identical to every other similarly-defined *reptile* object. This of course can never be the case for actual reptiles of the same species.

These concepts and rules are general to all OO systems, nevertheless (perhaps surprisingly) consensus regarding the concept of the *object* within OO design is far from agreed upon. For example, after introducing some key controversies observed in the OO literature, Date and Darwen ask, "So what exactly *is* an object? Is it a value? Is it a variable? Is it both? Is it something else entirely?" Due to the alleged ambiguity they go on to assert, "As a matter of fact, it is largely because of this confusion over what objects really are that we prefer … not to use object terminology at all, except in a few very informal contexts" (2000, p. 10). Instead Date et al., prefer to rely upon a vocabulary that draws upon predicate logic and set theory to explicate their model of data representation.

Whilst the sheer variety and volume of OO programming and design literature available no doubt contribute to the confusion, even across a small sample we see inconsistent points of view emerge. Booch for example (1994, p. 35) coins a simple truism, "What we can agree upon is that the concept of an object is central to anything object-oriented." Martin (1992, p. 241) perhaps in the tradition of Berkeley,[7] prefers, "An 'object' is anything to which a concept applies", and "A concept is an idea or notion we share that applies to certain objects in our awareness". On a first reading James Rumbaugh, one of the founding OO methodologists, appears to get closer to the mark with, "We define an object as a concept, abstraction or thing with crisp boundaries and meaning for the problem at hand" (1991, p. 21). Whilst this approach is certainly useful when it comes to defining an object class model for a well-defined problem domain to be implemented as an OO application or database, it seems problematic as the basis for a universally favoured ontology. Indeed quite the reverse seems to be the case. Concepts and ideas seem to have vague rather than crisp boundaries. This is particularly the case when the applicability of a predicate to its subject is tolerant, for example when does a child cease to be a child and begin being an adult? The fact is most concepts do not have easily-defined boundaries; reality is not crisp. On Rumbaugh's definition most of reality would be thrown in the too-hard basket with regard to object modelling. One might be tempted to argue; all we need is a set of clear *semantic rules* that apply to our artificial OO style language and we could by and large eliminate such vagueness and ambiguity. However this approach seems to point to a required preciseness in meaning which naturally gives way to appeal to definitions and hence does not appear to solve our problem.

---

[7] George Berkeley famously argued in his *Treatise Concerning the Principles of Human Knowledge* that material objects were merely ideas or concepts.

The confusion only deepens if we consider the historical discussion in philosophy regarding objects, classes and attributes. Whilst the Floridian notion of the information object posits attributes – collections of which constitute the nature or essence of an object and classes (following the OO paradigm) as "a named representation for an abstraction, where an abstraction is a named collection of attributes" (Floridi 2002, 2004b, 2008a) – Quine argues that the job of attributes can be adequately handled by classes. "Classes are on a par with attributes on the score of abstractness or universality, they serve the purposes of attributes so far as mathematics and certainly most of science are concerned; and they enjoy, unlike attributes, a crystal-clear identity concept" (1957, p. 19). I accept (and imagine many others to agree) that any identity problem for concrete objects can be generally cleared up by appeal to concepts of reference but there seems to be a lack of an identity concept for attributes. In the case of OO programming however there is a clear identity link between attributes belonging to certain object classes. Attributes in this case can only be exposed via the methods associated with each class using pointer mechanisms and the unique object ID construct; however this tells us nothing about how attributes can be identified by non-OO information objects. Thus talk of attributes in the OO sense seems to be quite different to the philosophical and ontological concept of attribute.

The point is this: in the OO world an object is a *referent*. Method invocation represents a layer of abstraction between some actual representation[8] that is implemented in some arbitrarily complex way on a computer disk (which is of no concern to an OO modeller) and the encapsulated object provides some kind of pointer reference to that data via the declared methods. Moreover object classes and their instantiated objects stand independent to reality and attempt simply to model reality in so far as *something* exists in time and space.

While we have introduced some controversies regarding Floridi's use of OO methodology and we have outlined, albeit briefly, the OO side of the story from a broad definitional perspective, we now need to tease out, a little more explicitly, how Floridi develops his concept of the informational object.

Floridi presents, via analogy, a picture of the information object using a well-known icon, the pawn chess piece (2002, 2008a). The identity of the pawn he argues is known not by its somewhat arbitrary properties as a physical object but rather by its function in the game of chess. In fact we could simply replace the pawn with any placeholder (Floridi suggests a cork) without any semantic loss. Alternatively we needn't have a placeholder at all – a pair of chess players with good enough memories and imagination could visualise an entire game thus suggesting that the real pawn is not really a material thing but a mental entity or an entity constituted by a bundle of properties (2008a, p. 30). Unsurprisingly the pawn makes for a very good analogy in the application of OO theory – its role is simple and its functional boundaries

---

[8] By *actual representation* I mean the physical or internal codification or implementation of the data as it exists on disk.

are conceptually crisp. Indeed there are a great many OO programs that have been written to represent electronic versions of the game of chess. Designing an object class with the appropriate attributes and methods that represent the physical pawn is a relatively trivial exercise from an OO programming perspective.

Such analysis according to Floridi relies upon another computational concept, that of levels of abstraction (LoA). Put simply, we can discuss computational systems at differing levels of abstraction. High conceptual levels often involve abstract diagrammatic models. At lower levels we might imagine some written computer code or logical statements in SQL or the like and at even lower levels, strings of scalar variables and combinations of bits and bytes. Thus according to Floridi, "The choice of LoA pre-determines the type and quantity of data that can be considered and hence the information that can be contained in the model" (2008a, p. 16). The entire notion of the information object thus is couched within the levels of abstraction concept.

Yet this reductionist construal still seems odd to me. I wonder what utility the use of a terminology specific to a particular LoA might have at other levels. For example the level where we might construe objects as OO-like informational objects and then naming related concepts, objects or structures and their relationships or linkages across varied levels of abstraction using OO terminology surely cannot give us any extra insight into the ontological nature of the corresponding real world objects. The OO-like model provides value insofar as it is a *representation* of its real world counterpart but OO-like concepts are usually reserved exclusively for the development of OO applications and are in part a matter of convention. Indeed there are various valid levels of abstraction used within the OO paradigm, these include unified modelling language (UML) constructs such as use-case diagrams, system sequence diagrams and state machine diagrams but each of these levels introduce structures and models that have very clear abstraction relationship rules linking them with domain class diagrams and their consequent computational implementations.[9] However, nothing in Floridi's literature suggests that informational objects exhibit similar relationships or rules across analogous levels of abstraction. This seems to be particularly the case between information objects and the real world objects that define them.

Thus, whilst I agree that a pawn (with all its requisite behaviours and attributes) may be imagined and that a pair of chess players with sufficient memories could somehow visualise an entire chess game, this is not the same thing as a pawn being represented in terms of OO theory, nor is it the same as a physical pawn. Of course there are certainly *some* properties that each pawn representation shares, but there are a great many differences also. Floridi, however, clearly takes a certain selected set of properties of the pawn quite seriously and these seem to be more definitive or significant to him. By way of example when I imagine moving a pawn on a chess

---

[9] I do not intend to rehearse the literature on the development of OO models and their abstraction relationships and rules across differing levels of abstraction, however, any review of UML OO modelling literature will suffice should the reader wish to read further. The UML Wikipedia page perhaps might be a good starting point.

board I visually imagine a three dimensional space, as well as a prototypical three dimensional image of a pawn. I fully expect this mental image to be quite different to a functional Magnetic Resonance Imaging (fMRI) of what is going on in my head at the time and we can be sure the fMRI will be slightly different on each consequent imagining. It is not clear that such different construals could be explained away as being merely different levels of abstraction of the same thing since they don't seem to be the same thing even though it could be argued that there are some tenuously shared properties.

To throw gasoline on the fire, consider the following: a Google image search of "pawn" returns a wide array of images of pawns not only within chess contexts but as an iconic image in its own right and these are all valid representations of pawns which bear little resemblance to the OO-like informationally austere pawn. There is no doubt at all that the pawn *can* be reduced to a set of behaviours and basic properties that can be modelled within an OO programming language but this (to me at least) only represents a minimally sufficient condition for an object to qualify as a pawn, in fact a *virtual pawn* as the current case may be.

The worry deepens when we consider imagined pawns. It is certainly not outside the realms of impossibility that mental representations of pawns between individuals vary greatly. Such representations of a "pawn" are not identical at all with the set of attributes, methods and other OO constructs that defines a pawn in an OO program. The same goes for physical pawns in physical chess sets. I cannot for example throw the information object that represents (or is) a pawn at my opponent when I'm losing thus suggesting that the set of pawns described purely in informational terms is not identical with the set of either physical pawns, nor with any mental representations of pawns whatever our normative notions about how pawns are meant to behave. Himma (2004) takes a similar tack as follows:

> Indeed, it is hard to see how a pawn could be identical with the information object that describes its properties and operations. If we conceive of the pawns as nothing more than information objects then all of the propositions in the set constituting the relevant information object are propositions that describe that set. Such an assumption would, of course render some of the sentences obviously false (information objects lack spatio-temporal location and hence can't be moved around) (p. 148)

This analysis I believe raises several questions. Firstly how do the information objects defined by things such as mental images relate, if at all, to all the other information objects that represent real world objects, which by ostension we'd agree are part of the same class or set other than by loose consensus? They surely relate at some level since they are all supposed to represent the same thing – but this, I contend is largely folk talk. What Floridi seems to be talking about with regard to pawns is a relativity of identity of type. The OO method of defining a pawn *necessarily* relies upon pawns being of the same type and thus sharing some well-defined properties. Floridi relies upon this methodology to clarify his information object and as such is quite serious about these particular properties. He takes it that these properties do exist and that they are constituent properties of pawns. Hence two different tokens, be it a cork or a carved piece of wood, have the same properties – such properties constitute the tokens' identity transcending the material properties, that identity presumably, in this case, being pawnhood. It seems that what Floridi is talking about with regard to the information object (taking "pawn" as the example), is something like the universal concept of pawnhood for which we already seem to have a theory, albeit a controversial one.[10]

However, it is clear that the cork pawn and the wooden pawn (as well as the imagined pawn and the OO pawn), whilst they share *some* properties, these properties are not identical across all pawns. Each set of properties relating to each pawn are particular to that pawn.

Another issue raised is this: although the LoA approach is well proven within OO and computer systems design, this is because there exist very explicit rules about how differing levels of abstraction are linked to one another. Such rules about how real world objects and their informational counterparts are linked via LoA do not seem to be addressed by Floridi. Instead he offers us a conceptual discussion regarding ontological commitment and levels of abstraction (2008a, p. 17).[11] The structural approach taken by Floridi works well for classes, and perhaps the abstract entities that are *information objects*, but we initially learn about pawns not through abstract structural discourse but via ostension. Thus while we might agree upon what qualifies as a pawn, my class of pawns can be quite different to yours.

Two categories of problems come to mind with regard to the Floridian Information Object. The first I will call the *Methodological Problem*. Whilst Floridi draws heavily upon OO programming and design terminology in order to explicate his informational realism (as well as supporting the role of the information object within his IE), the object concept itself, within OO programming or design, (issues of clarity aside) is heavily contextualised and specific. The rules linking different levels of abstraction from high level conceptual models to much lower level compiled object classes and programs and their consequent representation at the disk level are very explicit. To extract the OO object concept from its own theoretical environment leaves us wondering as to the explanatory value such discourse could have outside metaphor and analogy. The application of OO concepts is specific to their domain and use of them outside this domain requires considerable ad hoc addition and modification which mostly ends up in confusion and misunderstanding. This is even the case within computing circles, a clear example of which can be seen in recent attempts to apply OO concepts to the relational model of data.[12]

---

[10] The problem of Universals was originally discussed by Plato and Aristotle and as a topic has captivated philosophy ever since. Universals are generally considered repeatable or recurrent abstract entities that can be instantiated in individual objects, classic examples are considered to be qualities shared by entities, such as two green chairs sharing the quality of "greenness" and "chairness".

[11] Floridi does attempt to address ontological commitment to different LoA by attempting to reconcile epistemic and ontological structural realism. However, in this paper I am concerned with the relationships between information objects, OO concepts and real world objects, as such this is outside the scope of this particular paper.

[12] Date and Darwen (2000, p. 371) call this a "great blunder" arguing that it both dilutes OO concepts and undermines the conceptual integrity of the relational model.

The object concept seems to have such a wide range of applicability that it ends up somewhat ambiguous. We have noted that Date and Darwen (2000) dispense with OO terminology in favour of a vocabulary based on set theory and predicate logic in their discussions on data representation. Certainly on Rumbaugh's description it seems difficult to understand how OO-type objects could represent anything within the ranges of our normal understanding of language.

The second difficulty I am calling the *Identity Problem*. This issue relates to where and how a Floridian information object's data members are represented or manipulated. While Floridi bases his notion of the information object on OO concepts and terminology his goal for the information object is clearly quite different to that of an OO application or data model designer. It could be that a Floridian information object isn't meant to be a referent in the same way an OO object is. The issue seems to a problem related to identity or correspondence relations between the abstract information object and its real world counterparts.

The methodological problem is only a problem when OO concepts are used outside an OO context. OO programming for all intents and purposes, "works" and all the philosophical anxiety in the world over just what an *object* might be or whether it accurately addresses ontological problems doesn't really matter, certainly not to the OO designer or programmer who is simply solving what is usually an information management problem using a particular development/design environment. In other words OO theory, or at least parts of it, is instrumentally reliable with regard to the creation of "working" *Object Oriented* programs and their corresponding object structures. The question of whether they are (or not) ever directly representative or answerable to any external truth about the real world is not at issue. OO programs (or OO databases) are structured collections of relatively simple facts represented by sets of values and governed behaviourally by simple computational procedures and functions. However, the truth of such facts, or more precisely the correctness of the data representations depends entirely upon whether or not the values are *(a)* consistent with the rules (usually the "business rules") upon which the OO application has been designed, and *(b)* correspondent with the external environment, that is, not erroneous. Of course *a* is exclusively the responsibility of the OO designer, whereby *b* is almost certainly contingent.

Whilst there are philosophical disagreements regarding a precise definition of just what constitutes an object, the points above and the distinction between *a* and *b* is generally accepted within the OO community. An OO model is designed at a conceptual design level by an OO designer conceptualising object classes which are somewhat representative of the external (data) environment or as it is sometimes termed, "the problem domain". Moving through the requisite LoA appropriate to OO modelling, the OO model is implemented internally at a logical level as a set of definable structures specific to a particular vendor's database or programmatic development environment. The resulting OO application or database by and large serves some clear delimited business, engineering or scientific function.

Of course Floridi is *not* saying OO objects are the *same* as informational entities, the kind that he supposes reality is comprised of. He is not trying to directly co-opt the OO object concept into the service of ontology. Yet he does borrow and rely upon OO terminology and unless clarified, it seems reasonable to argue that the information object could only suffer the same ambiguous issues that the OO object endures. Thus it is not at all clear that OO terminology can play any pragmatic role in explicating a universal ontology other than providing some kind of loose model of the ontological components of his informational realism.

The Identity Problem is a more complex philosophical problem and concerns the issue of how we can have knowledge of abstract entities and the role their composite attributes play, as well as how and what such components reference. We shall consider this problem as part of the next section which considers the philosophical position with regard to abstract objects and some objections.

12.3 Abstract Objects and Their Problems

We have already stated that philosophy views "abstract objects" as those which are non-spatiotemporal in nature but it also often considers abstract objects to be causally inert, that is they generally have no direct ability to affect the "real world". Whilst there are varied theories about the nature of abstract objects, for the purposes of this essay we shall follow a rather generalist approach. Thus we acknowledge a general claim that all objects fall into two exclusive categories, either those which are concrete or those which are abstract. Whilst there is much debate surrounding the concrete/abstraction distinction following our general approach we will consider, for the most part, concrete objects to spatiotemporally extended, (like tables, chairs and mountains), and abstract objects not (like sets, prime numbers, predicates, fictional characters and informational objects). Further we assume no object can straddle the distinction. We also assume that whilst an information object itself is abstract, it can represent both concrete and abstract objects. Thus an information object is equally capable of representing Wonderland's Alice as it is Aoraki.[13]

The information object, however, seems to exhibit a special kind of abstractness. As discussed in the previous section, Floridi's use of OO theory to "clarify" the notion of the information object leads us to conclude that he must think there can be groups of things of the same *type* (pawns, for example), and these things are all members of the same class (which is something like an OO class), and that this sameness is taken in a strict sense – that is, these things share some important selected identical properties. This must be the case since OO programming requires strict relations of identity with regard to object construction via class methods. Whilst this notion of abstraction in the OO world is taken for granted, in philosophy it does not seem to be quite so simple. I tend to think (as per the illustrated case of pawns above) that notions of identity across groups of resembling objects are not strict at all, and whilst it is difficult to avoid talk of like-properties across individuals

[13] Aoraki is the indigenous Maori name for Mt Cook, New Zealand's highest mountain.

or particular objects such talk of similarity is largely an artifact of our language, culture and some innate human predisposition for organising and describing our world.

In philosophy the rejection of the notion of strict identity across objects is called *nominalism*. This form of nominalism amounts to a rejection of universals (see footnote 11). For Floridi's notion of the information object to work, particularly with regard to the OO concept of object it seems he needs to be a realist about universals, otherwise we cannot form a concept of pawnhood representing the identical properties shared across the class of pawns. Another form of nominalism rejects abstract entities just because the nominalist generally disagrees that non-spatiotemporal, causally inert objects exist at all. Part of the reason for this rejection is as follows. In science causality and explanation are closely linked and whilst there is an entire domain in philosophy devoted to this topic we will simply say that in many cases to explain a fact is to identify its cause.[14] In other words, the intellectual understanding we have about our world and all its systems and organisms and so on is often summarised in our scientific explanations, these explanations are often causal by nature (*cf.* Salmon 1998). The corollary to this is that it is difficult to explain how we could have knowledge or understanding about an object if there are no causal relations between it and us. Such causal relations could simply be the reflection of light from the surface of some such object impacting upon sensory apparatus (in that case our eyes). Quine argues, "Science itself teaches us that there is no clairvoyance; that the only information that can reach our sensory surfaces from external objects must be limited to two dimensional optical projections and various impacts of air waves on the eardrums and some gaseous reactions in the nasal passages and a few kindred odds and ends" (1974, p. 2). All explanations no matter how complex or convoluted ultimately stem back to these basic causal interactions.

Goodman and Quine do espouse a form of nominalism based upon a simple philosophical intuition, "What seems to be the most natural principle for abstracting classes or properties leads to paradoxes. Escape from these paradoxes can apparently be effected only by recourse to alternative rules whose artificiality and arbitrariness arouse suspicion that we are lost in a world of make believe" (1947, p. 105). Compared to the OO object conception whereby objects are derived from declared object classes the Floridian conception falls prey to at least one causally related paradox we alluded to above. That is, from an OO point of view an object *qua* referent points (literally and programmatically) to a physically implemented value (or variable). There is a logical difference between the appearance of that value exposed by the object's methods and the encoded version that is used to represent its appearance (via various layers of abstraction) internally (on a computer disk). But, and this is

[14] Generally the claim in this context refers to non-trivial or non-tautological facts. For example, 1 + 1 = 2 and, "unmarried men are bachelors" qualify as facts but they have no cause. (I thank Morgan Luck for pointing this out to me). We might note here as a by-line that Floridi and many others doubt the informative nature of tautologies or necessary truths.

the rub, while the referent doesn't point to the external thing it attempts to model, in the case of our example above, the OO pawn, it does point literally to some arbitrarily complex piece of data, and that is what the referent is physically referencing. Furthermore there are direct, explicit, and traceable causal links between the various levels of abstraction.

But what does an information object point to? Clearly the intention is that it either points to or emerges from the thing that it references. However in the case of an information object there is no implementable (as there is with an OO object) physical layer of abstraction which provides mappings between logical representations at higher levels of abstraction and physical, or the actual material data representation. Thus in the case of an information object the abstraction mapping process is either a behavioural or a cognitive process, relating some vague sensory input between the extant object and the agent or organism's experience of it. This experience will be different for each organism. Humans experience exposure to the sun for example in an entirely different way to the way algae *experience* it, if we can even call it that.

12.4 Natural Entities, Artifacts and Technology

There is a final related distinction I will make between OO objects and Floridian information objects. In this section I draw upon an argument made by Deborah G. Johnson (2006) regarding the distinction between natural phenomena or natural entities (which according to Floridi can be explained as "dynamically interacting informational entities") and human-made entities or what are often termed artifacts. I first outline Johnson's argument and then apply it to my general discussion.

Johnson's paper develops the thesis that computer systems can be moral entities but not moral agents. Although I don't intend to discuss the main thrust and conclusions of Johnson's argument regarding the moral status of computer systems in this paper, whilst developing her argument she highlights two important distinctions which are salient to the present discussion. The first distinction is between artifacts or what would normally be considered human-made entities and those entities which occur naturally. A second distinction is made between artifacts and technology. Although she concedes that these distinctions are inherently problematic in the sense that there is no sharp boundary, they are nevertheless significant. Any "rejection or re-definition of these distinctions obfuscates and undermines the meaning and significance of claims about morality, technology and computing" (2006, p. 196), Johnson asserts.

The challenges are illustrated as follows: a stick used by a tribesman as a spear to hunt an animal for instance, is a naturally-occurring object which has been used as a tool. Thus the stick, whilst seen as a natural entity, is also utilised as a form of technology. Newer technologies such as genetic modification and nano-molecular technology may also appear to blur the line between nature and technology. Just which parts are naturally occurring parts and which are human-made artifactual parts maybe difficult to assess. The only difference Johnson notes, between biotechnology and other types of technology such as computer systems and the like is the extent to which they manipulate, or the level at which the manipulation of nature occurs (2006, pp. 196–197). Thus while challenges can be made to the distinctions, Johnson argues it doesn't mean the distinctions are incoherent or untenable. On another level, according to Johnson, these distinctions allow us to make sense of the kinds of questions related to what kind of effect human behaviour has on the planet, versus something independent of human behaviour i.e. nature.

Although absolute definitions of *technology* are problematic, referencing Heidegger, Johnson attempts to avoid some of the debate by simply arguing that technology is a contrivance and inherently refers to human-made things. This of course includes computer systems at both logical and physical levels. While Johnson confines the term artifact to physical objects,[15] I think however, we can extend Johnson's definition of artifact. Computer systems are made up of both physical and logical components and the logical components are just as artifactual in nature as the physical or material components. This is clear since there are explicit design and development processes associated with the creation of all logical components of computer systems. The resulting objects have unambiguous designed functions, can be identified and map to specific arbitrarily complex data representations on disk or in the memory of the computer. They are indeed virtual artifacts.

Furthermore Johnson points out that "technology is a combination of artifacts, social practices, social relationships and systems of knowledge… sometimes called socio-technical systems" (2006, p. 197). Thus artifacts (as components of socio-technical systems) cannot exist according to Johnson without systems of knowledge, social practices and human interactions and relationships. Artifacts are created, distributed, utilised and have meaning only within the context of human social activity (*ibid.*). Whilst there are differences between logical and physical artifacts, they are both clearly the result of human behaviour and decision processes.

Object oriented objects are artifacts. They are nothing but the result of explicit design processes by humans. Alternatively informational objects seem to exhibit some of the characteristics evident in both natural entities and human-made artifacts. Information objects, however, may be just a way of talking about and interpreting the things and events around us, in other words, mental entities.

This points toward the contingency of information objects since the existence of such objects relies upon some vague correspondence between the object's internal structures, whether we describe such structures as "attributes" and "methods" and so on, and their relationship with the real world physical entity (table, chair, mountain etc.) they seek to represent. We should point out that Floridi's construal of the information object seems to differ from the role classes play in a similar discourse,

[15] Johnson (2006, p. 197) "A common way of thinking about technology – perhaps the layperson's way – is to think that it is physical or material objects. I will use the *artifact* to refer to the physical object."

where the information object seems to pick out individuals, the role of classes, to use Plato's metaphor, are our attempt to carve the beast of reality along its joints.

The problem becomes more complex when we include informational objects representing abstract entities into the mix, the kind that nominalism would typically reject. For example, there could be any number of imaginable information objects representing anything for any possible interpretation. To paraphrase Hayaki (2006, p. 81) who considers similar problems associated with contingent objects, we are not counting actual possible physical objects; we are counting the ways in which an object might be represented (by an information object) by any possible agent.

The information object suffers from an identity crisis. Following Johnson, in order to identify an information object as an independent entity requires us to separate the object from its context, however, in this process we are extracting it from the context that gives it its meaning and function. This appears to be a problem the analysis of information *qua* information suffers from in general.

Thus whilst I agree it does not make sense to ask the question, "Where are these information objects you talk of Luciano?" Abstract objects such as information objects do not exist in space, I do think it legitimate to ask, "How do everyday (concrete) objects map to their information object counterparts?" This question is answered in the OO case since there are clear and explicit abstraction relations between different levels of the model. This level of detail seems obscure with regard to informational objects. I hope I have made it clear that OO system entities are never isomorphic with any kind of external reality (but informational objects are supposed to be). OO objects are merely a more or less accurate *model* of reality. Indeed OO models are pragmatic by nature – the goal is to solve a business, engineering or scientific problem, that is, a problem that can be adequately solved with an OO application. The Floridian account however seems to suggest an object *qua* information object does indeed reference the real world object it purports to represent but just how this works is not explained. Floridi argues, "the ultimate nature of reality is informational, that is, it makes sense to adopt a level of abstraction at which our mind-independent reality is constituted by relata that are neither substantial nor material (they might well be but we have no reasons to suppose them to be so) but informational" (2004b, p. 5).

The problem is not that informational entities are not materially evident, neither are classes, but they are an essential part of the natural sciences. Quine's nominalism for instance admits of such abstract objects such as classes, numbers and sets and the like into his physicalist ontology because science simply could not proceed without them – this in essence reveals Quine's pragmatism. However, is such a comparison any argument for the admission of Floridian information objects as a legitimate ontological category? While most of us would agree that there are concrete objects in the world that are both substantial and material, the question remains as to how seemingly mind-independent, non-material, abstract relata causally interact with the material world. For this gentle nominalist at least Quine's (as well as Occam's) suspicions are aroused.

12.5 Conclusion

Whilst OO programming and design was in its infancy Gaifman (1975, p. 329) a philosopher, summed up our predicament, "objects are notoriously theory-laden; an informative discussion of objects of this or that kind presupposes already a whole conceptual scheme". The conceptual scheme which supports a notion of the information object isn't surprising, we live in the information age – our economy, social structure and culture are virtually defined by information and information technology. This technology is implemented in computer systems that are developed using OO development environments such as Java and C#. These environments are supported by a popular design methodology UML which explicitly lays out rules for mapping high level conceptual models to implementable OO class models and eventually computer code. The notion of the most basic primitive ontological category being informational by nature is somewhat attractive; it fits with our current and popular world view, it appeals to our natural desire to impose order upon our world.

Having said that, implementation of levels of abstraction within computing environments are explicit and functional. A fundamental and necessary property of OO objects is that they are referents and this referential property is implemented directly via the implementation of LoA between various conceptual and physical layers ending with bits mapped to hard disk or memory addresses. Identity relations with regard to OO objects are explicit and clear. OO objects of the same type, sharing the same attributes, are logically (and digitally) identical. It isn't the case that nature operates in a way anything like an OO application. Floridi doesn't say as much, but this does raise the question as to what extent a seemingly primitive natural kind, the information object, can be made clear with appeals to human-made artifacts. Careful derivation of semantic information from objects and states of affairs led to the development of the OO model. It doesn't then make sense that that the OO model can clarify nature in any way outside loose metaphor.

Bas Van Fraassen advises us that, "Theories with some degree of sophistication always carry some metaphysical baggage" (1980, p. 68), just like hidden variable theories in quantum physics, the hope is that carrying the baggage will eventually pay off. What I have done is presented a fallible argument as to how informational objects (if we at least entertain the possibility of such things) do not seem to be much like OO objects. The ontological argument for such objects and hence informational realism is ultimately metaphysical by nature but Floridi adopts a structuralist methodology thus what we can salvage with some conviction is the properties and relations that are part of these postulated entities. Whether these are captured by universals (as in the property of *pawnhood*) or are ultimately unique to instantiated objects (what Armstrong (1989) called *particulars*) has not been addressed by this paper but perhaps introduces a topic for future investigation. It should be evident (with respect to my own declared nominalistic tendencies) that my analysis of these problems has being cautiously parsimonious with regard to postulated entities. David Armstrong concludes his significant 1989 text *Universals* with a quote that seems so fitting I adapt it here for my own purposes. The topic of *informational realism* is most certainly an intellectually fascinating one for those interested in what D.C. Williams terms "grubbing around in the roots of being" (1966).

Acknowledgments I am indebted to both John Weckert and Morgan Luck who read earlier versions of this article and kindly provided many thoughtful and inspiring comments. I also graciously thank Skye Bothma for her indispensable editing and formatting assistance; of course any remaining errors or omissions are mine alone.

References

Armstrong, D.M. 1989. *Universals: An opinionated introduction*, Focus series. Boulder, Colorado: Westview Press.
Booch, G. 1994. *Object-oriented analysis and design with applications*. 2nd ed. Benjamin/Cummings Publishing Company, Inc: Redwood City, California.
Date, C., and H. Darwen. 2000. *Foundation for future database systems: The third manifesto*, 2nd ed. Boston/Reading, MA: Addison-Wesley.
Floridi, L. 2002. What is the philosophy of information? *Metaphilosophy* 33(1–2): 123–145.
Floridi, L. 2004a. Open problems in the philosophy of information. *Metaphilosophy* 35(4): 554.
Floridi, L. 2004b. Informational realism. In *IEG research report*, ed. G.M. Greco. Oxford: Information Ethics Group.
Floridi, L. 2008a. A defence of informational structural realism. *Synthese* 161(2): 219–253.
Floridi, L. 2008b. Modern trends in the philosophy of information. In *Philosophy of information*. Holland: Elsevier.
Gaifman, H. 1975. Ontology and conceptual frameworks, part I. *Erkenntnis* 9: 329–353.
Goodman, N., and W.V. Quine. 1947. Steps toward a constructive nominalism. *The Journal of Symbolic Logic* 12(4): 105–122.
Hayaki, R. 2006. Contingent objects and the Barcan formula. *Erkenntnis* 64: 75–83.
Himma, E. 2004. There's something about Mary: The moral value of things qua information objects. *Ethics and Information Technology* 6: 145–159.
Johnson, D.G. 2006. Computer systems: Moral entities but not moral agents. *Ethics and Information Technology* 8: 195–204.
Martin, J., and J. Odell. 1992. *Object oriented analysis and design*. Englewood Cliffs/Upper Saddle River, NJ: Prentice-Hall Gale.
Quine, W.V. 1953a. On what there is. In *From a logical point of view*, 1–19. Harvard: Harvard University Press.
Quine, W.V. 1953b. Two dogmas of empiricism. In *From a logical point of view*, 20–46. Harvard: Harvard University Press.
Quine, W.V. 1957. Speaking of objects. *Proceedings and Addresses of the American Philosophical Association* 31: 5–22.
Quine, W.V. 1974. *The roots of reference*. La Salle: Open Court.
Quine, W.V. 1992. Structure and nature. *The Journal of Philosophy* 89(1): 5–9.
Rumbaugh, J., M. Blaha, W. Lorensen, F. Eddy, and W. Premerlani. 1991. *Object-oriented modeling and design*. Upper Saddle River, NJ : Pearson Education.
Salmon, W. 1998. *Causality and explanation*. Oxford: Oxford University Press.
van Fraassen, B. 1980. *The scientific image*. Oxford: Oxford University Press.
Williams, D.C. 1966. The elements of being. In *The principles of empirical realism*. Springfield: Charles C. Thomas.
Wittgenstein, L. 1961. *Tractatus logico-philosophicus*. London/New York: Routledge.

Part V
Replies by Floridi

Chapter 13
The Road to the Philosophy of Information

Luciano Floridi

13.1 Introduction

There are places, like the small village where I live, that are difficult to find. They lay in remote locations, not well indicated on the map, few people have ever heard of them, and hardly anyone can tell you how to get there. There are places, like the university where I work, which are difficult to reach. They are so big that, if you are driving following a GPS, their postcodes actually take you miles away from the campus, to a mail deposit. Sometimes, I fear that the philosophy of information that I have been working on combines the geographical problems of my home and working places: difficult to find and hard to reach. This is why the invitation to contribute to this volume is not only a great honour, of which I am fully aware, but also a very welcome opportunity, for which I am deeply grateful. For it allows me to map some less tortuous paths that, if followed, should help the reader to get to the philosophy of information that I have in mind, and alert the same reader to some wrong turns, potential pitfalls and misleading road signs that have side-tracked more than a fellow traveller. Of course, being able to indicate more clearly how to reach a place does not mean that the place itself is worth visiting. I believe that the philosophy of information is the philosophy *of* our time properly conceptualised *for* our time, but then you might expect this level of commitment on my side. I also hope that the journey to reach it will be rewarding, but on this I can only rely on the traveller's experience. What I may say is that the view from here is very interesting and shows an immense conceptual space still virgin. If you join me, you will see.

L. Floridi (✉)
UNESCO Chair in Information and Computer Ethics, University of Hertfordshire,
de Havilland Campus, Hatfield, Hertfordshire AL10 9AB, UK

Faculty of Philosophy and Department of Computer Science, University of Oxford, Oxford, UK
e-mail: l.floridi@herts.ac.uk

H. Demir (ed.), *Luciano Floridi's Philosophy of Technology: Critical Reflections*, Philosophy of Engineering and Technology 8, DOI 10.1007/978-94-007-4292-5_13,

13.2 Reply to Dodig Crnkovic

I am indebted to Dodig Crnkovic for her very perceptive, informative and insightful chapter. In several cases, I doubt I could have put things any better. Her analysis of the method of levels of abstraction is remarkable, especially insofar as she correctly sees that

> *Some critics feel uneasy with Levels of Abstraction in fear of ethical relativism, but the fear is unfounded. Defining Level of Abstraction adds to our understanding of a model.*

Her analogy with the natural sciences is impeccable. As she writes:

> *Physics has specific models of the world on many different Levels of Abstraction: from elementary particles, atoms, molecules, solid state, classical mechanics and fluid dynamics, astrophysics to cosmological level. There is also a remarkable emerging field of complex systems which is not only about phenomena on specific levels of organization but it also deals with interactions among different levels. As a result, a complex system as a whole exhibits properties that are distinct from the properties of its individual parts. PI [philosophy of information] uncovers similar complex structures in epistemology and ontology while IE [information ethics] does the same for ethics. This makes IE a promising research programme, and its practical applications are already many and will surely increase in number and importance.*

Her further comparison between IE and molecular biology is equally enlightening:

> *One of frequent misunderstandings of IE is related to the intrinsic value of informational objects, which in its turn is connected to understanding of Levels of Abstraction of a model. A common misconception that follows this confusion is that IE will provide machinery for automatization of ethical decision-making. However, being on a fundamental level, IE will in the first place help us understand basic structures and underlying mechanisms. IE in relation to traditional ethical approaches is like molecular biology in relation to classical biology. We do not expect molecular biology to give us all answers on questions of the living world, but it provides a solid underpinning for the rest of biology. As in other research fields, diversity of ethical approaches is still equally valuable and it presupposes human judgment and interaction among theoretical structures.*

I could easily carry on (see for example her comparison between computational modelling in IE and the use of a microscope in medical diagnostics) but the reader will have grasped the point. This is a chapter from which I have learnt a lot, and it provides a very good introduction to information ethics as I understand it.

13.3 Reply to Wolf, Grodzinsky, and Miller

My shortest reply to the chapter by Wolf, Grodzinsky, and Miller is that I agree with them wholeheartedly. Their application of the method of levels of abstraction is clever and instructive. A slightly longer reply might include the clarification of a minor point.

In their chapter, the authors state that

We contend that the addition of LoAS to the method of levels of abstraction is consistent with Floridi's desire to formulate "an ethical framework that can treat the Infosphere as a new environment worth the moral attention and care of the human inforgs inhabiting it" (Floridi 2010c*:19).*

I agree. They then continue saying that

LoAS consolidates the concerns of those working on embedding values in design and those concerned with the effect of technology on society.

And this seems to be correct as well. However, they conclude by saying that

It expands Floridi's method beyond the levels of designer and user and includes society in the mix.

This is the point I am not sure that I fully grasp: in which sense is LoAS an "addition" or "expands" the method of LoA? The authors themselves clearly indicate that LoAS is just a third Level of Abstraction in which the "S" stands for "society." They are constituted by "the set of observables available to an observer of society." But this means that a LoAS does not really expand or extend or add anything to the method, it is just part of its application. Of the unlimited number of LoAs and combinations of LoAs into Gradients of Abstraction that are possible, the authors have chosen a "societal" one. This is correct, in terms of applicability of the method, and very useful, given the goals of their analysis in the chapter. Yet presenting it as an extension of it would be like describing $12-5=7$ as an addition or extension of the general method of subtraction.

As I wrote above, this is really a small clarification, which should not cast any doubt on my full agreement with their work and conclusion:

this method offers a usable framework in the analysis and development of software applications.

They are right. It seems to me the right method to approach the ethical and epistemological challenges emerging in our information society.

13.4 Reply to Lucas

While assessing Lucas' chapter, I encountered increasing difficulties in keeping under reasonable size the growing list of passages where I thought the text to be mistaken – e.g., how many times has it been clarified that Levels of Abstraction, or LoAs, come from computer science and hence can be quantitative, but can also be, and in fact often are, just purely qualitative? They are like interfaces, as I stressed *ad nauseam*, at least mine – misleading – e.g., the whole discussion about a "natural" LoA is largely meaningless – or (the inclusive or) grounded on misunderstandings – e.g., LoAs are indiscriminately confused with systems or closed-systems (which are their targets), models (which are their output), sets of observables (which are their constituents) and so forth, depending on the paragraph one is reading.

The longer the list grew, the worse the mess looked, the harder it seemed to me that I could even start explaining how flawed the chapter is, let alone putting things straight again. There is such a thing as fatal and irreversible conceptual damage, and I started wondering whether trying to improve the chapter at all costs might actually be a case of futile medical care. Luckily, this reminded me of a fundamental distinction, which I shall exploit, at the risk of disappointing the reader. Replying is a right, not a duty. As such, it does not have to be exercised. So, in this case, I hope the reader will accept my apology for being unable to engage with a text that I am unable to improve. Perhaps it is unfixable. Perhaps others more able than me will do better. They are welcome to try, but my suggestion is to follow Virgil's advice to Dante in the *Third Canto* of *Inferno*, verse 51:

Non ragionam di loro, ma guarda e passa.
Let us not talk (reason) about them, just look and move on.

Instead of fixing the chapter, I shall try to explain the method in simple terms. It seems to me to provide an intuitive and powerful approach. I hope the reader will agree. But just in case some cynic were to suspect that the problem lies with the hapless doctor, not with the hopeless chapter, let me invite anyone interested in understanding what the method of levels of abstraction is, and how it can be applied to ethical issues, to read the chapters in this book by Wolf, Grodzinsky, Miller and by Dodig Crnkovi. They are critical, but definitely worth your time. And now, here is the method again.

The latest formalisation of the *Method of Abstraction* can be found in Floridi (2011). The terminology has been influenced by an area of Computer Science, called *Formal Methods*, in which discrete mathematics is used to specify and analyse the behaviour of information systems. Despite that heritage, the idea is not at all technical and for present purposes no mathematics is required, for only the basic idea will be outlined.

Let us begin with an everyday example. Suppose we join Anne (A), Ben (B) and Carole (C) in the middle of a conversation. Anne is a collector and potential buyer; Ben tinkers in his spare time; and Carole is an economist. We do not know the object of their conversation, but we are able to hear this much:

A. Anne observes that it (whatever "it" is) has an anti-theft device installed, is kept garaged when not in use and has had only a single owner;
B. Ben observes that its engine is not the original one, that its body has been recently re-painted but that all leather parts are very worn;
C. Carole observes that the old engine consumed too much, that it has a stable market value but that its spare parts are expensive.

The participants view the object under discussion according to their own interests, which determine their conceptual interfaces or, more precisely, their own *levels of abstraction* (LoA). They may be talking about a car, or a motorcycle or even a plane, since any of these three systems would satisfy the descriptions provided by A, B and C above. Whatever the reference is, it provides the source of information and is called the *system*. Each LoA (imagine a computer interface) makes possible an analysis of the system, the result of which is called a *model* of the system (see Fig. 13.1). For example, one might say that Anne's LoA matches that of an

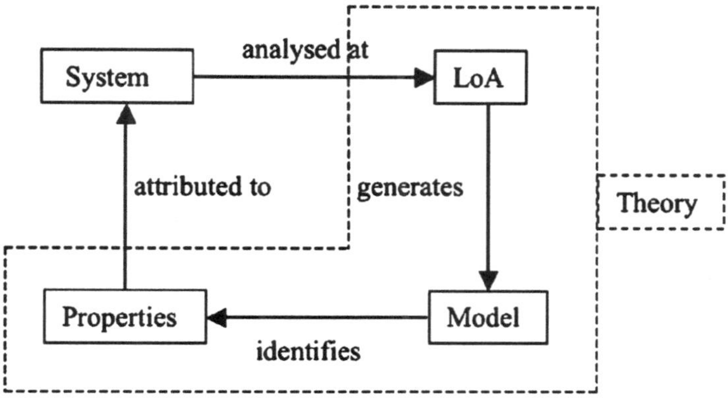

Fig. 13.1 The scheme of a theory

owner, Ben's that of a mechanic and Carole's that of an insurer. Evidently a system may be described at a range of LoAs and so can have a range of models.

A LoA can now be defined as a finite but non-empty set of *observables*, which are expected to be the building blocks in a theory characterised by their very choice. Since the systems investigated may be entirely abstract or fictional, the term "observable" should not be confused here with "empirically perceivable". An *observable* is just an *interpreted typed variable*, that is, a typed variable together with a statement of what feature of the system under consideration it stands for. It may be qualitative or quantitative, digital or analog, continuous or discrete. An interface (called a *gradient of abstractions*) consists of a collection of LoAs. An interface is used in analysing some system from varying points of view or at varying LoAs. In the example, Anne's LoA might consist of *observables* for SECURITY, METHOD OF STORAGE and OWNER HISTORY; Ben's might consist of observables for ENGINE CONDITION, EXTERNAL BODY CONDITION and INTERNAL CONDITION; and Carole's might consist of observables for RUNNING COST, MARKET VALUE and MAINTENANCE COST. The *gradient of abstraction* might consist, for the purposes of the discussion, of the set of all three LoAs.

The Method of Abstraction allows the analysis of systems by means of models developed at specific gradients of abstractions. In the example, the LoAs happen to be disjoint but in general they need not be. A particularly important case is that in which one LoA includes another. Suppose, for example, that Delia (D) joins the discussion and analyses the system using a LoA that includes those of Anne and Carole plus some other observables. Let's say that Delia's LoA matches that of a buyer. Then Delia's LoA is said to be more *concrete*, or *finely grained* or *lower*, than Anne's and Carole's, which are said to be more *abstract*, or more *coarsely grained* or *higher*; for Anne's or Carole's LoA abstract some observables which are still "visible" at Delia's LoA. Basically, not only has Delia all the information about the system that Anne and Carole might have, she also has a certain amount of information that is unavailable to either of them.

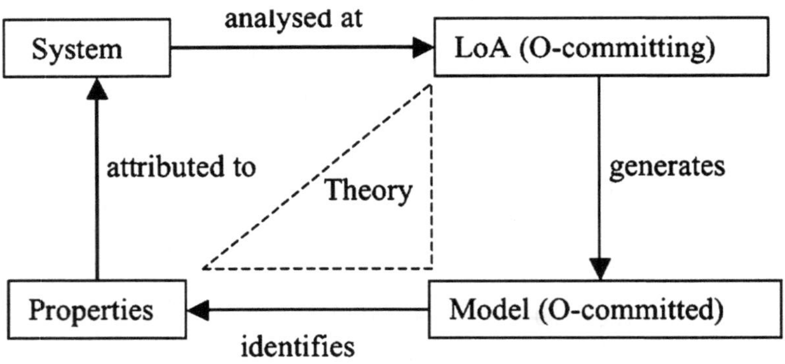

Fig. 13.2 The SLMS scheme with ontological commitment

It is important to stress that LoAs can be nested, disjoined or overlapping and need not be hierarchically related, or ordered in some scale of priority, or support some syntactic compositionality (the molecular is made by more atomic components).

We can now use the method of abstraction and the concept of LoA to make explicit the ontological commitment of a theory, in the following way.

A theory comprises at least a LoA and a model. The LoA allows the theory to analyse the system under investigation and to elaborate a model that identifies some properties of the system at the given LoA (see Fig. 13.1).

The ontological commitment of a theory can be clearly understood by distinguishing between a *committing* and a *committed* component, within the scheme.

A theory commits itself ontologically by opting for a specific LoA. Compare this to the case in which one has chosen a specific kind of car (say a Volkswagen Polo) but has not bought one yet. On the other hand, a theory is ontologically committed in full by its model, which is therefore the bearer of the specific commitment. The analogy here is with the specific car one has actually bought (that red, four-wheeled, etc. specific object in the car park that one owns). To summarise, by adopting a LoA a theory commits itself to the existence of specific types of objects, the types constituting the LoA (by deciding to buy a Polo Volkswagen one shows one's commitment to the existence of that kind of car), while by adopting the ensuing models the theory commits itself to the corresponding tokens (by buying that particular vehicle, which is a physical token of the type Polo Volkswagen, one commits oneself to that token, e.g., one has to insure it). Figure 13.2 summarises this distinction.

By making explicit the ontological commitment of a theory, it is clear that the method of abstraction plays an absolutely crucial role in any context in which we acquire and process information. In information ethics (IE), for example, different theories may adopt androcentric, anthropocentric, biocentric or ontocentric LoAs, even if this is often left implicit. IE is committed to a LoA that interprets reality – that is, any system – informationally. The resulting model consists of informational objects and processes.

13.5 Reply to Russo

The chapter by Russo is, quite frankly, impressive. She combines, in a coherent picture, several themes I developed in different writings, in a way that I can only admire. I would definitely recommend the reader to start with this text, if she wishes to have a clear, insightful, and at the same time critical and original analysis of topics such as the fourth revolution, the nature of inforgs, and the development of the infosphere. But enough of praise. Probably the best way to return the favour is to contribute one more idea to the coherent picture provided by Russo. The idea is that of *enveloping the world*. In order to explain it, I will need to introduce two concepts, that of *infosphere* and that of *re-ontologization* (Floridi 2007).

Infosphere is a neologism I coined years ago on the basis of "biosphere", a term referring to that limited region on our planet that supports life. It denotes the whole informational environment constituted by all informational entities (thus including informational agents as well), their properties, interactions, processes and mutual relations. It is an environment comparable to, but different from, cyberspace (which is only one of its sub-regions, as it were), since it also includes off-line and analogue spaces of information. It is an environment (and hence a concept) that is rapidly evolving.

Re-ontologising is another neologism that I have recently introduced in order to refer to a very radical form of re-engineering, one that not only designs, constructs or structures a system (e.g., a company, a machine or some artefact) anew, but that fundamentally transforms its intrinsic nature. In this sense, for example, nanotechnologies and biotechnologies are not merely re-engineering but actually re-ontologizing our world.

These two concepts are not indispensable – the reader is welcome to rely on any other useful shortcuts – but they are helpful to formulate the claim that digital ICTs are re-ontologizing the very nature of (and hence what we mean by) the infosphere, while the infosphere is progressively becoming the world in which we live. It follows that, while we are pursuing the development of digital technologies that can operate in the world, we are actually re-ontologising the world to fit them. Especially in recent years, the world as infosphere has been adapting to technologies' limited capacities increasingly well. Using a term from robotics, we have been *enveloping*[1] the world without fully realising it. The example of a dishwasher is elementary but still helpful in making the point. We do not build robots that wash dishes like us, we envelop micro-environments around simple robots to fit and exploit at best their limited capacities and still deliver the desired output. It is the difficulty of finding the right enveloping that makes ironing (as opposed to pressing) so time-consuming.

Enveloping used to be either a stand-alone phenomenon (you buy the robot with the required envelop, like a dishwasher or a washing machine) or implemented within the walls of industrial buildings (in a mundane context, think of the tunnel-like system

[1] In robotics, an *envelope* (also known as reach envelop) is the three-dimensional space that defines the boundaries that the robot can reach.

of the conveyorised, automatic car wash in which you drive). Nowadays, enveloping the environment into a technology-friendly infosphere has started pervading any aspect of reality and is visible everywhere, on a daily basis. If driverless vehicles can move around with decreasing trouble, this is not because AI has finally arrived, but because the "around" they need to negotiate has become increasingly suitable to AI applications.[2] We do not have semantically proficient technologies, but we have accumulated so much data, can rely on so many humans, and have such good statistical tools that purely syntactic technologies can bypass problems of meaning and understanding, and still deliver what we need: a translation, the right picture of a place, the preferred restaurant, the interesting book, the right answer, and so forth. The victory of Watson – IBM computer that answers questions posed in natural language – against two human players during a two-game, combined-point match of *Jeopardy!* is only the most recent episode in such trend. Indeed, some of the issues we are facing today, e.g., in e-health or in financial markets, already arise within highly enveloped environments in which all relevant (and sometimes the only) data are machine-readable, and decisions as well as actions may be taken automatically, by applications and actuators that can execute commands and output the corresponding procedures, from alerting or scanning a patient, to buying or selling some bonds. Examples could easily be multiplied. Enveloping is a trend that is robust, cumulative and progressively refining: everyday sees the availability of more tags, more humans online, more documents, more statistical tools, more devices that communicate with each other, more sensors, more RFID tags, more satellites, more actuators, more data collected on all possible transitions of any system, in a word, more enveloping.

This is good news for the future of smart technologies, which will be exponentially more useful and successful with every step we take in the expansion of the infosphere. Enveloping is a process that has nothing to do with some sci-fi singularity, for it is not based on some unrealistic (as far as our current and foreseeable understanding of AI and computing is concerned) speculations about some super AI taking over the world in the near future. But it is a process that raises some challenges. In order to express the one I have in mind, let me use a parody.

Two people T and H are married and they really wish to make their relationship work, but T, who does increasingly more in the house, is inflexible, stubborn, intolerant of mistakes and unlikely to change, whereas H is just the opposite, but is also becoming progressively lazier and dependent on T. The result is an unbalanced situation, in which T ends up shaping the relationship and distorting H's behaviours, practically, if not purposefully. If the marriage works, that is because it is carefully tailored around T. Now, AI and smart technologies play the role of T in the previous analogy, whereas their human users are clearly H. The risk we are running is that, by enveloping the world, our technologies might shape our physical and conceptual environments and constrain us to adjust to them because that is the best, or sometimes the only, way to make things work. New humans are born inside pre-existing technological environments and they plastically adapt to them. After all, T is the stupid

[2] See the progressive successes of the DARPA Grand Challenge.

but laborious spouse and humanity the intelligent but lazy one, who is going to adapt to whom, given that a divorce is not an option? The reader will probably recall many episodes in real life when something could not be done, or had to be done in a very cumbersome or silly way because that was the only way to make the technology in question do what it had to do. Here is a more concrete, trivial example (philosophically, things are way more complex). The risk is that we might end up building houses with round walls and furniture with sufficiently high legs in order to fit the capacities of a Roomba (http://www.irobot.com/) much more effectively. I certainly wish our house were more Roomba-friendly. The example is useful to illustrate not only the risk but also the opportunity represented by ICT's re-ontologising power and the enveloping of the world.

There are many "roundy" places in which we live, from igloos to medieval towers, from bow windows to public buildings where corners of the rooms are rounded for sanitary reasons. If we spend most of our time inside squarish boxes that is because of another set of technologies related to the mass production of bricks and concrete infrastructures, and the ease of straight cuts of building material. It is the mechanical circular saw that, paradoxically, generates a right-angled world. In both cases, squarish and roundy places have been built following the predominant technologies, rather than through the choices of their potential inhabitants. Following this example, it is easy to see how the opportunity represented by technologies' re-ontologising power comes in three forms: rejection, critical acceptance, and proactive design. By becoming more critically aware of the re-ontologising power of AI and smart ICT applications, we might be able to avoid the worst forms of distortion (rejection) or at least be consciously tolerant of them (acceptance), especially when it does not matter (consider the Roomba-friendly length of the legs of the furniture) or when this is a temporary solution, while waiting for a better design. In the latter case, being able to imagine what the future will be like and what adaptive demands technologies will place on their human users may help to devise technological solutions that can lower their anthropological costs. In short, intelligent design should play a major role in shaping the future of our interactions with forthcoming technological artefacts. After all, it is a sign of intelligence to make stupidity work for you.

13.6 Reply to Beavers

In his chapter, Beavers provides an interesting analysis of what I have called "the fourth revolution". He does so from the perspective afforded by the history of the technologies that have made possible the recording and transmission of data. The topic is immense and fascinating, and the chapter wisely highlights some of its most significant aspects. As I mentioned in my reply to Giardino, there are indeed many reasonable ways of interpreting the sort of radical information changes that we are witnessing in these decades. Among them, Beavers' approach is not only plausible, but also fruitful. Likewise, if one were to look for further perspectives, it might seem obvious to connect the information revolution to the agricultural and the industrial revolutions that preceded it. This would also make sense, and the reader keen on other ordinal numbers might wish to check the article "Lists of cultural, intellectual, philosophical and technological revolutions" provided by the usual Wikipedia. As for the "fourth revolution", in this brief reply I would like to clarify two points which may aid our understanding of what I mean by it, and why I believe it to be a useful perspective to conceptualise our time.

The first point is historical. Some people (not Beavers) seem to think that the origins of the fourth revolution can be dated back roughly to the invention of the first computing machines and the work of Alan Turing or perhaps Claude Shannon. This is fine, but it is not what I have been arguing. The information revolution understood as a fourth revolution dates back to the animals scratched by our ancestors on the walls of their caves and the rudimentary signs they used to communicate. Thus, the information revolution is not an episode in human history, but what makes history possible. The information revolution has always preceded us, for we are its children. The crucial difference is that it is only in the last decades that it has begun to be the most salient feature of our lives. And this leads me to the second point, which is hermeneutical. If the information revolution began such a long time ago, if it has been with us all along, why so much stress on it only now? And why call it a *fourth* revolution? Why not a third, or a fifth, or … you number it. Of course the number is not essential and other metrics are perfectly fine. What matters is that "the *fourth* (or *nth*) revolution" is an *interpretation* of the *information revolution* as a transformation whose greatest significance does not lie in the new ways in which we manage data, nor in what such new data management enables us to do in our interactions with the world, nor in how wealth and well-being is generated by such interactions, but rather in the way in which we are rethinking our nature and role in the universe. In other words, whatever number you think best captures the information revolution (third, if you count the agricultural and the industrial, second, if you count analogue then digital, etc.) the "fourth" refers to how many times we already have been through this radical change in our self-perception. We have looked into the mirror of science and technology before and realised that we had changed. It has happened with Copernicus, Darwin and Freud (or neuroscience, if you prefer), and it is now happening again witch computer science and ICTs. So the fourth revolution is not a serial number to label a sequence of historical transformations in our technologies. It is a way of recalling that we find the transformations brought about by our information and communication technologies so radical today because they are now changing who we think we are and can be. And this is revolutionary.

13.7 Reply to Giardino

The chapter by Giardino captures well several ideas I have articulated in recent years, while providing insightful comments, some interesting suggestions and a wealth of very valuable, if difficult, questions.

One of the ideas discussed by Giardino, on which I am particularly keen, is that philosophy is the last stage where the semanticisation of (the process of giving meaning to) Being becomes self-conscious and questions itself. All attempts to make sense of reality in the most general way lead to fundamental, open questions – that is, questions about which we care most, but about which well-informed, reasonable and tolerant disagreement is ineradicable. Such open questions are posed, shaped and answered by philosophy. And their covariance – the way in which open questions and philosophical answers mutually determine each other – is driven by human history. When our predicament changes, so do our philosophical questions and answers. Change the music, and the couple will dance differently. As I have argued elsewhere (Floridi 2011a), this explains why philosophy is, in principle, neither immutable nor unquestionable, but timely and rationally interactive. Anyone who objects that philosophy never provides final and progressively accumulating and more refined solutions to anything may as well complain that culinary art never improved on mammoth steaks. He deserves to be still in the cave.

Regarding Giardino's suggestions, I like the one concerning the fourth revolution understood as the second information revolution. Of course, it is a matter of intellectual taste and taxonomical inclinations how we go about counting the ways in which science and technology have modified our self-understanding. My choice for the "fourth" revolution is dictated by a sense of intellectual respect towards Freud. Not just because it was his brilliant idea to put Copernicus, Darwin and indeed himself in the same category (the first three revolutions), but mainly because he had a clear criterion in mind whereby such a selection should be made. He did not allow Gutenberg to join the club, for example, not because the latter could not represent a revolutionary figure, for he did, but because the printing revolution did not, *by itself*, radically change our self-understanding, certainly, not in the way in which Copernicus, Darwin, Freud and, I suggest, Turing, have. The long history of the information revolution has many episodes. Gutenberg is only one of the most important. As I have argued in (Floridi 2010a, b), it took roughly six millennia, from the Bronze Age until the end of the second millennium AD, for the information revolution to bear its main fruit. During that time, Information and Communication Technologies evolved from being mainly recording systems – writing and manuscript production – to being also communication systems – especially after Gutenberg and the invention of printing – to being also processing and producing systems, especially after Turing and the diffusion of computers. So the fourth revolution has been in the making for a very long time. As Giardino remarks,

> *to some extent we have been living in an informational environment all along. In fact, our culture deals by nature with information and pursues the realization of newer and newer means to reach the world and the others around us.*

We have always been Darwinian creatures on a Copernican planet, Freudianly opaque to ourselves. Likewise, we have always been Turing informational organisms (inforgs). We just did not know it. Yet, precisely for this reason, I agree with Giardino when she writes that "we have been inforgs all along", but I would not be inclined to add previous or different numbers to the fourth revolution. The first information revolution that Giardino has in mind is literally vital, but it does not seem to me to belong to the same line of development through which scientific advancements about the world and how we interact with it indirectly, ended up telling us a very significant story about ourselves and our place in the universe. The beginning of life on our planet and the evolution of DNA did not make us radically re-address the question about our fundamental nature. They allowed us to pose such question in the first place, but that is a different story.

This leads me to a clarification that might be of interest to the reader. Giardino is right in drawing a neat distinction between different ways in which we speak about information. I share the same concern (Floridi 2011b, 2004a). Simplifying, one might be talking about semantic information *about* something (consider the BBC news), of ontic information *as* something (consider the fingerprints of an individual), or of procedural information *for* something (consider a recipe for a cake). In Floridi (2010a, b) I have provided an introductory map of these and other cognate concepts and stressed, like Giardino, that much care needs to be exercised in order to avoid misleading confusions. However, when talking about inforgs in the infosphere, one must be able to use all three dimensions, the semantic, the ontic and the procedural, or the analysis would be over-simplistic. Thus, I have argued both that human agents are informational organisms – who share many features with artificial, biological and hybrid agents – and that, to the best of our current and foreseeable knowledge, our informational condition is utterly unique (the proviso is due to the possible discovery of intelligent life elsewhere in the universe and to the logical, though implausible, possibility of engineering real AI one day). There is no contradiction. At a very reasonable level of abstraction, we are informational structures, which process inputs in order to deal with their environments successfully, and as such we are indistinguishable from other agents. Think of those cases when, in your email exchanges with an online service, you are not sure whether you are dealing with a person or a computer. Or consider how these days you might be asked to prove that you are not a piece of software by completing a CAPTCHA (Completely Automated Public Turing test to tell Computers and Humans Apart), a simple test often involving pattern recognitions, administered and evaluated by a computer, which presumably a machine would be unable to pass. However, we are also the only informational structures in the universe capable of intelligent, semantic structuring. Humanity has informational organism as *genus* and structuring structure as *species*. This, I hope, clarifies the apparent tension between similarity and uniqueness: we are inforgs, but our intelligent anti-entropic nature is what makes us a special kind of inforgs. There is, however, a further potential confusion that I would like to avoid. We might be a glitch in the infosphere, what I like to call the *wonderful mistake*. For as long as we are here, we realise and rightly boast that (to the best of our current scientific knowledge) we are the infosphere's only chance of having a mental, conscious life. Such responsibility is enormous. However, unless there is a divine plan (and I am happy to leave the answer to this question to the reader), we are that portion of the infosphere that merely won the mental lottery. There was no reason to be the owners of the lucky ticket, so amazement is more than justified (the exclamation mark effect, the "we won the mind lottery!" attitude) but puzzlement would be out of place (the question mark effect, the "why did we win the mind lottery?" attitude). There are so many other lotteries that we lost. The cheetah, for example, won the lottery for the fastest runner on earth, with its astonishing speed of 70 mph, but lost the climbing lottery. We simply won the (possibly only) lottery (ticket) in the universe that allows a justified sense of amazement and a (possibly, if the atheist is right) mistaken sense of puzzlement. There is no why from wow.

13.8 Reply to Pasquinelli

I enjoyed reading Pasquinelli's chapter. It addresses an issue of immense importance for the future of our information societies: the relation between the information revolution and our educational practices. Unfortunately, she is right in stressing that:

> *Despite the potential effects (and side-effects) of the massive introduction of ICT in our daily life, the information revolution has barely modified the way we teach and learn (at least in school). Rather than happening, the information revolution is invoked, funded, measured, asserted as a goal for the wealth of the nations.*

Her analysis of the current state of education and what the future might bring is both informative and enlightening. In this brief reply, I would like to offer one more example that seems to go in the (right) direction outlined by Pasquinelli's chapter, and a broader suggestion of what education may look like after the fourth revolution, to use a phrase from her chapter.

The example first. Through small appliances, known as clickers, Classroom Response Systems or CRS (also known as Classroom Communication Systems, Personal Response Systems, Electronic Response Systems, or Audience Response Systems) allow a variety of interactions between students and teachers, e.g., by conveying yes/no answers to questions shown on the board. Such IT-mediated interactions can increase participation or provide immediate feedback on whether the material delivered has been understood, for example. In different forms, CRSs have been available since the 1960s. Their increasing popularity today is due not only to advancements in technology – clickers may easily be replaced by mobile phones – but also to the synergy between the systems and students' ordinary habit of writing and sending SMSs while attending their classes. Instead of prohibiting the uses of any communication technology in the classroom as mere distractions, a better approach is to harness the relevant technology and the corresponding skills in order to improve the learning experience. This example converges on the same conclusion reached in the chapter about the use of Wikipedia. It is pointless to try to stop students from relying on it. It is immensely more fruitful to teach them how to use it critically, and ask them to contribute and edit new entries, or improve old ones.

Similar examples point in the direction of a substantial change in our educational practices. I still believe that the acquisition of some basic information and skills is crucial. Of course, I do not mean learning by heart lists of names, dates, facts, or grammatical rules and so forth, but possessing the sort of basic information that allows one to understand a decent newspaper. There is little one can do intellectually without some reliable and critically assimilated input. Which information needs to be privileged today poses a challenge, but this too is hardly new. The novelty is represented by the interpretation of information societies as neo-manufacturing organizations, in which the raw material is represented by a zettabyte (10^{21}) of data. In such societies, learning by making, as was the case with artisans before the Industrial Revolution, seems crucial. Informational goods require new skills that will be increasingly important and can be kept updated only if properly learnt at the right age. Many of such skills are "linguistic". By this I do not mean to refer to natural languages, which of course are fundamental – especially one's own mother tongue on which clear and precise thinking depends so heavily – but to the languages spoken by the information society: general mathematics, logic, statistics, ICT. Such languages enable the critical and creative handling of data, the open-ended acquisition of new skills and further information, and the intelligent production of informational goods. Unfortunately, I am not very optimistic. Not because our technologies are "making us stupid". This is ridiculous. But because such technologies are making increasingly clear that the old hurdles of *availability* and *accessibility* of information were merely eclipsing the real difficulty of *understanding*. Today, a good Wikipedia entry is trivially available and accessible, but it might still be impossible to grasp its contents, if one lacks the required competences, e.g., if one does not speak "chemistry". The truth is that once understanding is unveiled as the real difficulty, it becomes clear that only time, patience, resolve and intelligence can help. And these have always been scarce resources that no educational system can miraculously multiply.

13.9 Reply to Cohen-Almagor

The chapter by Cohen-Almagor explores the relations between information ethics and net neutrality. I completely agree with him about the premise:

> *The issue of responsibility of ISPs and host companies is arguably the most intriguing and complex. Their actions and inactions directly affect the information environment.*

And of course I can only concur with his conclusion:

> *our interest in the sound construction of the infosphere must be associated with an equally important, ethical concern for the way in which the latter affects and interacts with the physical environment, the biosphere and human life in general, both positively and negatively.*

As a contribution to the debate on net neutrality I only wish to provide a brief and general consideration.

Unfortunately, the debate on net neutrality has been affected, among other things, by a loaded terminology, which has made the ecological aspect of the issue less visible. If we had been speaking all along in terms of *net diversity* instead, we would have been able to appreciate more easily the fact that, in a complex infosphere, more nuanced and articulated rules about the various services that could be offered to end-users could increase, and not decrease, the opportunities for growth and development, as long as a fair entry-level is guaranteed to all participants. This holds true in many transport systems, with different tickets for different classes in trains and airplanes, and in postal services, where public and private providers compete and different tariffs apply. The point is of course more complex, but it has been well argued in a recent paper (Turilli et al. forthcoming), which I recommend to the reader. The take home message seems to me that in net neutrality what matters are the minimal conditions of fair equality, not the maximal imposition of unfair sameness.

13.10 Reply to Silva and Ribeiro

The chapter by Silva and Ribeiro provides a wealth of details about information science (IS) and the philosophy of information (PI), and the connections between the two disciplines. It deserves to be studied by anyone interested in the interactions between PI and IS. In this reply, I would like to contribute to their effort by briefly recalling the contents of two articles in which I argued that IS (or LIS, library and information science, as it is known in the States) might be understood as applied PI, which could provide its conceptual foundations.

In Floridi (2002) I analysed the relations between PI, IS and social epistemology (SE). In that context, I argued that there is a natural relation between philosophy and IS but that SE cannot provide a satisfactory foundation for IS. Rather, SE should be seen as sharing with IS a common ground, represented by the study of information, to be investigated by PI. In that context, I outlined the nature of PI as the philosophical area that studies the conceptual nature of information, its dynamics and problems, and then defined IS as a form of applied PI. The hypothesis supported was that PI should replace SE as the philosophical discipline that can best provide the conceptual foundation for IS. In the conclusion, I suggested that the "identity" crisis undergone by IS has been the natural outcome of a justified but precocious search for a philosophical counterpart that has emerged only recently, namely PI. The development of IS should not rely on some borrowed, pre-packaged theory.

In a later contribution (Floridi 2004b), I defended the suggestion that, as applied PI, IS can fruitfully contribute to the growth of basic theoretical research in PI itself and provide its own foundation. We often hear about the differences between the information worker, busily involved in managing and delivering information services, and the information scientist or the IS expert, deep in theoretical speculations. The line of reasoning here seems to be that a foundation for IS should satisfy both, and that this is something that PI cannot achieve, hence the objection that PI is not "social" enough. I accept the inference, but I disagree on the premise. For I think we should distinguish as clearly and neatly as possible between three main layers. There is a first layer where we deal with information contents and services. Compare this with the accountant's calculations and financial procedures. One may wish to develop a theory of everyday mathematics and its social practices – surely this would be a worthy and interesting study – but it seems impossible to confuse it with the study of mathematics as a formal science. The latter is a second layer. It is what IS amounts to, what one learns, with different degrees of complexity, through the university curriculum that educates an information specialist. There is then a third layer, in which only a minority of people is interested. We call it foundational. For mathematics, it is the philosophy of mathematics. I suggested PI for IS. My point here is that it is important to acknowledge and respect the distinction between these three layers; otherwise one may criticize *x* for not delivering *y* when x is not there to deliver *y* in the first place anyway. When checking whether the bank charged you too much for an overdraft, you are not expected to provide an analysis of the arithmetic involved in terms of Peano's axioms. Likewise, a scientist may be happy with a clear understanding of statistics without ever wishing to enter into the philosophical debate on the foundations of probability theory. So it seems to me that IS could be provided with an equally theoretical approach, capable of addressing issues that the ordinary practitioner and the expert would deem too abstract to deserve attention in everyday practices (mind that I am talking about layers, not people; one can wear different hats in different contexts; this is not the issue here). In the end, I agree that PI should seek to explain a very wide range of phenomena and practices. I would add that this is precisely the challenge ahead. The scope of PI spans a whole variety of practices, precisely because the aim of PI is foundationalist. IS seems to be well posed to benefit enormously from the development of a sound philosophy of information.

13.11 Reply to Kerr and Pritchard

I welcome and indeed advocate the informational approach to knowledge to be found in the chapter by Kerr and Pritchard. It seems high time that we focus more on information than on beliefs, as the basis of our epistemic interactions with the world (Floridi 2011a). I discussed informational scepticism elsewhere (Floridi 2010a, b) so, in this reply, I would like to deal with a key issue tackled by Kerr and Pritchard, namely the closure principle and its relation with sceptical doubts.

The reader may recall that, according to Kerr and Pritchard:

> *It is because of the possibility of deceptive environments like this that Dretske denies that information alone could ever answer a skeptical doubt.*

They hold that this is so because

> [...] *on Dretske's view I can have an informational basis for believing that I am in Edinburgh but I can have no informational basis for believing that I am not a BIV* [brain in a vat] *on Alpha Centauri (a skeptical hypothesis which entails that I am not in Edinburgh), even whilst I know that if I am a BIV on Alpha Centauri then I am not in Edinburgh. It is for this reason that Dretske denies epistemic closure.*

I find their analysis convincing. What I wish to discuss is whether Dretske's position is actually defensible once we move from a hybrid context of doxastic, epistemic and informational concepts to a purely informational one. I shall argue that it is not: the principle of information closure is perfectly acceptable within a modal logic of being informed (Floridi 2006).

Dretske's argument has the form of a *modus tollens*. The first step, which requires some patient refinement, consists in formulating the principle of closure in informational terms. This is not as straightforward as it might seem because there is a zoo of alternative formulations of the principle of epistemic closure (PEC),[3] each with some interesting if subtle mutations. Luckily, the informational translation makes our task less daunting. Minimalism does help to declutter our conceptual space. Let us see how.

A standard way of formulating PEC under *known* entailment is:

K) *If, while knowing that* p, S *believes that* q *because* S *knows that* p *entails* q, *then* S *knows that* q.

Suppose we avoid any reference to beliefs and knowledge. After all, we are seeking to formulate a principle of information closure (PIC) that should apply to human and artificial agents – including computers that may be able to hold information physically (say in their RAMs) – or to hybrid agents like banks or online services, which might hold information in their files, or in the memories of their employees. Neither artificial nor hybrid agents can literally believe or know that *p*, for they lack the required mental states or propositional attitudes. In this case, **K** becomes the principle of known information closure:

PKIC) *If, while holding the information that* p, S *holds the information that* q *because* S *holds the information that* p *entails* q, *then* S *holds the information that* q.

Trivial, isn't it? **PKIC** just states in a very verbose way that *S* holds the information that *q*. This will not do. It would be interesting to understand better why the translation deprives **K** of its conceptual value, but this would go well beyond the scope of this reply, so let us not get side-tracked but check whether we can obtain PIC by adapting another version of PEC, known as the *straight principle* of epistemic closure. This states that:

SP) *If* S *knows that* p, *and* p *entails* q, *then* S *knows that* q.

The informational translation gives us:

SPIC) *If* S *holds the information that* p, *and S holds the information that* p *entails* q, *then* S *holds the information that* q.

Note that **SPIC** treats *p* entails *q* as another piece of information held by *S*. The advantage is that, in this way, **SPIC** becomes the philosophical counterpart of the axiom of distribution in epistemic logic: $\Box(\varphi \rightarrow \psi) \rightarrow (\Box\varphi \rightarrow \Box\psi)$, which is the source of the debate on PEC in that context. And there seems to be no disadvantages. Adding that *S* holds the information that *p* entails *q* seems to be unproblematic (it is in **K**).

[3] The interested reader is referred to the excellent (Luper 2010). I use **K** and **SP** for consistency with the literature.

SPIC is not trivial, or at least not in the sense in which **PKIC** is. And it seems exactly what we need to revise Dretske's argument informationally, depending on how we *handle* the entailment occurring in it. Mind, I do not say *interpret* it, for this is another matter. In what follows, I shall simplify our task by assuming that the entailment is interpreted in terms of material implication.

The entailment in **SPIC** can be handled in several ways. I shall mention two first, for they provide a good introduction to a third one that seems preferable for our current purpose.

A modest proposal is to handle *p* entails *q* in terms of *feasibility*. *S could* obtain the information that *q*, if only *S* cares enough to extract it from the information that *p* and the information that *p* entails *q*, both of which *S* already holds. Consider: the bank holds the information that Peter, one of its customers, is unemployed. As a matter of fact, the bank also holds the information (endorses the entailment) that, if a customer is unemployed then that customer does not qualify for an overdraft. So the bank can (but might not) do something with or about the entailment. Peter might keep enjoying his overdraft for as long as the bank fails to use the information at its disposal to generate the information that he no longer qualifies.

A slightly more ambitious proposal, which has its roots in work done by Hintikka (1962), is to handle *p* entails *q normatively*: *S should* obtain the information that *q*. In our example, the bank should reach the conclusion that Peter no longer qualifies for an overdraft; if it does not, that is a mistake, for which someone (an employer) or something (e.g., a department) might be reprimanded.

A further alternative, more interesting because it bypasses the limits of the previous two, is to handle *p* entails *q* as part of a *sufficient procedure for information extraction* (data mining): in order to obtain the information that *q*, it is sufficient for *S* to hold the information that *p* entails *q* and the information that *p*. This third option, which captures better the formulation of the closure principle based on the distribution axiom, leaves unspecified whether *S* will, can or even should extract *q*. One way for the bank to obtain the information that Peter does not qualify for an overdraft is to hold the information that if a customer is unemployed that customer does not qualify for an overdraft, and the information that Peter is unemployed. Handling the entailment as part of a sufficient procedure for information extraction means qualifying the information that *q* as obtainable independently of further experience, or empirical evidence or factual input, that is, it means showing that *q* is obtainable without overstepping the boundaries of the available database. This is another way of saying that the information in question is obtainable *a priori*.

SPIC, with the entailment embedded in it handled in terms of *a priori* information extraction, provides the necessary translation of the first step in Dretske's revised argument. The second and third step are very simple, for they consist in providing an interpretation of the information that *p* and of the information that *q* such that *p* entails *q*. Following Kerr and Pritchard, we have:

p:=S is in Edinburg
e:=if S is in Edinburg then S is not a brain in a vat on Alpha Centauri.

The fourth and final step is a negative thesis, already formulated by Dretske in an informationally suitable vocabulary:

NT) *information alone could never answer a skeptical doubt.*

NT seems very plausible: I agree with Dretske that one cannot solve sceptical doubts of a Cartesian nature by piling up information. One of the reasons for raising them is precisely because they block such possibility. We would have stopped discussing sceptical questions a long time ago if this were not the case.

We can now reformulate Dretske's argument informationally thus:

(i) if **SPIC**, *p* and *e*
(ii) then *S* can generate the information that *q*;
(iii) but *q* is sufficient for *S* to answer the sceptical doubt (in the example, *S* holds the information that *S* is not a brain in a vat on Alpha Centauri);
(iv) and (iii) contradicts **NT**;
(v) but **NT** seems unquestionable;
(vi) so something is wrong with (i)–(iii): in a Cartesian scenario, *S* would simply be unable to discriminate between being in Edinburgh or being a brain in a vat on Alpha Centauri, yet this is exactly what has just happened;
(vii) but (iii) is correct;
(viii) and the inference from (i) to (ii) is correct;
(ix) and *e* in (i) seems innocent;
(x) so the troublemaker in (i) is **SPIC**, which needs to be rejected.

It all sounds very convincing, but I am afraid **SPIC** has been framed, and I hope you will agree with me, once I show you by whom.

Admittedly, **SPIC** looks like the only suspicious character in (i). However, consider more carefully what **SPIC** really achieves; that is, look at *e*. The entailment certainly works, but does it provide any information that can answer the sceptical doubt? Not by itself. For *e* works even if both *p* and *q* are false, of course. This is exactly as it should be, since valid deductions, like *e*, do not generate new information, a scandal (D'Agostino and Floridi 2009) that, for once, it is quite useful to expose. **NT** has a logical counterpart: deductions alone could never answer a sceptical doubt, either. If *e* did generate new information, we would have a case of synthetic *a priori* reasoning (recall the handling of the entailment as a sufficient procedure for information extraction), and this seems a straightforward *reductio*. The fact is that the only reason why we take *e* to provide some anti-sceptical information about *S*' location is because we also assume that *p* in *e* is true. *Ex hypothesis*, not only *S* is actually in Edinburgh, but *S* holds such information as well. So, if **SPIC** works anti-sceptically, it is because *q* works anti-sceptically, but this is the case because $e + p$ work anti-sceptically, but this is the case only if *p* is true. Now, *p* is true. Indeed it should be true, and not just in the chosen example, but in general, or at least for Dretske and anyone else, like me, who subscribes to the veridicality thesis, according to which *p* qualifies as information only if *p* is true. But then, it is really *p* that works anti-sceptically. All the strength in the anti-sceptical interpretation of (i)–(iii) comes from the truth and informativeness of *p*. This becomes obvious once we realise that no shrewd sceptic will ever concede *p* in the first place, because she knows that, if you concede *p*, then the sceptical challenge is over, as Descartes correctly argued. Informationally (but also epistemically), it never rains, it pours: you never have just a bit of information, you always have a lot more, Quine was right about this. Allow a crack in the sceptical dam and the epistemic flooding will soon be inevitable. This is why, in the end, local or circumscribed scepticism is either just critical thinking or must escalate into global scepticism of a classic kind, e.g., Pyrrhonian or Cartesian. So it is really the initial input quietly provided by *p* that is the real troublemaker and **SPIC** is only following orders, as it were. For **SPIC** only exchanges the higher informativeness of a true *p* (where *S* is located) into the lower informativeness of a true *q* (where *S* is not located, being located where he is). This is like exchanging a 20 lb banknote into many 1 $ bills. It might look like you are richer, but of course you are just a bit poorer, in the real life analogy because of the exchange rate and the commission charged, and in Dretske's argument because you moved from a positive statement (where you actually are located) to a negative one (one of the infinite number of places where you are not, including places dear to the sceptic). If you do not want the effects of *q*, do not blame **SPIC**, just never concede *p* in the first place.

It follows that the informational answer to the sceptical doubt, which we agreed was an impossibility, is provided not by *q*, but by *p*, and this disposes of Dretske's objection that **SPIC** is untenable because information can never provide an answer to sceptical doubts. It never does because you may never be certain that you hold it (you cannot assume *p*), not because, if you hold it, it does not.

One may object that all this leaves the last word to the sceptic. I agree, it does, but it does so only in this context, and this is harmless. **SKIP** was never meant to provide an anti-sceptical argument in the first place. It was the alleged accusation that it did in a mistaken way that was seen to be the problem. So what happens next? If being in Edinburgh means that I may not be sure that I am there, then we are talking about a scenario in which no further information, no matter how far-reaching, complex, sophisticated or strongly supported, will manage to eradicate once and for all such Cartesian doubt. I believe it is this the proper sense in which all the information in the world will never meet the sceptical challenge. For information is a matter of empirical facts, and sceptical doubts are based on logical possibilities. The former just cannot cure the latter. Is this, then, finally a good reason to reject **SPIC**? The answer is again in the negative. **SPIC** was not guilty when we were assuming to have a foot in the door, a piece of information about how the world really is, namely *p*. It is not guilty now that we are dealing with a web of information items that might turn out to be a complete fabrication. On the contrary, in the former case it is **SPIC** that helps us to squeeze some (admittedly rather useless) further bits of information from *p*. In the latter case, it is **SPIC** (though of course not only **SPIC**) that makes the coherence of the whole database of our information tidy and tight. But if **SPIC** is to be retained in both cases, what needs to be discharged? Either nothing, if we are allowed a foot in the door, because this is already sufficient to defeat the sceptical challenge; or the value of absolute scepticism as a weapon of total information destruction, if all that this can ever mean is that the logically possible is empirically undefeatable. Once made fully explicit and clarified in detail, radical informational scepticism, with its fanciful scenarios of possible worlds, can be proved to be entirely redundant informationally (Floridi 2010a, b), so it can be disregarded as harmless. Wondering whether we might be dreaming, or living in a Matrix, or might be butterflies who think they are humans, or might be characters in a sci-fi simulation created by some future civilization, and so forth, are interesting speculations that may be intellectually stimulating or simply amusing, but that make no significant difference to the serious problem of how we acquire, manage, and refine our information about the world.

13.12 Reply to Brenner

Brenner's chapter offers an articulated discussion of the method of levels of abstraction (LoA). As a contribution to the debate, I would like to offer a clarification about the ontological commitment entailed by the method.

That the same reality might be analysed in different ways is a truism hardly worth stating. That such ways might be understood as levels of abstraction (Floridi 2008b), and then be subject to some formal assessment also seems uncontroversial. The real bone of contention, if I might be allowed to anticipate a pun, is the interpretation of such analytic procedure, and hence of its outcome. Plato is usually credited to have been the first to theorise it explicitly. In a famous passage in the *Phaedrus* (265d–266a), Socrates discusses "a pair of procedures", which might be called *synthesis* and *analysis*. Through the former "we bring a dispersed plurality under a single form". Today, one may say that this is the ability to draw sound and fruitful connections between different bits of information in order to compose a more satisfactory account. Through analysis, "the reverse of the other",

> *we are enabled to divide into forms, following the objective articulation; we are not to attempt to hack off parts like a clumsy butcher.*[4]

This is what happens when we apply the method of abstraction, but Plato's well-known metaphor of "carving nature at the joints" is unfortunate. Not merely because we have acquired a different sensitivity about animals, but mainly because, through his metaphor, a form of ontological interpretation quietly sneaks in. From the old debate on universals to the more recent debate on natural kinds, the metaphor might easily lead one to presuppose as uncontroversial the view that the structures, invariants, patterns, types, universals, natural kinds and so forth, identified by our

[4] Translation from Plato (1989).

analytic procedures, are entirely intrinsic to the system. They are discovered, in the same way as we carefully discover where to carve a body. This is a mistake. It would be like saying that, given the contents of the fridge, we simply discover the dishes we can cook. Meaningless. The ingredients provide affordances and constraints, but there are different ways in which we can take advantage of the former while respecting the latter. The most important things to remember are, first, that the choice of the level of abstraction and hence the purpose which orients such choice, make a significant difference in the way in which we analyse the structure of any system, be this biological, artificial, chemical, physical, social and so forth. And, second, that the system under observation is – or, more cautiously, that it would be safer to assume that it most likely is – a unity, in which articulations, organizing patterns and so forth are still aspects of a single whole. As it has been repeatedly and convincingly argued[5] reality does not come in well-organised bits, all properly disjoint in non-overlapping pigeonholes, which only need to be collected and catalogued. Taxonomy is a teleological science of design (not invention nor discovery), based on levels of abstractions.

> *There is nothing wrong with talking about carving nature at the joints as long as one bears in mind that nature's joints are not always disjoint.* (Khalidi 1993, p. 112)

For all these reasons, I would be much happier to use either the cooking metaphor, introduced above, or adapt from Leibniz a different carving analogy, which he used in a related context, the debate on innate ideas (Leibniz 1996). A system, anything from the smallest and simplest element to the all-encompassing universe, is like a block of veined marble (Leibniz's metaphor) or better, a gemstone. We carefully carve and polish a cameo according to our goals (the purpose of the analysis), skills (the specific level of abstraction chosen in view of the purpose) and the veins or contrasting colour (the ontological constraints and affordances) in the gemstone. The patterns in the gemstone encourage some outcomes but not others. They allow for several different images to be carved, but not any: for some will be impossible, some unlikely and require *ad hoc*, *virtuoso* solutions, and some others will be much more feasible, say the picture of an eagle. Like gemstones, systems are inhomogeneous, not disjoint. This does not undermine the scientific realism and reliability of our analyses: the eagle is no less real just because we could have carved a phoenix. The omelette is no more a fiction of our imagination than the zabaglione we could have obtained from the same eggs. How we structure the world depends both on us and on the world. It is a mistake to underestimate either side of this interactive relation. I hope this clarifies why I am so reluctant to accept any theory that postulates ontological levels of organization existing independently of the level of abstraction at which they are conceptualised.

[5] For a great article with plenty of further references and scientific examples of the criss-crossing nature of reality see Khalidi (1993).

13.13 Reply to McKinlay

The interesting chapter by McKinley tackles a very important issue, namely the informational nature (or I should say conceptualization, but see below) of worldly objects. If I do not misunderstand him, McKinlay quotes Quine *approvingly*, when the latter states that:

> *the very notion of object, or of one and many, is indeed as parochially human as the parts of speech; to ask what reality is really like, however, apart from human categories, is self-stultifying.* (Quine 1992, p. 9)

If so, then I could hardly agree more. Quine, McKinlay and I are on the same side of the river, the other bank being populated by all those who hold that the world is *really* made of objects, with the latter being pretty much the sort of objects with which we deal in our kitchens. There is much more on which I agree with McKinlay. In his chapter, for example, he provides a clear and perfectly shareable analysis of the nature of objects as understood in Object Oriented Programming (OOP). The reader who does not know much about this topic will find that part helpful. He also spends a considerable amount of time arguing that

> *informational objects (if we at least entertain the possibility of such things) do not seem to be much like OO objects.*

Maybe (see the two problems discussed below), but the important fact is that this is irrelevant. For I actually suggested that

> *OOP provides us with a rich concept of informational objects that can be used to conceptualize a structural object as a combination of both data structure and behaviour in a single entity, and a system as a collection of discrete structural objects. Given the flexibility of the conceptualization, it becomes perfectly possible, indeed easy, to approach the macroworld of everyday experience in a structural-informational way.* (Floridi 2008a)

I thought using *x* to conceptualise *y* (this is what I intended in the passage above) was something very different from, and hardly mistakable for, saying that *x* is like *y* (but this is what McKinlay seems to be taking me to say). For example, one may use a chess game to conceptualise a battle, or a social conflict, or a duel, or ... you get the picture. But a chess game is a chess game, as I suspect McKinlay would insist, and we must all agree. I think it can be used to cast light and make sense of a battle, a social conflict or a duel, but McKinlay might point out the endless number of differences, on which it would be impossible to disagree. Yet ultimately the real point is whether a conceptualization is helpful. This of course depends on many factors, but above all on whether it makes things clearer and helps one to grasp them better. So if the link between informational objects and objects in OOP does not help the reader, the invitation is to drop the comparison immediately. It is only a tool to help make sense of a philosophical thesis, if it is unhelpful you would do better to avoid it. If the way in which I discussed the relation between informational object and objects in OOP confused other people apart from McKinlay, I am grateful to him for clarifying it.

Let me now turn to some more important problems with the chapter. McKinlay highlights a crucial point when he writes that:

> *We are obliged to point out that Floridi does limit the scope of his adoption of OO concepts and theory by saying "OOP is not a viable way of doing philosophical ontology, but a valuable methodology to clarify the nature of our ontological components."* (2004a, b, p. 5)

Yet here I wish he had treated such "obligation" seriously, and used it to inform the whole chapter, instead of relegating it to a footnote and then forgetting about it. If he had taken his own advice seriously, then in the chapter we would have encountered at least a passing reference to what I actually mean by informational objects. They are the structural objects discussed in structural realism, and

> *A straightforward way of making sense of these structural objects is as informational objects, that is, as cohering clusters of data, not in the alphanumeric sense of the word, but in an equally common sense of differences de re, i.e. mind-independent, concrete points of lack of uniformity.* (Floridi 2008a)

In that article, I attempt to clarify such a definition, so it is not worth rehearsing that line of reasoning here. The interesting point is that, if one starts from this, admittedly complex, description of informational objects as differences *de re*, then it is easy to highlight the two major limits of the chapter, which undermine its hermeneutic value.

First, the chapter interprets informational objects as if they were abstract objects, a bit like classes or natural numbers (at least in some interpretations of the latter). Here is a typical sentence among many:

> *Abstract objects such as information objects do not exist in space.*

Second, the chapter further interprets such abstract informational objects as if they were some sort of catalogues of the properties qualifying their physical referents or counterparts in the world.

While the first mistake is not common, the semantic fallacy behind the second (understanding information as if it were information *about* something, not information *as* something, compare the difference between a train timetable and some fingerprints) has affected other interpreters. This probably means that there is something wrong in the way I have explained what I mean by informational objects. So let me try again and, in order to avoid misleading echoes, let us talk of *i-objects* (assuming Apple has not trademarked them already). Your couch is an *i-object*, and so is your car, you and me, and the moon, and the tree in front of my house. If you find this counterintuitive, you are in good company. If you find it unintelligible, then compare *i-objects* to Leibniz's monads, or Berkeley's ideas and you might no longer be alone. Let me hasten to specify that *i-objects*, monads and Berkeleian ideas only share a family resemblance, namely the ontological direction in which they should be interpreted. Given that your fridge is an *i-object*, you will understand why I cannot accept McKinlay's offer to agree with me when he writes

> *Thus whilst I agree it does not make sense to ask the question, "Where are these information objects you talk of Luciano?"*

For that is a perfectly sensible question to ask, which *ultimately* (this clause requires some philosophical work, I admit) has a perfectly sensible answer: *i-objects* are in the world (well, the world as experienced on this side of our human cognitive interfaces, i.e. our levels of abstraction) you are sitting on one, and driving another. It is his other question that makes no sense to me, namely:

> *I do think it legitimate to ask, "How do everyday (concrete) objects map to their information object counterparts?"*

What does he mean "map"? This is like asking how everyday (concrete) water maps to its chemical object counterpart, H_2O. Water does not *map*, it *is* H_2O, at the chosen chemical level of abstraction. In OOP, a button does not "map" a button, it *is* a button. Everyday concrete objects are (aggregates of) *i-objects*, at the informational-structuralist level of abstraction. I hope this also clarifies why I find the following remark utterly puzzling:

> *The Floridian account however seems to suggest an object qua information object does indeed reference the real world object it purports to represent but just how this works is not explained. Floridi argues, "the ultimate nature of reality is informational, that is, it makes sense to adopt a level of abstraction at which our mind-independent reality is constituted by relata that are neither substantial nor material (they might well be but we have no reasons to suppose them to be so) but informational."* (2004b, p. 5)

I disagree, for *i-objects* most definitely do not purport to represent anything, they are not bits of semantics, they are the real thing (pun intended), otherwise my statement quoted at the end by McKinlay about the informational nature of reality would be simply idiotic rather than controversial.

Clarifying the previous two mistakes leads me to a final comment. McKinlay seems to share with Quine a naturalistic approach to epistemology, and there is a fundamental thesis on which I agree with them:

> *Natural science tells us that our ongoing cognitive access to the world around us is limited to meager channels. There is the triggering of our sensory receptors by the impact of molecules and light rays. Also there is the difference in muscular effort sensed in walking up or down hill. What more? Even the notion of a cat, let alone a class or number, is a human artifact, rooted in innate predisposition and cultural tradition. The very notion of an object at all, concrete or abstract, is a human contribution, a feature of our inherited apparatus for organizing the amorphous welter of neural input. [...]*
>
> *Reification and implication are the key principles by which that organizing proceeds. [...]*
>
> *The reification of bodies comes in stages in one's acquisition of language, each successive stage being more clearly and emphatically an affirmation of existence. The last stage is where the body is recognized as identical over time, despite long absences and interim modifications. Such reification presupposes an elaborate schematism of space, time, and conjectural hidden careers or trajectories on the part of causally interacting bodies. Such identifications across time are a major factor in knitting implications across the growing fabric of scientific hypotheses.* (Quine 1992, p. 6)

Change Quine's talk of "innate predisposition", "cultural tradition", and "schematism of space, time [etc.]" with the equivalent levels of abstraction, and what you get is a position quite close to the one I defend in my philosophy of information. There is, however, a major and substantial difference. The process whereby our meagre data are transformed in what we experience epistemically as the outside world is one of *construction* not one of mimetic representation. How could it be otherwise? Think of the processes whereby a computer interface transforms the even more meagre flow of zeros and ones into a colourful, dynamic, interactive, noisy game, for example. The world sends signals, which we interpret through our bodily hard-wired and mentally soft-wired interfaces (LoAs). What we make of such signals or data is, partly, up to us as informational organisms. Not anything goes, but it is a *poietic* interaction. It baffles me why Quine (and perhaps McKinlay as well, but I might be wrong) never acknowledged the constructionist nature of our epistemic interactions with the world (Floridi 2011c). Perhaps some forms of empirical addiction can seriously damage your critical thinking. The point could not be simpler: it is like saying that what we have as chefs is a constant and abundant reserve of very meagre ingredients, from which we can cook quite elaborate meals and then try to convince you that the meals represent the ingredients. That is where the naturalization of epistemology stops making sense and needs to be supplemented by a constructionist one.

References

D'Agostino, Marcello, and Luciano Floridi. 2009. The enduring scandal of deduction. Is propositional logic really uninformative? *Synthese* 167(2): 271–315.

Floridi, Luciano. 2002. On defining library and information science as applied philosophy of information. *Social Epistemology* 16(1): 37–49.

Floridi, Luciano. 2004a. *The Blackwell guide to the philosophy of computing and information*. Malden/Oxford: Blackwell.

Floridi, Luciano. 2004b. LIS as applied philosophy of information: A reappraisal. *Library Trends* 52(3): 658–665.

Floridi, Luciano. 2006. The logic of being informed. *Logique et Analyse* 49(196): 433–460.

Floridi, Luciano. 2007. A look into the future impact of ICT on our lives. *The Information Society* 23(1): 59–64.

Floridi, Luciano. 2008a. A defence of informational structural realism. *Synthese* 161(2): 219–253.

Floridi, Luciano. 2008b. The method of levels of abstraction. *Minds and Machines* 18(3): 303–329.

Floridi, Luciano. 2010a. *Information – A very short introduction*. Oxford: Oxford University Press.

Floridi, Luciano. 2010b. Information, possible worlds, and the cooptation of scepticism. *Synthese* 175(1): 63–88.

Floridi, Luciano. 2010c. Ethics after the information revolution. In *The Cambridge handbook of information and computer ethics* (Chapter 1), ed. L. Floridi, 3–19. Cambridge: Cambridge University Press.

Floridi, Luciano. 2011a. *The philosophy of information*. Oxford: Oxford University Press.

Floridi, Luciano. 2011b. Semantic conceptions of information. In *The Stanford encyclopedia of philosophy*, ed. E. N. Zalta.

Floridi, Luciano. 2011b. A defence of constructionism: Philosophy as conceptual engineering. *Metaphilosophy* 42(3): 282–304.

Hintikka, Jaakko. 1962. *Knowledge and belief: An introduction to the logic of the two notions, contemporary philosophy*. Ithaca: Cornell University Press.

Khalidi, Muhammad Ali. 1993. Carving nature at the joints. *Philosophy of Science* 60(1): 100–113.
Luper, Steven. 2010. The epistemic closure principle. In *The Stanford encyclopedia of philosophy*, ed. E. N. Zalta.
Plato. 1989. *The collected dialogues of Plato: Including the letters*, 14th ed, ed. Edith Hamilton and Huntington Cairns, with introduction and prefatory notes. Princeton: Princeton University Press.
Quine, Willard Van Orman. 1992. Structure and nature. *The Journal of Philosophy* 89(1): 5–9.
Turilli, Matteo, Antonino Vaccaro, and Mariarosaria Taddeo. forthcoming. Network neutrality: Ethical issues in the internet environment. *Philosophy & Technology*.
von Leibniz, Gottfried Wilhelm Freiherr. 1996. *New essays on human understanding*. Cambridge: Cambridge University Press.

Index

H. Demir (ed.), *Luciano Floridi's Philosophy of Technology: Critical Reflections*, Philosophy of Engineering and Technology 8, DOI 10.1007/978-94-007-4292-5,

MIX
Papier aus verantwortungsvollen Quellen
Paper from responsible sources
FSC® C105338
FSC
www.fsc.org

If you have any concerns about our products,
you can contact us on
ProductSafety@springernature.com

In case Publisher is established outside the EU,
the EU authorized representative is:
Springer Nature Customer Service Center GmbH
Europaplatz 3, 69115 Heidelberg, Germany

Printed by Libri Plureos GmbH
in Hamburg, Germany